General and Applied Aspects of Halophilic Microorganisms

NATO ASI Series

Advanced Science Institutes Series

A series presenting the results of activities sponsored by the NATO Science Committee, which aims at the dissemination of advanced scientific and technological knowledge, with a view to strengthening links between scientific communities.

The series is published by an international board of publishers in conjunction with the NATO Scientific Affairs Division

A	**Life Sciences**	Plenum Publishing Corporation
B	**Physics**	New York and London
C	**Mathematical and Physical Sciences**	Kluwer Academic Publishers
D	**Behavioral and Social Sciences**	Dordrecht, Boston, and London
E	**Applied Sciences**	
F	**Computer and Systems Sciences**	Springer-Verlag
G	**Ecological Sciences**	Berlin, Heidelberg, New York, London,
H	**Cell Biology**	Paris, Tokyo, Hong Kong, and Barcelona
I	**Global Environmental Change**	

Recent Volumes in this Series

Volume 196—Sensory Abilities of Cetaceans: Laboratory and Field Evidence
edited by Jeanette A. Thomas and Ronald A. Kastelein

Volume 197—Sulfur-Centered Reactive Intermediates in Chemistry and Biology
edited by Chryssostomos Chatgilialoglu and Klaus-Dieter Asmus

Volume 198—Selective Activation of Drugs by Redox Processes
edited by G. E. Adams, A. Breccia, E. M. Fielden, and P. Wardman

Volume 199—Targeting of Drugs 2: Optimization Strategies
edited by Gregory Gregoriadis, Anthony C. Allison, and George Poste

Volume 200—The Neocortex: Ontogeny and Phylogeny
edited by Barbara L. Finlay, Giorgio Innocenti, and Henning Scheich

Volume 201—General and Applied Aspects of Halophilic Microorganisms
edited by Francisco Rodriguez-Valera

Volume 202—Molecular Aspects of Monooxygenases and Bioactivation
of Toxic Compounds
edited by Emel Arınç, John B. Schenkman, and Ernest Hodgson

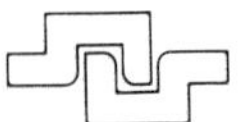

Series A: Life Sciences

General and Applied Aspects of Halophilic Microorganisms

Edited by

Francisco Rodriguez-Valera

University of Alicante
Alicante, Spain

Plenum Press
New York and London
Published in cooperation with NATO Scientific Affairs Division

Proceedings of a FEMS-NATO Advanced Research Workshop on
General and Applied Aspects of Halophilic Microorganisms,
held September 17–22, 1989,
in Alicante, Spain

Library of Congress Cataloging-in-Publication Data

FEMS-NATO Advanced Research Workshop on General and Applied Aspects of
    Halophilic Microorganisms (1989 : Alicante, Spain)
    General and applied aspects of halophilic microorganisms / edited
  by Francisco Rodriguez-Valera.
      p.   cm. -- (NATO ASI series. Series A, Life sciences ; v.
  201)
      "Proceedings of a FEMS-NATO Advanced Research Workshop on General
  and Applied Aspects of Halophilic Microoganisms, held September
  17-22, 1989, in Alicante, Spain"--T.p. verso.
      "Published in cooperation with NATO Scientific Affairs Division."
    Includes bibliographical references and index.
    ISBN 0-306-43816-X
    1. Halophilic microorganisms--Congresses.   I. Rodriguez-Valera,
  Francisco.   II. Federation of European Microbiological Societies.
  III. North Atlantic Treaty Organization.   Scientific Affairs
  Division.   IV. Title.   V. Series.
    [DNLM: 1. Halobacterium--congresses.   2. Water Microbiology-
  -congresses.   QW 80 F329g 1989]
  QR97.S3F46   1989
  576--dc20
  DNLM/DLC
  for Library of Congress   91-3015
                                                                CIP

© 1991 Plenum Press, New York
A Division of Plenum Publishing Corporation
233 Spring Street, New York, N.Y. 10013

Printed in the United States of America

PREFACE

During recent years the subject of extreme environments and extremophiles has become a central topic in modern Biology. The capability of some microorganisms to withstand, and often prefer, the harsh conditions found in such environments is helping to define the physicho-chemical limits of life and in consequence its essential nature. Halophiles are one of the most representative types of extremophiles, requiring high concentrations of inorganic salts, mostly sodium chloride, to grow and survive. They inhabit hypersaline environments, the distribution and abundance of which during geological eras are attested by the vast amounts of evaporite rocks present in the Earth crust and by their role in the generation of petroleum deposits. The conditions of high osmolarity and ionic strength that are concomitant with concentrated salt solutions challenge the stability of lipid bilayers and the structure of proteins forcing halophilic microbes to develop specialized molecules and physiological mechanisms to cope with this environmental stress. Even so, halophilism is a widespread trait in the microbial world. All the major groups of eucaryotic microbes, two groups of archaeobacteria and most phylogenetic branches of eubacteria have halophilic representatives. Therefore, the study of halophilic microorganisms is indeed a highly heterogeneous and extense topic.

The present volume contains the contributions to the FEMS-NATO Advanced Research Workshop on "General and Applied Aspects of Halophilic Microorganisms" held at Alicante, Spain, September 17-22, 1989. At this meeting 46 invited speakers and 51 other participants presented their latest results through lectures and posters dealing with this diverse subject. I think we succeeded in getting a remarkably comprehensive view of the state of the art in the field. Obviously there were some areas that had to be treated shallowly, for example, the biophysics of the halobacterial retinal proteins, which has become a vast and highly specialized field with regular and frequent meetings dedicated specifically to it. On the other hand the spectacular advances in the field of haloadaptation and the molecular biology of halobacteria (Sessions 2 and 4) have been widely covered. The rapid progress in these areas make them particularly appropiate for revision in a meeting such as this.

As editor of the book I have to thank the authors for their contributions, the extremely high standard of which made my job very simple. I am very grateful to both NATO and the Federation of European Microbiological Societies (FEMS) for their sponsorship. Without the joint support from both parties this meeting could not have been held. Other organizations made substantial contributions to financing the meeting, particularly the Spanish Ministry of Education and Science and the Valencian Government (Generalitat). All the staff and students of the Department of Molecular Genetics and Microbiology, University of Alicante worked hard and long helping in the organization of the meeting and I am very grateful to all of them. Finally a special acknowledgment is required of the invaluable contribution made by Kathleen Hernandez, the secretary of our Department, to both the organization of the meeting and the typing and preparation of this book, thank you.

June 1990   F. Rodriguez-Valera

CONTENTS

PART II

PHYSIOLOGY

PART III

COMPOSITION AND STRUCTURE

PART IV

MOLECULAR BIOLOGY AND GENETICS

PART V
APPLIED ASPECTS OF HALOPHILIC ORGANISMS

PART I

TAXONOMY, ECOLOGY AND PHYLOGENY

HALOPHILY, TAXONOMY, PHYLOGENY AND NOMENCLATURE

Hans G. Trüper, Jörg Severin, Axel Wohlfarth, Ewald Müller and
Erwin A. Galinski

Institut für Mikrobiologie & Biotechnologie
Rheinische Friedrich-Wilhelms-Universität
Meckenheimer Allee 168
D-5300 Bonn 1
Federal Republic of Germany

ABSTRACT

On a first view there is apparently little correlation between taxonomy and halophily. The distribution of different compatible solutes in the eubacterial phyla Proteobacteria and Firmacutes follows a pattern, however, that at least follows certain regularities: (1) All eubacteria that gain energy from photosynthesis or respiration and are capable of haloadaptation are able to accumulate and/or synthesize compatible solutes. (2) When a eubacterium can grow in a non-complex medium, it usually can synthesize ectoine. Growth on yeast extract does not necessarily exclude ectoine synthesis. (3) Extreme halophily in eubacteria is always accompanied by glycine betaine synthesis. The complete synthesis of glycine betaine from CO_2 or simple carbon compounds has only been proven for cyanobacteria, Ectothiorhodospiraceae and *Actinopolyspora halophila*. (4) A not yet identified compound "Y" occurs preferably in Firmacutes; *Bacillus* species accumulate and synthesize proline and glutamate as compatible solutes.

HALOPHILY - TAXONOMY - PHYLOGENY

Halophily and halotolerance are cellular properties that require metabolic and structural adaptations in order to withstand hyperosmotic stress. Halophily and halotolerance are gradual properties ranging from slightly via moderately to extremely halophilic or halotolerant.

Therefore one should not be surprised that bacterial species which need less than seawater salinity have been given the epithet *halophilus/a*, as well as such that prefer saturation concentrations.

General and Applied Aspects of Halophilic Microorganisms
Edited by F. Rodriguez-Valera, Plenum Press, New York, 1991

Table 1. Compatible solutes of Proteobacteria

Species	% NaCl	be-taine	ectoine	tre-halose	suc-rose	others
Alpha subdivision						
Rhodospirillum salinarum	20	⊕	+	−	−	
Rhodospirillum salexigens	15	⊕	−	−	−	
Rhodobacter sulfidophilus	10	−	⊕	−	−	GG
Rhodopseudomonas marina	7.5	−	−	⊕		A
Paracoccus halodenitrificans	10	−	⊕	−	−	G
Gamma subdivision						
Chromatium salexigens	7.5	⊕	−	−	+	
Chromatium purpuratum	7.5	+	−	−	⊕	
Ectothiorhodospira halophila	20	⊕	+	+	−	
Ectothiorhodospira halochloris	20	⊕	+	+	−	
Ectothiorhodospira abdelmalekii	20	⊕	+	+	−	
Ectothiorhodospira marismortui	15	⊕	−	−	+	B
Halomonas elongata	10	−	⊕	−	−	G, Y
Halomonas halmophila	10	−	⊕	−	−	G, Y
Vibrio costicola	10	−	⊕	−	−	G
Vibrio alginolyticus	10	−	⊕	−	−	G
Deleya halophila	10	−	⊕	−	−	G, Y

A, B, Y as yet unidentified, GG glycosylglycerol, G glutamate.
Predominant compatible solutes are encircled.

Throughout the phylogenetic system of the prokaryotes in the modern sense [1] there exist halophilic and halotolerant genera or species in close (phylogenetic) relationship with non-halophilic and non-halotolerant ones. Therefore halophily and halotolerance as such may be of little value for the study of phylogenetic relations. It is, however, interesting to compare the mechanisms bacteria in the different phylogenetic branches have developed to overcome salt stress.

It is a well established fact that the extremely halophilic archaeobacteria of the family Halobacteriaceae conteract the high osmotic pressure and ionic strength of their environments by accumulating potassium ions in their cytoplasm and excluding sodium ions at the same time [2]. The high ionic strength in the cytoplasm of these bacteria affords special structural adaptations of all macromolecules that are potentially affected [3, 4]. Thus, e.g., the proteins of halobacteria contain a substantially higher percentage of acid amino acids, as compared with those of non-halophilic bacteria.

The occurrence of elevated cytoplasmic salinities in methanogenic bacteria as well as that of extremely halophilic species in this group [1] permit speculation that the potassium mechanism of osmoregulation in the Halobacteriaceae may have evolved in their ancestral methanogens.

Table 2. Compatible solutes of Firmacutes

Species	% NaCl in media	be-taine	ectoine	tre-halose	pro-line	others
Low GC subdivision						
Sporosarcina halophila	10	+ !	–	–	–	Z
Planococcus citreus	10	+ !	–	–	–	Z
Bacillus species (several)	10	–	–	–	+	Z
High GC subdivision						
Micrococcus halobius	10	–	⊕	+	–	Y
Micrococcus halobius var. *halophilus*	10	–	⊕	+	–	Y
Streptomyces griseolus	5	+ !	–	+	–	Y
Actinopolyspora halophila	20	⊕	–	–	–	–
"*Actinomycete A5-1*"	10	–	⊕	+	–	Y
Marinococcus halophilus	10	–	⊕	–	–	Y, G

G glutamate, Y and Z as yet unidentified, ! only on complex media. Predominant compatible solutes are encircled.

Apparently the situation in halophilic and halotolerant eubacteria is principally different. They rely upon organic osmolytes that must be compatible with their cytoplasmic metabolic functions, i.e., enzymes, and are therefore called "compatible solutes" [5]. We have screened a considerable number of eubacterial strains and identified their compatible solutes by HPLC and NMR techniques (for methods cf. Wohlfarth et al. [6]) These studies revealed certain regularities. Table 1 shows the distribution of compatible solutes within the Proteobacteria studied.

In the alpha subdivision of the Proteobacteria the predominant compatible solutes are clearly glycine betaine or ectoine, as minor components glutamate, glycosylglycerol and the unidentified "A" occur.

In the gamma subdivision the phototrophic species contain betaine as the predominant solute (except *C. purpuratum*). They all have three components namely betaine, ectoine and a disaccharide (except *E. marismortui*), the *Chromatium* species lack ectoine, however. The Halomonadaceae (*Halomonas* and *Deleya* uniformly possess ectoine as the predominant solute and glutamate and the unidentified "Y" as minor components. A detailed study of the compatible solute content within the species of this family has been done by Wohlfarth et a. [6]. Table 2 shows the distribution of compatible solutes within the Firmacutes (Gram-positive bacteria) studied.

The unidentified solute "Z" appears to be typical for the low GC subdivision, whereas "Y" and ectoine appear to be the most common solutes in the high GC subdivision. *Actinopolyspora halophila* is so far the only non-phototrophic bacterium able to synthesize betaine from a simple carbon compound. Other bacteria in this group can synthesize betaine only when suitable immediate precursors are offered them in complex media.

We further studied *Spirochaeta halophila* and found another as yet unidentified compound, "X", as the dominant compatible solute besides betaine, which also here can only be formed from complex media.

The strictly anaerobic bacterial genera *Sporohalobacter, Haloanaerobium* and *Halobacteroides*, all species of which are extremely halophilic [7, 8], do not contain any detectable compatible organic solute. They must rely on inorganic ions to maintain osmotic equilibrium with their environment. Thus this property cannot be considered a principal difference between archaeobacterial and eubacterial halophiles!

For the organic compatible solutes occurring in cyanobacteria the reader is referred to Reed et al [9]. Glycine betaine, trehalose, glycosylglycerol, sucrose, glutamate, betaine were reported from halophilic cyanobacteria.

When we critically summarize the data reported above we arrive at some regularities that rather bear reference to physiology than to phylogeny:

(1)   All eubacteria that gain energy from photosynthesis or oxygen-respiration and are capable of haloadaptation, are able to accumulate and/or produce compatible solutes.

(2)   Archaeobacteria and anaerobic fermenting eubacteria are incapable of synthesizing organic compatible solutes.

(3)   Most eubacteria that can grow in non-complex media are capable of ectoine biosynthesis.

(4)   Extreme halophily in non-fermenting eubacteria is usually accompanied by glycine betaine biosynthesis. The complete de novo synthesis of betaine from carbon dioxide or simple organic carbon compounds has only been proven for cyanobacteria, *Ectothiorhodospira* species and *Actinopolyspora halophila*.

The regularities observed may, however, not stand for long, as there are still those unidentified solutes which may change the whole picture. Also, we are quite convinced that further search will present more unknown compatible solutes.

6

Obviously the biosynthesis of ectoine requires much less energy than that of glycine betaine. Therefore ectoine rather than glycine betaine (as postulated by Imhoff and Rodriguez-Valera [10]) is the most common compatible solute in eubacteria. As most heterotrophic bacteria are usually grown in complex media they are used to accumulating betaine or its precursors (e.g. choline) from the media.

NOMENCLATURE

Under these aspects the term halophily needs to be reconsidered. If the ability to survive and grow in the presence of elevated salt concentrations depends on special ingredients in the media, these acquire a vitamin-like status, although - quite different from vitamins - they have to be supplied in up to molar concentrations. Perhaps we should call such bacteria "facultatively halotolerant".

This fact also relates to nomenclature, because there are certainly some bacteria carrying the name *halophilus/a*, that belong in this category.

ACKNOWLEDGEMENT

This work was sponsored by a grant from the Deutsche Forschungsgemeinshaft.

REFERENCES

[1]   C. R. Woese, Bacterial evolution, *Microbiol. Rev.,* 51: 221 (1987)

[2]   J. K. Lanyi, Salt dependent properties of proteins from extremely halophilic bacteria, *Bacteriol. Rev.* 38: 272 (1974)

[3]   R. Reistad, On the composition and nature of the bulk protein of extremely halophilic bacteria. *Arch. Microbiol.* 71: 353 (1970)

[4]   L. P. Visentin, C. Chow, A. T. Matheson, M. Yaguchi and F. Rollin, *Halobacterium cutirubrum* ribosomes. Properties of the ribosomal proteins and ribonucleic acid. *Biochem. J.* 130: 103 (1972)

[5]   A. D. Brown, Microbial water stress. *Bacteriol. Rev.* 40: 803 (1976)

[6]   A. Wohlfarth, J. Severin, and E. A. Galinski, The spectrum of compatible solutes in heterotrophic halophilic eubacteria of the family Halomonadaceae. *J. Gen. Microbiol. (in press)* (1989)

[7]   A. Oren, Intracellular salt concentrations of the anaerobic halophilic eubacteria *Haloanaerobium praevalens* and *Halobacteroides halobius. Can. J. Microbiol.* 32: 4 (1985)

[8]   S. Rengpipat, S. E. Lowe and J. G. Zeikus, Effect of extreme salt concentrations on the physiology and biochemistry of *Halobacteroides acetoethylicus. J. Bacteriol.* 170: 3065 (1988)

[9]   R. H. Reed, L. J. Borowitzka, M. A. Mackay, J. A. Chudek, R. Foster, S. C. R. Warr, D. J. Moore, and W. D. P. Stewart, Organic solute accumulation in osmotically stressed cyanobacteria. *FEMS Microbiol. Rev.,* 39: 51 (1986)

[10]   J. F. Imhoff and F. Rodriguez-Valera, Betaine is the main compatible solute of halophilic eubacteria. *J. Bacteriol.* 160: 478 (1984)

THE MICROBIOTA OF SALINE LAKES OF THE

VESTFOLD HILLS, ANTARCTICA

P. D. Franzmann

Deutsche Sammlung von Mikroorganismen und Zellkulturen GmbH
Mascheroder Weg 1B
D-3300 Braunschweig
Federal Republic of Germany

THE VESTFOLD HILLS

The Vestfold Hills is a 400 km^2, coastal, ice free area in east Antarctica. The region is geologically young, formed in the last 7000 years after ice retreat. It contains in excess of 300 lakes and ponds which were formed from relic seawater trapped in valleys and depressions after isostatic readjustment [1]. Eleven of the lakes are meromictic [2], with monimolimnia ranging in salinity from 1.4 % to 22.8 %. A number of the lakes are hypersaline; the most saline has a total salt concentration up to 28%. Water column temperatures can fall as low as -17°C over winter, but summer temperatures in the limnetic region of the lakes are above 0°C, and in some cases as high as +14°C. The biota of two of the lakes have been examined.

ORGANIC LAKE, PHYSICAL CHARACTERISTICS

Some physical and chemical characteristics of Organic Lake are given in Table 1. The chlorinity of the lake increased with depth and varied between *ca.* five and six times the chlorinity of seawater. After melting of the winter ice cover, surface water salinity was 0.83%, salinity of the bottom water was *ca.* 22.8% [3]. The lake was meromictic; an ice cover excluded wind induced turbulence in the water column throughout winter. The summer thermal profile of the lake and the increasing salinity of the water column with depth prevented lake turnover in the ice free, summer period. As a result of water column stability and microbial activity, the bottom waters have become depleted of oxygen and the products of fermentation were detected. Hydrogen sulfide or other strong reductants were not present and the redox potential remained positive.

General and Applied Aspects of Halophilic Microorganisms
Edited by F. Rodriguez-Valera, Plenum Press, New York, 1991

Table 1. Physical and chemical characteristics of Organic Lake.

Depth (M)	Temp.[a] (°C)	Temp.[b] (°C)	O_2 (mg l^{-1})	E_h (mV)	Cl^- (g l^{-1})	Acetate (mmol l^{-1})	Bacteria ($\times 10^6$ ml^{-1})
0	+3.6	ND	12.8	+378	ND	Tr	0.3
1	+8.9	-3.9	9.8	+363	99	Tr	1.4
2	+2.6	-4.9	6.2	+370	107	Tr	5.1
3	-3.7	-6.1	3.2	+367	107	Tr	4.4
4	-6.3	-6.5	1.1	+208	110	0.22	8.5
5	-6.3	-5.9	0.0	+120	129	0.32	8.4

[a]= summer temperatures. [b]= winter temperatures. ND= not determined due to ice cover. Tr= present in trace amounts.
Data derived from reference [4].

ORGANIC LAKE, BIOTA

A combination of high salinity and low temperatures has restricted the diversity of the biota of Organic Lake. No multicellular organisms inhabited the lake. Eucaryotic populations were dominated by a diflagellated green alga, *Dunaliella* sp., which reached maximal numbers, 2.8 10^4 cells ml^{-1}, above the oxycline at a depth of 4m. Similar numbers of a *Chaetoceros* sp., a diatom, occurred in the upper water column and were generally restricted to the low salinity water formed from the melting of surface ice. A further three species of eucaryote occurred in the lake, a flagellated green alga, *Pyramimonas* sp., a diatom, *Navicula* sp., and a choanoflagellate, *Acantheocopsis unguiculata*, but their abundance did not exceed 500 cells ml^{-1}.

Bacterial numbers (determined as direct counts by fluorescence microscopy after staining with acridine orange) increased with depth to 4m. A high proportion of the bacteria in the bottom waters may have been cells sedimenting from the oxic zone and preserved in the cold brine. No bacteria capable of anaerobic growth were isolated from the lake, however, anaerobic methodology was limited to the use of anaerobic jars. The accumulation of acetate (Table 1) and other fermentation products in the waters of the monimolimnion suggests fermentative bacteria were present. More rigorous investigation using Hungate techniques would undoubtedly yield anaerobic strains.

The concentration of sulfate in the monimolimnion was a high as 78 mmol l^{-1}. A survey of meromictic lakes of the Vestfold Hills by Gibson et al. [5] determined that lakes with salinity above 19% and temperatures below *ca.* -3°C contained no H_2S in the water column, although H_2S was ubiquitous in the anaerobic monimolimnion of meromictic lakes of lower salinity. The combined effects of low temperature and high salinity in Organic Lake are probably beyond the tolerance limits of species of sulfate reducing bacteria.

Table 2. Species of bacteria described from Organic Lake.

Species	Colony colour	DNA mol % (G+C)	Salt Range (%)
Halomonas subglaciescola	White	61 - 63	0.5 - 20.0
"Halomonas meridiana"[a]	White	59	0.01 - 20.0
"Flavobacterium gondwanense"[b]	Orange	39	1.0 - 16.0
"Flavobacterium salegens"[b]	Yellow	41	0.01 - 20.0

[a] = proposed name [7]. [b] = proposed name [8].

Bacterial strains were isolated from the water column of Organic Lake with the Skerman Micromanipulator [6]. Single cells were selected from microcolonies which formed after aerobic incubation at 10°C on a solid substrate, composed of the lake water solidified with 1.5% agar. after inoculation with a water sample. Some characteristics of the taxa thus far isolated from Organic Lake are given in Table 2.

All of the strains isolated from Organic Lake were psychrotrophic, capable of growth at 0°C and at temperatures over 20°C. The effect of temperature on the growth rates of five strains of *Halomonas subglaciescola* [9] were tested by the Ratkowsky Model [10]. The minimum temperature at which each strain was grown and the predicted minimum, optimum and maximum temperature for growth as determined from the model are given in Table 3. Salinity and temperature were especially variable in the mixolimnion of Organic Lake where salinity ranged from 0.83% to 17% and temperatures ranged from -14°C to +9.5°C. There must be a period in each winter

Table 3.[a] The observed minimum temperature[b] for growth and the predicted minimum, maximum, and optimum growth temperatures of five strains of *Halomonas subglaciescola*.

Strain	Observed T_{Min}	Predicted temperature range		
		T_{Min}	T_{Opt}	T_{Max}
ACAM 12^T	0.0	-3.3	23.4	32.3
ACAM 30	-5.4	-9.9	20.2	29.8
ACAM 11	-5.2	-9.2	20.0	29.7
ACAM 15	-5.4	-9.2	22.3	31.6
ACAM 19	-3.4	-7.9	23.1	32.1

[a] = Data derived from reference [9]
[b] = Temperature in °C

when *Halomonas subglaciescola* could not grow in Organic Lake, as temperatures in the lake fall below the T_{Min} of strains of the species. Although temperature data is currently not available for the other three bacterial taxa isolated from Organic Lake, the salinity range for growth of these taxa would suggest the species are well adapted for growth in the lake.

Organic Lake contains species which seem well adapted to high salinity and low temperatures. Antarctic sea ice would seem to be the most likely colonization source of Organic Lake by the halotolerant strains. Brine channels within the sea ice can reach salinities greater than 15% and considerable uridine and thymidine incorporation occurs in the sea ice microbial community at 7 to 8% [11]. Comparison of 16S rRNA catalogues has shown that members of the genus *Halomonas* are related to members of the genus *Deleya* [12], a genus originally composed of marine species [13].

DEEP LAKE, PHYSICAL CHARACTERISTICS

Deep Lake contains water of *ca.* ten times the salinity of sea water with ratios of ions roughly similar to the ratios as they occur in sea water [14]. Due to its high salinity, an ice cover does not form on the lake. The lake is essentially monomictic; the water column turns over in winter when the water column is - 16 to -17°C throughout [14]. Bottom temperatures rarely warm to - 14°C and temperatures exceeding 0°C are only recorded in the upper 10-15 m of the water column and only for three to four months each year. The highest temperature recorded from two years monitoring of the lake was +11.5°C [15].

DEEP LAKE, BIOTA

Deep Lake contains a single species of eucaryote, a diflagellated green alga, *Dunaliella* sp. Although direct counts of bacterial cells in the lake estimated a population of bacteria of up to 10^6 cells ml^{-1}, many attempts to isolate bacteria on media prepared with undiluted Deep Lake water were unsuccessful, however bacteria could be readily isolated on media that contained less than 25% Deep Lake water [16]. For some time it was considered that the large population of bacteria in the lake was composed only of cells which had been washed into the lake and preserved in the cold brine. Attempts to isolate the *Dunaliella* sp. from Deep Lake at 18°C led to the enrichment of pleomorphic rod-shaped bacteria. Isolated strains formed red colonies on agar composed of undiluted Deep Lake water, 0.1% yeast extract and a vitamin supplement [17]. The strains were described as the new species *Halobacterium lacusprofundi* [17]. Further proof that *H. lacusprofundi* grew in Deep Lake came from diether lipid analysis of the lake sediment which contained a lipid fingerprint very similar to the diether lipid fingerprint of pure cultures of *H. lacusprofundi* (J. K. Volkman, unpublished data).

12

Amongst members of the genus *Halobacterium*, analysis of 16S rRNA catalogues [17] and polar lipids [18] has shown *H. lacusprofundi* was most closely related to *Halobacterium saccharovorum* and *Halobacterium sodomense*. Unlike all other species of *Halobacterium* described to date, *H. lacusprofundi* grew at temperatures less than 10°C. The predicted minimum temperature for growth for *H. lacusprofundi* ACAM 34^T was +1°C; determined from the application of the Ratkowsky Model to temperature and growth rate data [9]. As all regions of the Deep Lake had a temperature below the T_{Min} of *H. lacusprofundi* for at least eight to nine months of the year, growth of this species must have been limited to the short summer period. Application of the Ratkowsky Model and knowledge of the temperature fluctuations in Deep Lake suggested that *H. lacusprofundi* is restricted to about 8 generations per year in the most temperature favourable region of Deep Lake [9]. This estimate was based on temperature limitation of growth alone; other factors such as limited nutrient supply may decrease the number of generations achieved each year.

Analysis of the glyceryl ether lipids of *H. lacusprofundi* has shown the presence of unsaturated isoprenoid glyceryl ethers [17]. This may be a result of adaptation to growth at low temperatures, since presence of unsaturated lipids has been suggested as a means by which organisms maintain membrane fluidity at low temperatures.

The total number of bacteria in Deep Lake was *ca.* 10^{18} cells (determined from the number of cells per ml and the approximate volume of the lake). It is probable that a significant proportion of these cells was not *H. lacusprofundi*. If cells remained intact throughout the overwinter period of non growth, the species could achieve this population in 8 years after inoculation of the lake with a single cell. As Deep Lake was separated from the sea *ca.* 6000 years B.P. and has become progressively more saline since then [1] *Halobacterium lacusprofundi* has had considerable time in which to colonize the lake.

CONCLUSIONS

Most strains of bacteria isolated from Antarctic saline lakes have represented new species, distinct from species inhabiting warmer hypersaline environments. The bacterial taxa of Antarctic saline environments display an ability to grow at reduced temperatures when compared with their taxonomic counterparts isolated from tropical or temperate environments. The saline lakes of the Vestfold Hills are young, and strains have not been isolated which can exploit these environments in winter when temperatures fall below minimum temperatures for growth. Given further evolution, organisms which can exploit their environment more fully may develop.

ACKNOWLEDGEMENTS

The work reported here was conducted with the support of the German

Collection of Microorganisms (DSM), the Australian Research Council, the Antarctic Science Advisory Committee, and the University of Tasmania.

REFERENCES

[1]   J. A. Peterson, B. L. Finlayson and Z. Qingson, *in* "Biology of the Vestfold Hills, Antarctica", pp. 221-226. Kluwer Academic Publishers, Dordrecht (1988)

[2]   J. A. E. Gibson, R. C. Garrick, P. D. Franzmann, P. P. Deprez and H. R. Burton, *ANARE Research Note No. 67*, Antarctic Division, Kingston (1989)

[3]   P. D. Franzmann, H. R. Burton and T. A. McMeekin, *Int. J. Syst. Bacteriol.* 37: 27 (1987)

[4]   P. D. Franzmann, P. P. Deprez, H. R. Burton and J. van den Hoff, *Aust. J. Mar. Freshw. Res.* 38: 409 (1987)

[5]   J. A. E. Gibson, R. C. Garrick, P. D. Franzmann, P. P. Deprez and H. R. Burton, *Paleogeography, Paleoclimatology, Paleoecology* (In Press).

[6]   V. B. D. Skerman, *J. Gen. Microbiol.* 54: 287 (1968)

[7]   S. R. James, S. J. Dobson, P. D. Franzmann, T. A. McMeekin, *Syst. Appl. Microbiol.* (Submitted)

[8]   S. J. Dobson, S. R. James, P. D. Franzmann, T. A. McMeekin, *System. Appl. Microbiol.* (Submitted)

[9]   T. A. McMeekin and P. D. Franzmann, *Polar Biology* 8: 293 (1988)

[10]   D. A. Ratkowsky, R. K. Lowry, T. A. McMeekin and A. N. Stokes, *J. Bacteriol.* 154: 1222 (1983)

[11]   S. T. Kottmeier and C. W. Sullivan, *Polar Biology*, 8: 293 (1988)

[12]   P. D. Franzmann, U. Wehmeyer and E. Stackebrandt, *Syst. Appl. Microbiol.* 11: 16 (1988)

[13]   L. Baumann, R. D. Bowditch and P. Baumann, *Int. J. Syst. Bacteriol.* 33: 793 (1983)

[14]   J. M. Ferris and H. R. Burton, *Hydrobiol.* 165: 115 (1988)

[15]   R. J. Barker, *ANARE Scientific Rep. Pub. No. 130*, Aust. Govt. Publ. Service, Canberra (1981)

[16]   R. M. Hand, *in* "Biochemistry of Ancient and Modern Environments. pp. 123-129. Australian Academy of Science, Canberra (1980)

[17]   P. D. Franzmann, E. Stackebrandt, K. Sanderson, J. K. Volkmann, D. E. Cameron, P. L. Stevenson, T. A. McMeekin and H. R. Burton, *Syst. Appl. Microbiol.* 11: 20 (1988)

[18]   B. J. Tindall, *FEMS Microbiol. Lett.* (In Press).

HALOPHILIC ORGANISMS AND THE ENVIRONMENT

Alberto Ramos-Cormenzana

Department of Microbiology
University of Granada
Campus Universitario de Cartuja
Granada
Spain

INTRODUCTION

The halophilic bacteria are usually found in habitats where salt concentration
is higher than the concentration of salt in the sea, and also in numerous environments
with different salinities. Halophiles have been described in a wide diversity of
environments, including those from museum cultures [1]. But although most of the
research has been done on hypersaline environments, they have also been isolated from
other non-saline habitats such as freshwaters [2].

The description of halophiles in saline habitats suggests an important role of
these organisms in nature; and a wide correlation between the different types described
has been taken into account. The determination of saline spectrum is very important
to the characterization of the type of microorganisms to be regarded as halotolerants
or extreme, moderate and slight halophiles. I am going to consider all these kinds of
organisms because they perform important activities in the saline environment, but
more properly those organisms that need more than 3% NaCl for their optimum
growth; or in another sense those able to grow at 10% NaCl. So I am excluding the
non-tolerant, those which tolerate only a small concentration of salt (about 1% w/v)
[3], and also those that although they could be included in other groups, grow better
under non-halophilic conditions. When the halophiles were isolated for the first time
nobody realised the great impact that these organisms would have in the future.

TYPES AND DIVERSITY OF HALOPHILES

Halophiles constitute a wide group of organisms which includes a great
diversity of protozoa, algae and bacteria. Together with the eukaryotic halophiles
where the algae, such as *Dunaliella*, seem to play an important role as primary

producers, the halophilic bacteria could be assigned to two categories: the archaeobacterial and the eubacterial halophiles. The ecology and taxonomy of both groups of organisms has been studied [4, 5].

Many different groups of bacteria have been described. Phototrophic bacteria, both oxygenic and anoxygenic are common in marine and hypersaline environments. Their distribution depends on their different response to oxygen: cyanobacteria usually inhabit the upper part of the photic zone in saline lakes and also in the sea; in contrast the anoxygenic phototrophic green and purple bacteria, which are restricted to the lower anoxic part of the photic zone [6]. Phototrophic bacteria were also found in marine salterns, where *Chromatium* species and *Rhodospirillum salexigens* were predominant [7].

At first only two genera were considered in the archaeobacterial halophiles, but after the discovery of *Natronobacterium* and *Natronococcus* [8] the number of new genera increased, such as *Haloferax* and *Haloarcula* [9]. Similar facts happen in relation to eubacterial halophiles, among them there are numerous groups, both Gram-positive and Gram-negative, and new genera are constantly being proposed, such as the two new genera *Gibbonsiella* and *Volcaniella* (Results not yet published). Some reviews have recently been done in relation to the study of archaeobacterial [10] and eubacterial halophiles [11].

Although there was belief in the banal influence of halophiles in pathogenesis [12] and there are scarce reports on halophilic pathogens, perhaps they should be taken into account, at least there are some descriptions of halophilic vibrios associated with infections in man [13, 14].

Bacteriophages specific for halobacteri ., both moderate and extreme, have been shown to occur and may play an interesting ecological role, although we are still far from a thorough understanding of the ecology of such habitats.

Procedures to study extreme and moderate halophiles have been described [12]. The methods clearly influence results, as I will show later on, this could be the reason why different data are obtained by different authors. Recently, new isolates of halophilic bacteria have demonstrated the specific adaptations to particular environments. It could be easy to select some types of halophiles according to the methodology used. The maximum recovery of halophilic archaeobacteria was obtained on media with natural brines and a whole cell extract of *Halobacterium* [15]. The isolation of pigmented *Bacillus* or *Sporosarcina* strains could be done from heat-treated saline samples [16]. Under anaerobic conditions novel microorganisms have been characterized and described, such as *Halobacteroides*, bacteria which could be selectively enriched when samples were first heated to 80-100°C [17]. Halophilic leptospirae have been isolated from sea water by culture at 20°C and were examined at seven-day intervals by dark-field microscopy [18].

I consider the use of methods to study these organisms *in situ* of great importance, although there are some difficulties in relation to the observation in the laboratory. Scarce papers have been published on field studies and very few studies have been carried out with the scanning electron microscope and/or with the epifluorescent microscope. The distribution and temporal fluctuations in the density of bacteria, ranging from about 1 to 19 millions of bacteria/ml, in the water covering a high-salinity marsh, were investigated employing epifluorescence, scanning electron microscopy and plate counts [19]; there was a peculiar correlation between salinity and density of bacteria, on increasing salinity the direct count of bacteria first decreases and then increases, although this was not discussed by the authors. I would like to encourage such types of work which may tell us more properly the activities of halophiles in the natural environments. To date most of the papers are laboratory studies which in some way show us some perspectives of these organisms, although they are less realistic. An example is the geomicrobial activity of these microorganisms.

The distribution of halophilic microorganisms is conditioned with the places or habitats where studies have been done but I do not want to emphasize this subject because it has been well studied in hypersaline soils [20 - 22], marine salterns [23, 24], inland salterns [25], bottom sediments [26] or other different habitats [27, 28].

After the isolation of an extreme halophile from sea water [29], two questions were of great interest to me: do these organisms survive in environments where there is a low salt concentration and why does this happen?; and the second, do eubacterial halophiles come from non-halophilic eubacteria by an adaptation process or do eubacteria evolve from halophiles by an adaptation mechanism to non-halophilic conditions. Perhaps both the hypotheses could be possible.

Part of the questions could be solved by a paper presented in this meeting on myxobacterial adaptation to saline conditions.

To solve the other question it is important to know if the moderate and extreme halophiles can possibly survive in environments where there is a low salt content, in the same way as non-halophiles were isolated from hypersaline environments [30, 24]. The first thing was to try to isolate halophilic organisms from places where, at least in my mind, scarce or no previous studies have been done to date. I have selected three habitats, rivers, water reservoirs and sewage, to isolate these organisms. To simplify the experiments I have also selected the region of Granada (Spain), where previous studies have been done in saline soils and waters [31].

On studying the habitats of these bacteria in the Granada zone, it was found that they could be found in the three places studied, rivers, reservoirs, and waste waters. Until now the number of halophiles isolated are 16% from waste waters, 38% from reservoirs, and 46% from rivers. I do not think it is important to show the numbers of isolates belonging to the different categories of halophiles, nor the types of the isolates; but what I think is really important is the fact of isolating extreme halophiles of habitats where there is not only no opportunity of development but neither the theoretical possibility of survival. How may these organisms exist in those non-halophilic habitats? How could these be isolated at a high salt concentration and afterwards no growth could be obtained at the same concentration? At the moment these are questions that I cannot answer. But there is a general conclusion that halophilic bacteria, both moderate and extreme, are present in non-saline habitats, in the same manner as non-halophilic bacteria are present in hypersaline environments [20, 24, 30]. So the halophilic and non-halophilic bacteria coexisted in saline and non-saline environments. The halophilic bacteria, both moderate and extreme, present in non-saline habitats are a still relatively unexplored subject.

ENVIRONMENTAL APPLICATIONS

As the main subject of this Advanced Research Workshop is on general and applied aspects of halophilic microorganisms, it seems logical to make some reference to the applied aspects of these organisms in the environment. The environmental applications of these types of organisms could be directed to understand microbial activity in natural habitats. But, as was pointed out before, the main problem is the difficulty in carrying out controlled and reproducible experiments.

Several hypotheses have been postulated and I want to suggest others based on experimental facts, perhaps some of them could be considered speculative, however I am interested to discuss some of them, which are briefly mentioned, these are:

Reclamation of saline soils.
Possible use as a decontaminant in saline or alkaline industrial waters.
Making use of agricultural wastes.
Use as bioindicators.
Potential use in enhanced oil recovery.

One of the problems with hypersaline soils which affects vegetal productivity, is the low microbial activity. Therefore knowledge of the microbial activity in a saline soil will allow the use of these bacteria in the reclamation of saline soils. The existence of toxic ions, and frequently high pH are reasons which most likely inhibit the microbial development. The existing references on this topic, show that salt addition can inhibit microbial processes in soil, such as carbon mineralization, nitrification, total heterotrophic counts and soil enzyme activities [32, 33, 34, 35]. It could be concluded that microbial activity decreased, when salts increased. However, the influence of salts depends on the nature and level of the salt soil properties, and type and diversity of microorganisms capable of adapting to salinity stress [36].

Nitrogen fixation and cellulolitic activity are particularly sensitive to this fact. Salinity and alkalinity of the soil were found to diminish the population of *Azotobacter* and its capacity to fix nitrogen [37]. More recently on a paper about the salt inhibition of free-living diazotroph population, it was found that salts inhibited diazotroph numbers; although the survival of these organisms was evident even at "the highest salt levels" [38]. Our experience about this shows that in saline soils, it is easy to isolate a high number of bacteria able to grow in a nitrogen-free medium, although we have not found nitrogenase activity. However the nitrogen fixation in salt marsh has been proved [39] and the cyanobacteria seemed to play an important role [40]. The effect of salinization and desalinization on the uptake, distribution and assimilation of

18

fertilizer nitrogen by nodulated soybeans has been studied [41]. Nitrogen absorption, the incorporation of nitrogen into the leaves, was depressed by salinization, even after salt removal. Authors suggested that the irrigation of soybean plants with fresh water after salty water may alleviate the depressed effect of salinity on nitrogen nutrition.

Other microbial activities are affected such as ammonification and nitrification, but in a lesser degree. It seems that ammonification is a non-specific process that could be possible under several diverse conditions. If both processes could work properly, the addition of organic matter into the soil would improve not only the soil structure, but also the assimilable nitrogen. Denitrification, on the other hand, is often increased in relation to fertile soils; and this is a clear index of the terrible physical conditions of these soils, in which the anaerobic conditions should be more favoured than usual.

The knowledge of the microbial activity in a saline soil, will allow us to deduce the response to fertilization directed to favour the growth of resistant vegetables to the salinity. According to a previous paper on the influence of CO_2 on pH of alkaline soils [42] a study of the reclamation of alkaline soils has been done [43]. Of course although it is not so easy as one would like it to be, these previous studies however proved this could be a good method to reclaim saline soils, which could be used in the near future by man. There are reasons for and against its use.

Another possible use is as a decontaminant in saline or alkaline industrial waters. It is known that sewage engineers exploit the denitrification process for removal of nitrate, it seems evident [44] that the use of halophilic denitrifying bacteria could be the best way when high saline concentrations were found in waste waters. Other possible applications of these halophilic bacteria could be to remove metals and other elements, such as phosphate (too expensive by chemical methods) in waste water treatment. The ability of some halophiles to precipitate struvite has been proved, now it is important to know the conditions to enable this process to be used in the purification stations. These could be the reasons in favour of use of these organisms in sewage treatment.

One interesting application of these microorganisms could be in making use of agricultural wastes to obtain other useful microbial products. My personal suggestion is to use halophilic organisms to produce biopolymers, from agricultural wastes. The ecological and economic considerations of microbial exopolymers have been described [45]. The interest of extreme halobacteria (*Haloferax*) able to produce considerable amounts of poly-ß-hydroxybutyrate, as biodegradable plastic [46], and an extracellular polysaccharide [47], was described. One important problem in our region is the water contamination by the alpechin [48]. And there are two facts that could support this hypothesis: first to be considered is the great number of strains isolated from the alpechin [48] that need or tolerate saline concentration of potassium ions; the alpechin has a relatively high amount of potassium ions. The second one is that as you know we have been working with moderate halophiles and up until now some of them are able to form a great amount of exopolysaccharide. It is important to determine the nature of these compounds and to determine their rheological properties. I do not think it is

19

difficult to develop the techniques to make use of agricultural wastes, but certainly there are important reasons for its use.

Another environmental application of selected halophilic bacteria is their use as bioindicators of saline wells. This could be easily done, by studying the saline spectrum of moderate and some extreme halophiles. Among the moderate halophilic bacteria there is a wide range of NaCl for growth and some of them do not require much salt. The results in liquid medium validate this application, it is very important to make the saline spectrum determinations in liquid media (not in solid media). One of the strains assayed, which does not show a very great saline requirement in order to grow, has the important property that it does not grow under 0.1 (g/l) NaCl saline concentration. This fact makes it easily possible to produce a kit to be used by farmers. I know how easy it is to determine NaCl in water, but this is not so easy to the farm workers who need to analyse the water that they are using on the crops introduced recently in the subtropical area of Almería and Granada, such as the avacado tree (*"Persea americana"*). The farmers are using the water from wells close to the sea, so step by step the water is increasing in salinity up to levels too toxic for the avocado tree. Actually a saline concentration above 100 mg/l prevents the growth of the tree. A freeze-dried sample of the strain should grow when the saline concentration is up to the above mentioned concentration, which shows the toxicity of water used to irrigate the avocado trees.

The last application that I am going to analyse is the potential use of halophiles in enhanced oil recovery processes. There are two main ways known to use microorganisms in enhanced oil recovery: the use as emulsifiers (surfactants) or as viscofiers (biopolymers). There are excellent papers on this subject, the pioneering paper of Clark et al [49] and others in which halophiles seem to have possessed an important role such as the recent ones of Pfiffner et al. [50] and Post and Al-Haram [51]. It has been clearly exposed that the success of a waterflood or an enhanced oil recovery process depends on the sweep efficiency of that process and one of the most important factors affecting sweep efficiency is permeability variation. In consequence thereof an oil reservoir which contains regions with different permeabilities, the injected fluid will preferentially move through the more permeable regions, and the oil entrapped in the less permeable regions will be bypassed and unrecovered. The production of a microbial product for field use involves the standard procedures of finding a suitable microorganism, determining the optimal environmental conditions for production, growing the culture, isolating or partially purifying the product, and delivering the product to the well site for injection. But many of these steps could be eliminated and the process made more economically feasible if the culture could be grown in the reservoir and the product produced *in situ*. It is necessary to find the proper microorganism which can produce the desired product under the environmental conditions which exist in the reservoir. An initial screening program has been done by Clark et al. [[49], they also described the conditions of the reservoirs. The search for organisms producing biopolymers could be used in situ as selective plugging agents. As many oil reservoirs have high salinities, this was the success of Pfiffner et al. [50] which isolated and characterized bacteria that grow rapidly and produced extracellular polymer up to 12% (wt/vol) NaCl, and with other properties of growing in anaerobic

20

conditions and at a high temperature (50°C). One of these biopolymers (SP018) does have many properties that make it suitable for enhanced oil recovery.

More recently the biodegradation of xanthan by salt-tolerant aerobic microorganisms has been discovered [52], since then the means for its prevention has become an important area of research in avoiding the loss of solution viscosity under enhanced oil recovery operation conditions.

I would like it if this talk serves to stimulate young scientists in the development of this area of knowledge, because I am sincerely convinced about the applicability of the halophiles to solve some environmental problems.

REFERENCES

[1]   L. I. Lenova and E. V. Borisova, Bacteria accompanying some halophilic algae, *Microbiol. J.* 45: 39 (1983) (in Russian).

[2]   H. Larsen, Halophilism. *in* "The Bacteria", I. C. Gunsaulus and R. Y. Stanier (eds), Vol. IV, pp 297-342, Academic Press, New York (1962)

[3]   H. Larsen, Halophilic and halotolerant microorganisms, an overview and historical perspective, *FEMS Microbiol. Rev.*, 39: 3 (1986)

[4]   W. D. Grant and H. M. N. Ross, The ecology and taxonomy of halobacteria, *FEMS Microbiol. Rev.*, 39: 9 (1986)

[5]   F. Rodríguez-Valera, The ecology and taxonomy of aerobic chemo-organotrophic halophilic eubacteria. *FEMS Microbiol. Rev.*, 39: 17 (1986)

[6]   J. F. Imhoff, Halophilic Phototrophic Bacteria, *in* "Halophilic Bacteria", F. Rodriguez-Valera (ed), Vol. 1, 85-108, C. R. C. Press, Boca Raton, Florida.(1988)

[7]   F. Rodriguez-Valera, A. Ventosa, G. Juez and J. F. Imhoff, Variation of environmental features and microbial populations with salt concentration in a multi pond saltern. *Microb. Ecol.* 11: 107 (1985)

[8]   B. J. Tindall, H. N. M. Ross and W. D. Grant, *Natronobacterium* gen. nov. and *Natronococcus* gen. nov., two new genera of haloalkaliphilic archaebacteria. *System. Appl. Microbiol.* 5: 41 (1984)

[9]   M. Torreblanca, F. Rodríguez-Valera, G. Juez, A. Ventosa, M. Kamekura and M. Kates, Classification of non-alkaliphilic halobacteria based on numerical taxonomy and polar lipid composition, and description of *Haloarcula* gen. nov. and *Haloferax* gen. nov. *System. Appl. Microbiol.* 8: 89 (1986)

[10]   G. Juez, Taxonomy of Extremely Halophilic Archaebacteria, *in* "Halophilic Bacteria, Vol. 2", F. Rodríguez-Valera, ed, CRC Press, Boca Ratón, Florida (1988)

[11]   A. Ventosa, Taxonomy of moderate halophilic heterotrophic eubacteria, *in* "Halophilic Bacteria, Vol. 1", F. Rodríguez-Valera, ed., CRC Press, Boca Ratón, Florida (1988)

[12]   N. E. Gibbons, Isolation growth and requirements of halophilic bacteria, *in* "Methods in Microbiology Vol. 3B", J. R. Norris and ? Ribbons, eds. pp.169-183, Academic Press, New York.

[13]   D. B. Craig and D. L. Stevens, Halophilic *Vibrio* sepsis, *South. Med. J.*, 73: 1285 (1980)

[14] J. L. Larsen, A. F. Farid and I. Dalsgaard, A comprehensive study of environmental and human pathogenic *Vibrio alginolyticus* strains, *Zbl. Bakt. Hyg., I.Abt. Orig.* A. 251: 213 (1981)

[15] A. C. Wais, Recovery of halophilic archaebacteria from natural environments, *FEMS Microbiol. Ecol.*, 53: 211 (1988)

[16] D. Claus, H. Fahmy, J. Rolf and N. Tosunoglu, *Sporosarcina halophila*, sp. nov., an obligate, slightly halophilic bacterium from salt marsh soils. *System. Appl. Microbiol.* 4: 496 (1983)

[17] A. Oren, A procedure for the selective enrichment of *Halobacteroides halobius* and related bacteria from anaerobic hypersaline sediments. *FEMS Microbiol. Lett.* 42: 201 (1987)

[18] M. Cinco, M. Tamaro and L. Cociancich, Taxonomic, cultural and metabolic characteristics of halophilic leptospirae. *Zbl. Bakt. Hyg., I.Abt. Orig.* A 233: 400 (1975)

[19] C. A. Wilson and L. H. Stevenson, The dynamics of the bacterial population associated with a salt marsh. *J. Exp. Mar. Biol. Ecol.* 48: 123 (1980)

[20] Y. Henis and J. Eren, Preliminary studies on the microflore of a highly saline soil. *Can. J. Microbiol.*, 9: 902 (1963)

[21] E. Quesada, A. Ventosa, F. Rodriguez-Valera and A. Ramos-Cormenzana, Types and properties of some bacteria isolated from hypersaline soils, *J. Appl. Bacteriol.*, 53: 155 (1982)

[22] E. Quesada, A. Ventosa, F. Rodríguez-Valera, L. Megías and A. Ramos-Cormenzana, Numerical taxonomy of moderately halophilic Gram-negative bacteria from hypersaline soils, *J. Gen. Microbiol.* 129: 2649 (1983)

[23] R. H. Vreeland, C. D. Litchfield, E. L. Martin and E. Elliot, *Halomonas elongata*, a new genus and species of extremely salt tolerant bacteria, *Int. J. Syst. Bacteriol.* 30: 485 (1980)

[24] M. C. Márquez, A. Ventosa and F. Ruiz-Berraquero, A taxonomic study of heterotrophic halophilic and non-halophilic bacteria from a solar saltern. *J. Gen. Microbiol.* 133: 45 (1987)

[25] A. Del Moral, B. Prado, E. Quesada, T. García, R. Ferrer and A. Ramos-Cormenzana, Numerical taxonomy of moderately halophilic Gram-negative rods from an inland saltern. *J. Gen. Microbiol.* 134: 733 (1988)

[26] A. Oren, W. G. Weisburg, M. Kessel and C. R. Woese, *Halobacteroides halobius* gen. nov., sp. nov., a moderately halophilic anaerobic bacterium from the bottom sediments of the dead Sea. *System. Appl. Microbiol.*, 5: 58 (1984)

[27] E. Quesada, V. Bejar, M. J. Valderrama, A. Ventosa and A. Ramos-Cormenzana, Isolation and characterization of moderately halophilic non-motile rods from different saline habitats. *Microbiología* 1: 89 (1985)

[28] K. J. Miller and S. B. Leschine, A halotolerant *Planococcus* from Antarctic Dry Valley soil. *Curr. Microbiol.* 11: 205 (1984)

[29] F. Rodríguez-Valera, F. Ruiz-Berraquero and A. Ramos-Cormenzana, Isolation of extreme halophiles from sea water. *Appl. Environ. Microbiol.* 38: 164 (1979)

[30] J. Brisou, D. Courtois and F. Fenis, Microbilogical study of a hypersaline lake in French Somaliland. *Appl. Microbiol.* 27:819 (1974)

[31]   A. Del Moral, A study of Halophilic Bacterial Microflora from an inland saltern. Thesis. University of Granada (in Spanish)

[32]   D. D. Johnson and W. D. Guenzi, Influence of salts on ammonium oxidation and carbon dioxide evolution from soil. *Soil Sci. Soc. Am. Proc.*, 27: 663 (1963)

[33]   P. Heilman, Effects of added salts on nitrogen release and nitrate levels in forest soils of the Washington Coastal area. *Soil Sci. Soc. Am. Proc.* 39: 778 (1975)

[34]   W. T. Frankenberger and F. T. Bingham, Influence of salinity on soil enzyme activities. *Soil Sci. Soc. Am. J.* 46: 1173 (1982)

[35]   G. McClung and W. T. Frankenberger, Soil nitrogen transformations affected by salinity. *Soil Sci.*, 139: 405 (1985)

[36]   R. O. Laura, Effects of neutral salts on carbon and nitrogen mineralization of organic matter in soil. *Plant Soil*, 4: 113 (1974)

[37]   L. A. Cervantes and J. Olivares, Microbiological studies of a saline soil in El-Salitre, Colombia, *Rev. Latinoam. Microbiol.* 18: 73 (1976)

[38]   M. M. El-Shinnawi and W. T. Frankenberger, Salt inhibition of free-living diazotroph population density and nitrogenase activity in soil. *Soil Sci.* 146: 176 (1988)

[39]   K. Jones, Nitrogen fixation in a salt marsh. *J. Ecol.* 62: 553 (1974)

[40]   M. E. Casselman, W. H. Patrick, Jr, and R. D. DeLaune, Nitrogen fixation in a Gulf Coast Salt Marsh. *Soil Sci. Soc. Am. J.* 45: 51 (1981)

[41]   R. K. Rabie and K. Kumazawa, Effect of salinization and desalinization on the uptake, distribution and assimilation of fertilizer nitrogen by nodulated soybeans. *Soil Sci. Plant Nutr.*, 34: 493 (1988)

[42]   R. S. Whitney and R. Gardner, The effect of carbon dioxide on soil reaction. *Soil Sci.* 55: 127 (1943)

[43]   C. Sauberán and J. S. Molina, Recuperación de bajos alcalinos, *Ciencia e Investigación* 16: 337 (1960)

[44]   G. Denariaz, W. J. Payne and J. Le Gall, A halophilic denitrifier *Bacillus halodenitrificans* sp. nov. *Int. J. Syst. Bacteriol.* 39: 145 (1989)

[45]   G. G. Geesey, Microbial exopolymers: ecological and economic considerations. *ASM News* 48: 9 (1982)

[46]   R. Fernández-Castillo, F. Rodríguez-Valera, J. Gonzalez-Ramos and F. Ruiz-Berraquero, Accumulation of poly-(ß-hydroxybutyrate) by halobacteria. *Appl. Environ. Microbiol.* 51: 214 (1986)

[47]   J. Anton, I. Meseguer and F. Rodríguez-Valera, Production of an extracellular polysaccharide by *Haloferax mediterranei*, *Appl. Environ. Microbiol.* 54: 2381 (1988)

[48]   E. Moreno, J. Quevedo-Sarmiento and A. Ramos-Cormenzana, Antibacterial activity of waste waters from olive oil mills. *Adv. Environm. Technol. Management* 1: (in press) (1990)

[49]   J. R. Clark, D. M. Munnecke and G. E. Jenneman, *In situ* Microbial Enhancement of oil production. *Dev. Ind. Microbiol.* 22: 695 (1980)

[50]   S. M. Pfiffner, M. J. NcInerney, G. E. Jenneman and R. M. Knapp, Isolation of halotolerant, thermotolerant, facultative polymer-producing bacteria and

characterization of the exopolymer. *Appl. Environ. Microbiol.* 51: 1224 (1986)

[51] F. J. Post and F. A. Al-Harjan, Surface activity of halobacteria and potential use in microbially enhanced oil recovery. *System. Appl. Microbiol.* 11: 97 (1988)

[52] C. T. Hou, N. Barnabe and K. Greaney, Biodegradation of xanthan by salt-tolerant aerobic microorganisms. *J. Ind. Microbiol.*, 1: 31 (1986)

ESTIMATION OF THE CONTRIBUTION OF ARCHAEBACTERIA AND EUBACTERIA TO THE BACTERIAL BIOMASS AND ACTIVITY IN HYPERSALINE ECOSYSTEMS: NOVEL APPROACHES

Aharon Oren

The Division of Microbial and Molecular Ecology
The Institute of Life Sciences
The Hebrew University of Jerusalem
Jerusalem 91904
Israel

ABSTRACT

In order to evaluate the contribution of archaebacteria (the *Halobacterium* group) and halophilic eubacteria to the bacterial biomass and activity in hypersaline environments I used bile salts, which in very low concentrations cause lysis of halobacteria, while eubacteria and halococci remain intact. The decrease in total bacterial numbers upon bile salt addition is a measure for the community size of bacteria of the *Halobacterium* group. Bile salts can be used as specific inhibitors for halobacterial activity in measurements of heterotrophic activity of the community. In addition, protein synthesis inhibitors specific for archaebacteria (anisomycin) and eubacteria (chloramphenicol) were used to differentiate between the groups in natural communities.

INTRODUCTION

During the past decades our knowledge on the extremely halophilic bacteria has increased dramatically. Many new organisms have been isolated and characterized, and much is known about their physiology, biochemistry, molecular biology and taxonomy. However, our understanding of the ecology of the *Halobacterium* group and of the halophilic eubacteria remains extremely limited. Even data on bacterial community sizes in hypersaline environments are scarce. Most studies on the subject deal with viable counts on plates, and these cannot yield a reliable picture as colony yields are in general two or more orders of magnitude lower than the real bacterial numbers. Thus, for example, numbers of moderately halophilic bacteria of only up to 4×10^4 colony-forming bacteria/ml were reported in saltern ponds of intermediate salinity, and in crystallizer ponds numbers of colony - forming halobacteria did not exceed $2x10^4$/ml [1,2]; in crystallizer ponds coloured red by halobacteria, bacterial numbers of 10^7/ml and higher should be expected [3].

In the present study several novel approaches to the study of the ecology of hypersaline ecosystems are proposed and tested. The approaches used exploit the profound differences that exist between the halophilic archaebacteria (*Halobacterium*, *Haloferax*, *Haloarcula*) and eubacteria, differences that can be used to assess the contribution of the different groups to the bacterial community sizes and activities in hypersaline environments. To test the developed methods two hypersaline environments were studied: saltern ponds of different salinity values (at a solar saltern facility near Eilat), and surface waters of the Dead Sea.

THE USE OF BILE SALTS IN THE ESTIMATION OF HALOBACTERIAL NUMBERS IN HYPERSALINE ECOSYSTEMS

Already more than 30 years ago it was reported that halobacteria are lysed by very low concentrations of bile salts [4], and already then it was suggested that bile salts can be used for the differentiation of halobacteria from other halophilic microorganisms, albeit in pure cultures [5]. A high content of bile salts was recently identified as the cause of lysis of halobacterial cultures by Bacto-Peptone [6].

A differential microscopic enumeration procedure for halophilic archaebacteria (not including the halococci, which are not lysed by bile salts) was developed [7]: 10 ml portions of brine are centrifuged for 15 min at 12,000 x g. Approximately 9.5 ml of the supernatant is removed, whereafter the contents of the

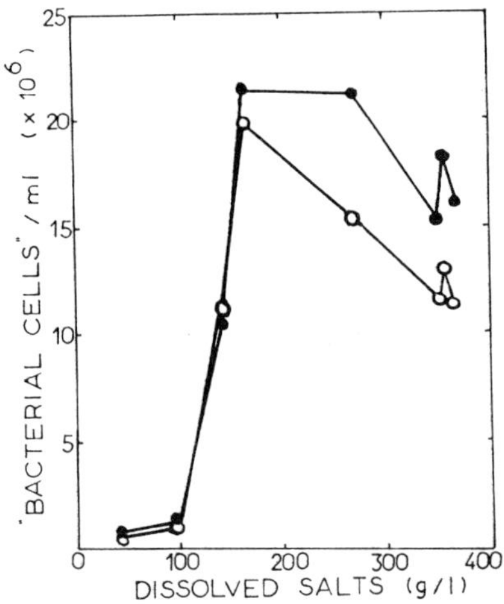

Figure 1.   Microscopic enumeration of bacteria in saltern brines of increasing salinity, December 28, 1988, before (●) and after (o) treatment with taurocholate.

tubes are thoroughly mixed, and the volume of the remaining concentrate measured. The bacterial density of the final suspension is then determined microscopically in a counting chamber. Prior to centrifugation part of the samples are supplemented with 0.003% Na-deoxycholate or 0.025% Na-taurocholate, and incubated for 5 min at room temperature. Deoxycholate was preferentially used with Dead Sea samples; when used in saltern brines precipitates formed upon addition of deoxycholate, and therefore taurocholate was used instead.

Total microscopic bacterial counts of samples from saltern ponds (Figure 1) show bacterial community sizes that generally increase with salinity. In the red crystallizer ponds bacterial numbers encountered were between 1.5×10^7 and 2.1×10^7 microscopically recognizable cells/ml (December 28, 1988). These numbers are somewhat higher than those reported from another similar pond system (Exportadora de Sal, Mexico), using a similar enumeration procedure [3].

To estimate the contribution of bacteria of the *Halobacterium* group to the total bacterial counts, the enumeration procedure was repeated after treatment of the samples with 0.025% Na-taurocholate. It is shown (Figure 1) that, while at the lower salinities the addition of taurocholate did not significantly change cell numbers, at salinities above 250 g/l counts in the presence of bile salts were up to 30 - 50% lower than in the untreated controls.

THE USE OF BILE SALTS TO ESTIMATE THE CONTRIBUTION OF HALOBACTERIA TO OVERALL HETEROTROPHIC ACTIVITY IN HYPERSALINE ECOSYSTEMS

If indeed the low bile salt concentrations that cause lysis of halobacteria do not affect halophilic eubacteria, bile salts can be used as specific inhibitors abolishing activity of halobacteria, while other bacteria remain unaffected. To test this assumption I determined amino acid incorporation rates as a measure of the heterotrophic activity of the bacterial community. Preliminary experiments with pure cultures of halophilic eubacteria (e.g. *Vibrio costicola*) showed no significant inhibition of amino acid incorporation by the bile salt concentrations used in the experiments.

To test the contribution of the bile salt-sensitive bacteria of the *Halobacterium* group to heterotrophic bacterial activity in ponds of different salinities, the incorporation rates of labelled amino acids were determined in the absence and in the presence of 0.005% Na-taurocholate [7]. Portions 25 ml) of the brines examined were introduced into sterile 100 ml Erlenmeyer flasks, and incubated with shaking at 32°C in the presence of 25 μl [U-14C]-protein hydrolysate (50 μCi/ml, 57 mmol C/mCi), and with bile salts as indicated (added 5 min prior to the start of the experiment). After different incubation periods samples were withdrawn, filtered on Millipore filters, the filters were washed with prefiltered water from the same pond or with cold 10% trichloroacetic acid, and the radioactivity retained on the filters was determined.

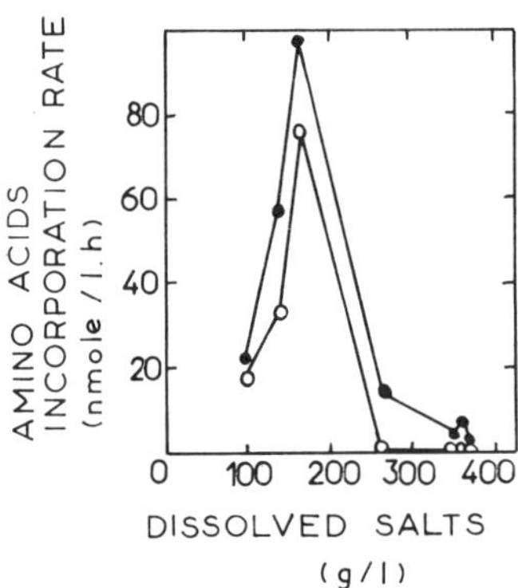

Figure 2. Rates of amino acid incorporation by saltern samples of different salinities, collected December 28, 1988, in the absence (●) and presence (o) of 0.005% Na-taurocholate at 32°C. Data on bacterial numbers present in the same samples were given in Figure 1.

At the lower salinities (up to 200 g salt/l) taurocholate only slightly inhibited amino acid uptake (inhibition from 10 to up to 50%), at a salinity of 267 g/l the residual activity in the presence of taurocholate amounted to only 9% of the control, while at the highest salinities activity was altogether abolished by the bile salt (Figure 2).

The use of bile salts enabled us to obtain a more reliable picture on the presence of viable halobacteria in the Dead Sea. Our studies on the biology of the Dead Sea have shown the development of a dense bloom of halobacteria (up to 2.2×10^7 cells/ml) in the lake's surface waters in the summer of 1980 [8]. This community declined slowly, and almost disappeared with a new overturn of the lake at the end of 1982; no new bacterial bloom has developed since. In recent years bacterial numbers have been low - in the sample used (December 12, 1988) total microscopic "bacterial" counts were $8.4 \pm 2 \times 10^5$ "cells"/ml, and $7.4 \pm 2 \times 10^5$/ml in the presence of taurocholate. Thus, above counts probably represent for the greater part debris and inorganic particles. However, a low amino acid incorporation activity persisted in Dead Sea surface water sampled at the end of 1988 (about 0.03 nmol amino acids/l.h), and this activity was abolished by the addition of 0.003% deoxycholate or 0.005% taurocholate. Comparison of the activity measured with uptake activities by Dead Sea halobacteria suspended in Dead Sea water enabled an estimate of the presence of about $2-4 \times 10^4$ active halobacterial cells/ml only.

THE USE OF SPECIFIC PROTEIN SYNTHESIS INHIBITORS TO ESTIMATE THE ACTIVITY OF HALOPHILIC ARCHAEBACTERIA AND EUBACTERIA IN HYPERSALINE ECOSYSTEMS

The profound differences in the protein synthesis machinery of archaebacteria, eubacteria and eukaryotes are reflected in highly different sensitivities to different antibiotics affecting the ribosomes. To test whether these differences can

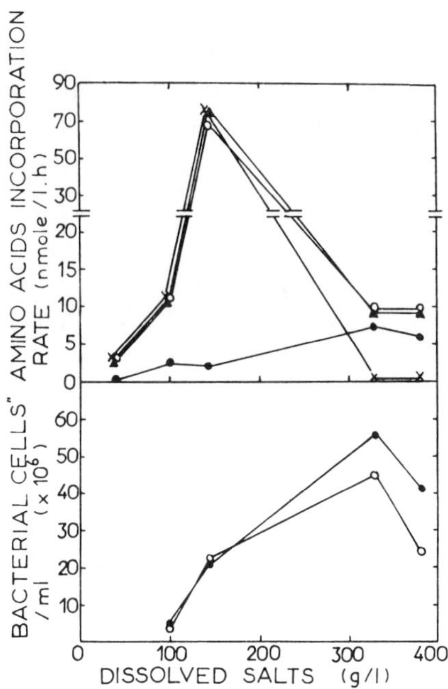

Figure 3. Rates of amino acid incorporation by saltern samples of different salinities, collected May 14, 1989, at 32°C, in the absence (o) and presence of different protein synthesis inhibitors: chloramphenicol (20 μg/ml) (●), anisomycin (40 μg/ml) (x), and cycloheximide (40 μg/ml) (▲). The lower part of the figure shows the bacterial numbers present as determined microscopically, in the absence (●) and presence (o) of 0.025% Na-taurocholate.

be used in ecological studies, amino acid incorporation studies as described above were performed in the presence of chloramphenicol (inhibiting eubacterial protein synthesis, but also possibly causing some inhibition of halobacteria [9]), anisomycin (an inhibitor of eukaryotic ribosomes, being also a potent inhibitor of halobacterial protein synthesis [9]), and cycloheximide (affecting eukaryotes, but neither eubacteria nor archaebacteria). Samples from saltern ponds were preincubated with the antibiotics for 30 min prior to addition of the labelled amino acids mixture. As shown in Figure 3, anisomycin quantitatively inhibited amino acid incorporation at the highest salinities; the same samples were also completely inhibited by taurocholate (not shown). The inhibition by anisomycin cannot likely be attributed to the activity of eukaryotes, as cycloheximide did not significantly affect the rates. With chloramphenicol almost quantitiative inhibition was obtained in the lower salinity range, consonant with the dominance of eubacteria in these waters, but even at the highest salinities some (28 - 42%) inhibition was observed; this inhibition is probably due to the action of chloramphenicol on halobacteria [9], rather than to the presence of active eubacteria, a conclusion based e.g. on the complete inhibition of activity by anisomycin in these samples. Similar to in the high-salinity saltern ponds, anisomycin completely inhibited amino acid uptake in Dead Sea surface water samples, while chloramphenicol caused a 40% inhibition, and cycloheximide did not affect uptake rates. Anisomycin thus turned out to be an effective inhibitor of halobacterial activity, which can be used in ecological studies to yield information on the contribution of halophilic archaebacteria to the total bacterial activity in hypersaline waters.

DISCUSSION

The present study shows that bile salts and specific protein synthesis inhibitors can be used to obtain information on the contribution of halophilic archaebacteria of the *Halobacterium* group to bacterial biomass and activity in hypersaline ecosystems.

The microscopic counting procedure for halobacteria, using bile salts, may yield more relevant information than either plate counts or simple microscopic enumerations as: (1), plate count methods are generally inefficient; (2), part of the bacteria observed microscopically may not be viable, and may have originated from lower salinity ponds, and accumulated in the high salinity ponds due to evaporation; (3), the inclusion of inorganic particles, including crystals of different salts, in the total microscopic "bacterial" counts [3].

The effect of bile salts on activity parameters of halobacterial communities is much more pronounced than the effect on microscopic counts. No interference is encountered here by inorganic particles, or by dead or inactive bacteria, not adapted to the actual salinity of the environment. It was shown that both in salterns at the highest salinities tested, as well as in the Dead Sea, all amino acid incorporation activity was abolished by low concentrations of bile salts, suggesting that only halobacteria were active at these high salt concentrations.

From a theoretical point of view, the use of bile salts does not allow the differential measurement of halophilic archaebacterial and eubacterial activities: it was shown that members of the genus *Halococcus* are insensitive to bile salts [5,6]. Practically, though, halococci cannot be expected to interfere in the proposed differentiation between archaebacterial and halophilic eubacterial activities, as halococci have never been shown to be of great quantitative importance in hypersaline ecosystems. A very good correlation was found between inhibition by bile salts and by anisomycin, which also inhibits halococci.

The studies described demonstrate that the use of antibiotics and other specific inhibitors enables the exploitation of the profound differences between archaebacterial and eubacterial halophiles, not only in physiology and biochemistry, but in ecological studies as well.

ACKNOWLEDGEMENTS

I thank the Israel Salt Co., Ltd. for allowing access to the Eilat salt ponds, and the staff of the Interuniversity Institute of Eilat for laboratory facilities and logistic support.

This study was supported by grants from the Israeli Ministry of Energy and Infrastructure and the Hebrew University of Jerusalem Mutual Fund.

REFERENCES

[1]   F. Rodriguez-Valera, F. Ruiz-Berraquero and A. Ramos-Cormenzana, Characteristics of the heterotrophic bacterial populations in hypersaline environments of different salt concentrations, *Microb. Ecol.* 7:235 (1981).

[2]   F. Rodriguez-Valera, A. Ventosa, G. Juez and J. F. Imhoff, Variation of environmental features and microbial populations with salt concentration in a multi-pond saltern, *Microb. Ecol.* 11: 107 (1985).

[3]   B. J. Javor, Planktonic standing crop and nutrients in a saltern ecosystem, *Limnol. Oceanogr.* 28: 153 (1983).

[4]   H. P. Dussault, Study of red halophilic bacteria in solar salt and salted fish: I. Effect of Bacto-oxgall, *J. Fish. Res. Bd. Canada,* 13:183 (1956).

[5]   H. P. Dussault, Study of red halophilic bacteria in solar salt and salted fish: II. Bacto-oxgall as a selective agent for differentiation, *J. Fish. Res. Bd. Canada,* 13:195 (1956).

[6]   M. Kamekura, D. Oesterhelt, R. Wallace, P. Anderson and D. J. Kushner, Lysis of halobacteria in Bacto-Peptone by bile acids, *Appl. Environ. Microbiol.* 54: 990 (1988).

[7]]   A. Oren, Estimation of the contribution of halobacteria to the bacterial biomass and activity in solar salterns by the use of bile salts, *FEMS Microbiol. Ecol.,* 73: 41 (1990).

[8]   A. Oren, The microbial ecology of the Dead Sea, in "Advances in microbial ecology", Vol. 10, K. C. Marshall, ed., Plenum Publishing Company, New York (1988).

[9]   T. Pecher and A. Böck, In vivo susceptibility of halophilic and methanogenic organisms to protein synthesis inhibitors, *FEMS Microbiol. Lett.* 10:295 (1981).

SOME PRIMARY AND SECONDARY METABOLITES OF

HYPERSALINE MICROBIAL MATS AND ASSOCIATED SEDIMENTS

B. J. Javor[1,2] and J. Porta[1]

[1]Scripps Institution of Oceanography
A-002 University of California
San Diego La Jolla
CA 92093
United States of America

[2] Department of Natural Science San Diego State University
San Diego
CA 92182
United States of America

ABSTRACT

Microbial mats and associated sediments of hypersaline ponds are rich sources of halophilic microorganisms and their primary (major) and secondary (minor) metabolites. An assessment of these metabolites in solar salterns was made in order to 1) understand the chemical conditions under which these microorganisms thrive, 2) identify and isolate novel halophiles, 3) establish the nature and role of chemical interactions between microorganisms, and 4) identify potentially interesting natural products in these unique environments. Porewaters of mats and sediments contained up to millimolar concentrations of sulfide and other dissolved thiols, total alkalinity, total amino acids (as primary amines), and total carbohydrates.

Mats, overlying brines, and associated organisms, including bacterial isolates, were extracted and tested for possible antimicrobial activity and novel metabolites. Several bacteria produced antibiotics, including aerobic, moderately halophilic spore-formers. Extracts from hypersaline microbial mats as well as from brine shrimp (*Artemia salina*) and their cysts contained substances that were inhibitory to Gram-positive bacteria. The nature of these secondary metabolites is currently under investigation.

General and Applied Aspects of Halophilic Microorganisms
Edited by F. Rodriguez-Valera, Plenum Press, New York, 1991

INTRODUCTION

Microbial mats are compact and complex, layered communities typically dominated by a bilayer of phototrophs in the top several mm: cyanobacteria on top, underlain by a layer of anoxygenic phototrophic bacteria. Beneath the phototrophic layer is an anaerobic zone of decomposition. The exact heterotrophic bacterial composition of microbial mats is poorly known but most likely includes a variety of aerobes in the top few mm and diverse anaerobes both within and below the photic zone [1].

Several studies have focused on the non-water-soluble organic constituents of hypersaline microbial mats (e.g. [2, 3]) but relatively little is known about the exact composition of the brines within the mats (porewaters). Such characterizations are necessary to define the major types and concentrations of dissolved organic compounds produced and consumed by the microbial community, the buffering capacity of the porewaters, sulfide and other thiols produced as a result of anaerobic processes within the mats, and possible secondary metabolites produced by some of the microbes.

Chemical characterization of the water-soluble constituents of porewater brines may provide a basis for isolating novel halophilic bacteria which are known to be present but which have never been isolated and maintained in pure culture. Some of these elusive halophilic bacteria include representatives from such physiological groups as sulfate reducers and nitrifiers [1]. In addition, microscopic examination of hypersaline microbial mats has revealed the presence of distinctive bacteria such as filamentous phototrophic bacteria, *Beggiatoa*, and *Achromatium*, but these microorganisms have not been successfully cultured [4, 5, 6].

Some bacteria that thrive in microbial mats may best compete by producing low concentrations of secondary metabolites that suppress the growth of other bacteria and possibly eucaryotes. Antibiotic activity produced by extreme halophiles that is active against other extreme halophiles has been documented [7, 8, 9], but these halophiles were not isolated from microbial mats. No systematic analysis of secondary metabolites of microbial mats or of halophilic microorganisms has been published. The study of hypersaline microbial mats and their associated brines and organisms provides a basis to introduce halophiles as potentially interesting sources of natural products.

MATERIALS AND METHODS

Microbial mats, brines and *Artemia salina* were collected from the Israel Salt Co. saltern (Eilat, Israel) and the Western Salt Co. saltern (Chula Vista, California, U.S.A.). Porewaters were extracted from plastic cores or from blocks of mat by carefully cutting layers. Sediments were centrifuged at 10,000 x g for 15 min at 10 - 15°C. For organic and alkalinity analyses, porewaters were transferred to 1.5-ml Eppendorf tubes and spun again for 5 min at 16,000 x g. A few crystals of sodium azide were added to retard bacterial development in the stored porewaters. Porewaters were stored at -10°C or 4°C until analysis. For thiol determinations, porewaters from sediments collected in cores were extracted under N_2 and analyzed immediately after the first centrifugation.

Salinity was determined with an Atago S-28 refractometer. Sulfide was determined by the methylene blue method of Gilboa-Garber [10] and by the monobromobimane method of Vetter et al. [11] which also separates and measures other thiols. Total alkalinity was determined by titration of 2.0 or 5.0 ml porewater with 0.10 N HCl to the CO_2 inflection point which was determined graphically [12].

Complex carbohydrates were analyzed with hydrazine sulfate reagent with glucose standards according to Strickland and Parsons [13]. Analysis of monosaccharides followed the procedures of Johnson and Sieburth [14] and Johnson et al. [15] using glucose standards. Dissolved free amino acids were estimated with fluorescamine reagent using glycine standards according to North [16], except the measurement was made colorimetrically at the absorption peak of the reacted reagent (385 nm). Dissolved combined amino acids were determined with fluorescamine after first hydrolyzing the porewaters in 6 N HCl overnight at 110°C in sealed ampoules. The samples were cooled and neutralized with 6 N NaOH. Total dissolved organic carbon was determined with a Dohrmann Envirotech organic analyzer after first acidifying the samples to pH 2 with HCl and sparging them with N_2.

In vivo spectra were determined in sonicated samples in 50 mM Tris-HCl buffer, pH 7.6, after removing cell debris by centrigation and filtration (GF/C filter). Photomicroscopy and epifluorescence microscopy were performed on a Zeiss microscope. Spectrophotometric determinations were made on either a Beckman DU-6 or a Gilford Stastar II spectrophotometer.

Bacteria were isolated and grown in a seawater-based medium supplemented with NaCl (10-20% total NaCl), 0.4% yeast extract, 0.5 or 1.0% starch, 0.2% choline chloride, and adjusted to pH 7.5. Solid biological materials (mats, brine shrimp and cysts) were freeze-dried and extracted in chloroform-methanol (1:1). Brines and liquid cultures were extracted in ethyl acetate. Concentrated extracts were tested for antibiotic activity by disc assay. Extracts were separated by thin layer chromatography on silica gel plates with ethyl acetate. Proton nuclear magnetic resonance (NMR) analysis was performed on a WP200SY Bruker NMR spectrometer.

RESULTS

Microbiology

The microbial mats of pond 203 of the Israel Salt Co. were investigated in detail for their content of total dissolved organic carbon (DOC), dissolved primary amines, dissolved carbohydrates, total alkalinity, and sulfide. Salinity varied between 13.2% and 17.8% during the course of study (November, 1986 to June, 1987). The top 2 mm of the mats was dominated by *Aphanothece halophytica*, with abundant filaments of *Spirulina* sp. and *Oscillatoria* sp. Between 2 and 4 mm depth in the mat, the same

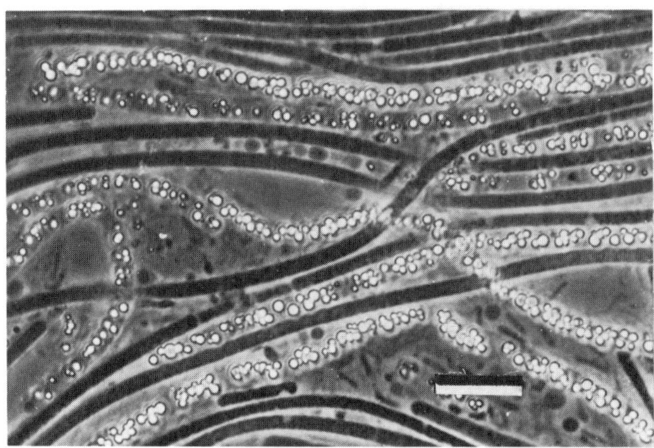

Figure 1.   Photomicrograph of *Chloroflexus*-like filaments with and without sulfur globules in Pond 203 microbial mats. Bar = 10 μm.

cyanobacteria were found in association with thin, S^o-containing filaments (Figure 1). At the base of the photosynthetic zone at 4 mm depth was a cream-coloured layer of the S^o-containing filaments associated with a 1 mm thick band of sand-sized calcite crystals (determined by x-ray diffraction, Hebrew University). Below the calcite layer, the sediments were black and they lacked cohesiveness below 8 mm depth.

The in vivo spectrum of the cream-coloured layer dominated by the S^o-containing filaments showed a pronounced peak of bacteriochlorophyll c (748-752 nm). The filaments were thin (1.7 μm wide) with barely visible crosswalls. They appeared green when they were concentrated under the cover slip. Epifluorescence microscopy showed they lacked chlorophyll a and phycobilins. The internal globules were determined to be elemental sulfur because they disappeared after treatment with pyridine. Due to their apparent pigments and their lack of resemblance to known species of *Beggiatoa*, they are presumed to be related to the green filamentous bacteria *Chloroflexus*.

Attempts to culture the filaments were not successful. However, enrichments of the same mats in Winogradsky columns produced pronounced populations of benthic and planktonic phototrophic bacteria in 15% salinity brine. In vivo spectra of these microorganisms showed obvious peaks of bacteriochlorophyll a (795, 831, 851 and 876 nm) and bacteriochlorophyll c (748-752 nm) as well as a distinct peak at 712 nm (bacteriochlorophyll e). In vivo spectra of the plankton also showed a 776 nm peak which could not be correlated with a known bacteriochlorophyll. The 712 nm and 776 nm peaks in these spectra indicate that a variety of undescribed halophilic phototrophic bacteria have yet to be isolated and characterized from these environments.

Porewater Chemistry

Porewaters in the photosynthetic zone of the pond 203 mats were extremely

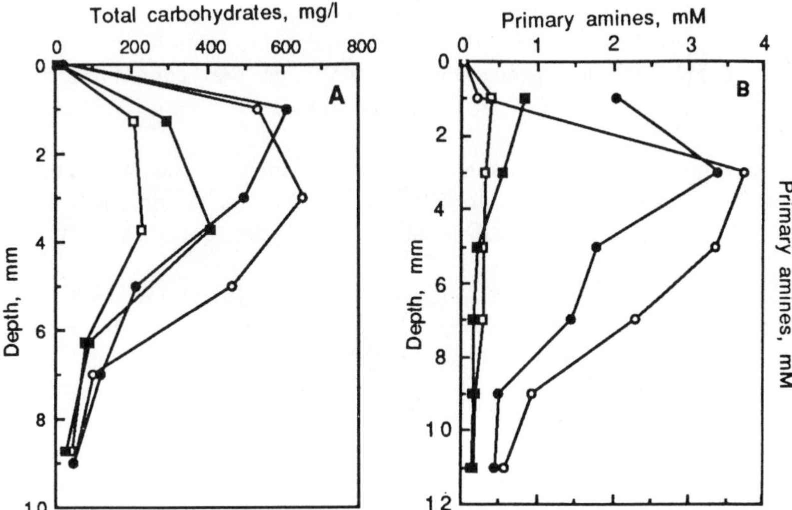

Figure 2. Dissolved organic constituents in Pond 203 mat porewaters. (A) Total carbohydrates in porewaters of light- (open symbols) and dark- (closed symbols) incubated mats. Circles: Jan. 27, 4.5 h incubation. Squares: June 11, 8 h light, 5 h dark. (B) DFAA (circles) and DCAA (squares) measured as primary amines in light- (open symbols) and dark- (closed symbols) incubated mats, June 23, 4 h light, 4.5 h dark.

viscous due to the extracellular gelatinous sheaths of *A. halophytica*. In some cases it was impossible to remove all the cyanobacterial cells by centrifugation or filtration. About 200 ppm total DOC was found in the porewater of the photosynthetic zone of these mats but less than 50 ppm was detected in the porewater below 1 cm depth in the sediment. Only 14.7 ppm total DOC was measured in the brines overlying the mats in the saltern.

Total dissolved carbohydrates that accumulated in and just below the photosynthetic zone of the pond 203 mats are shown in Figure 2a. There were significantly higher concentrations in the photosynthetic zone of the mats during summer as compared to winter. About 200 mg/l was found in a January sample while up to 600 mg/l was detected in the summer. The concentration decreased sharply to relatively low levels by 1 cm depth in the sediments. Surface brines had only 5-10 mg/l. When collected mats were incubated in sunlight and darkness to determine the effects on carbohydrate excretion, differences were found between the winter and summer. In the winter, higher levels of dissolved carbohydrates were detected in the photosynthetic zone of mats incubated for 4.5 h in darkness vs. those incubated at the same time under sunlight. The results are probably significant because 5 or more cores were pooled for porewater analysis at the end of the incubations. In the summer experiment using mat collected from the same site (incubations of 8 h sunlight or 5 h dark), higher concentrations were detected in the dark-incubated mat only in the top

2 mm of the photosynthetic zone. It is unknown whether the seasonal differences may be due to temperature effects on the metabolism of the microorganisms or whether there is a change in the composition of the biota that cannot be detected by microscopy.

The monosaccharide content of the porewater of the pond 203 mats was relatively low, ranging from 6 to 10 mg/l in the top 4 mm. Monosaccharides constituted about 1-3% of the total dissolved carbohydrates in these mats.

A comparison of the dissolved free amino acids (DFAA) and the dissolved combined amino acids (DCAA) in porewater profiles of the pond 203 mats is shown in Figure 2b. In summer mats incubated under sunlight and darkness, no significant differences were noted between treatments. In the photosynthetic zone, free amino acids were in much lower concentrations than total combined amino acids. In the summer mat, 2 mM glycine equivalents of DCAA were detected in the top 2 mm of the mat while 3-4 mM were found at the base of the photosynthetic zone. In a mat collected in March, only 0.7-0.8 mM glycine equivalents of DCAA were detected in the top 2 mm of mat with decreasing concentrations below (data not shown). Surface brines had 9 μm DFAA and 70 μm DCAA. Like the porewater profiles of dissolved carbohydrates, the DFAA and DCAA profiles show there were higher concentrations in the photosynthetic zone in summer.

Sulfate-reduction in the mats resulted in the accumulation of millimolar concentrations of sulfide in freshly collected sediments. Sulfate reduction also results in the accumulation of bicarbonate in the porewater which can be determined by

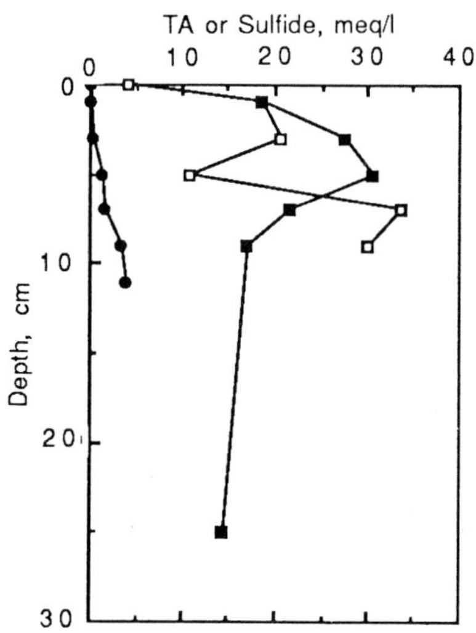

Figure 3. TA (June 11) and twice the sulfide concentration (May 12) of Pond 203 mat porewaters. (From Lazar et al., 1989). Open squares = TA, 4 h light incubation; solid squares = 4 h dark incubation; circles = sulfide.

Table 1. Thiols (μM) in porewaters of the Western Salt Co. saltern sediments (from [11]).

Depth cm	Salinity %	Sulfide (a)	Sulfide (b)	Thio-sulfate (b)	Sulfite (b)	Total thiols (b)
			CORE I			
0-1	9.6	70	0	48	0	48
1-2	9.3	169	83	51	5	139
2-3	8.9	198	153	79	8	240
3-4	8.5	328	116	152	26	293
6-7	8.2	1430	1360	270	37	1660
10-11	8.1	2620	1340	443	64	1850
14-15	8.0	2340	1980	227	36	2250
19-20	8.3	2660	2160	228	42	2430
			CORE II			
0-1	12.8	459	289	57	7	350
1-2	12.6	930	764	57	13	843
2-3	12.2	1160	681	80	11	771
3-4	11.7	713	390	130	8	528
6-7	11.0	1890	678	156	16	850
10-11	10.8	702	155	279	14	448
14-15	10.6	1740	706	236	14	955
19-20	10.3	1100	319	237	16	571

(a) Determined by the methylene blue method.
(b) Determined by the monobromobimane method.

measurements of total alkalinity (TA). Porewater profiles of TA and twice the sulfide concentration (to correspond to the TA generated by bacterial sulfate reduction) in pond 203 summer mats are shown in Figure 3. While the surface brines had only 4 meq/l TA, porewaters were very well buffered. The greatest TA was found at the base or just below the photosynthetic zone where more than 30 meq/l was detected. This is also the 1 mm-thick zone where calcite precipitates, indicating that an even greater alkalinity would accumulate at this horizon if Ca^{2+} did not precipitate it. The peak in alkalinity does not correspond to the highest concentrations of sulfide, but both biological and non-biological processes may be consuming the sulfide in this sediment horizon.

Even higher TA levels were found in the porewaters of pond 203 mats that had been incubated in Winogradsky columns for 6 months. Some layers had over 100 meq/l with correspondingly high concentrations of sulfide (60-80 mM). However, when these data were normalized to the salinity of the porewaters and compared to the

39

TA of normal seawater, it was found that TA increased to a maximum level (about 10 x the TA/salinity ratio of normal seawater) while dissolved sulfide continued to increase (data not shown). These data indicated that the missing TA could be accounted for by carbonate precipitation. X-ray diffraction analyses showed relative increases in $CaCO_3$ as calcite in those sediments.

Sulfide and other thiols were also measured in porewater profiles of two different hypersaline mats and sediments from the Western Salt Co. saltern (Table 1). These data also show that millimolar concentrations of sulfide can accumulate within hypersaline sediments. In addition, significant concentrations of sulfite and especially thiosulfate accumulated in the anaerobic zone of the sediments. In some cases, thiosulfate concentrations exceeded those of sulfide. Similar profiles taken in nearby intertidal bay sediments of seawater salinity showed up to 80 μM thiosulfate could accumulate in the anaerobic zones while 200-400 μM was found in the hypersaline sediments. Such high concentrations of this partially oxidized thiol indicate that the microbial sulfur cycle of hypersaline sediments may be qualitatively different from normal marine sediments. The sulfide values determined by the monobromobimane technique were always lower than those determined by the methylene blue technique in the strong brines. It may be that sulfide-positive particles contributed to the sulfide analysis by the colorimetric technique (methylene blue) or more likely that the high salinity inhibited the bimane reaction. Porewaters that were first diluted to seawater salinity with buffer gave correspondingly higher values, but led to a loss of sensitivity for trace thiols [11].

Antimicrobial Activity

Preliminary studies of secondary metabolites were undertaken to determine whether natural materials from solar salt ponds as well as halophilic bacterial isolates might produce antimicrobial compounds or other interesting natural products. Natural material from the Western Salt Co. ponds analyzed included microbial mats and brines from a range of salinities between 8.5% and NaCl saturation, as well as from *Artemia salina* adults and fresh cysts. Bacterial strains analyzed included those isolated from *A. salina* adults and fresh cysts as well as isolates from cysts and mats that were first treated for 10 min at 80°C to select for aerobic spore-formers. Extracts were tested for activity against *Candida albicans*, *Escherichia coli* and *Bacillus subtilis*. Several of the extracts gave positive results against *B. subtilis* (Table 2). A strain of *Marinococcus halophilus* isolated from *Artemia* cysts gave a positive result in a biological induction assay against *E. coli*. Preliminary analysis by thin-layer chromatography and proton-NMR spectroscopy indicate that some of these extracts contain potentially interesting compounds that are currently under further study.

DISCUSSION

High concentrations of primary amines and carbohydrates were found in hypersaline *Aphanothece* mats, especially in the photosynthetic zone. These compounds were subject to strong vertical zonation, seasonal differences, and in some cases daily

40

Table 2. Natural materials and bacterial isolates that showed antimicrobial activity against *Bacillus subtilis*

Sample	Description	TLC (a)	NMR (b)
Microbial mats	8.5-23% salinity, dominated by *Microcoleus, Oscillatoria, Spirulina*, or *Aphanothece*	-	-
Artemia	Extract of lysate of adults	+	-
	Extract of unwashed cysts	+	-
Strain A1b	Halotolerant motile rods isolated from *Artemia* adult	+	+
Strain 1b	Moderately halophilic, Gram-variable spore-former from 9% salinity	+	+
Strain 3a	Moderately halophilic, large yellow Gram (+) rod from 23% salinity	+	+
Strain 4a	Halotolerant Gram (+) coccus from *Artemia* cysts	+	-

(a) TLC = Thin-layer chromatography. + indicates polar fractions of potential interest detected under UV light or after charring.
(b) NMR = Proton nuclear magnetic resonance spectroscopy. + indicates potentially interesting downfield peaks in the region that aromatic and halogenated compounds give signals.

variations. Klug et al [17] also reported high concentrations of volatile fatty acids in saltern porewaters, especially acetate. Some of these low molecular weight acids could be responsible for the antibiotic activity observed in microbial mat extracts. High concentrations of these DOC compounds may be necessary for the growth of halotolerant sediment bacteria which must spend much of their biosynthetic energy on osmoregulatory processes.

It is unknown how the sulfur cycle of hypersaline sediments differs from that of normal marine sediments. Unidentified salt-tolerant, sulfide-oxidizing phototrophic bacteria were detected by microscopy and pigment analysis. Thiosulfate was a relatively important thiol in the hypersaline sediments investigated, but less so in normal marine sediments. The porewaters were well buffered by bicarbonate due to the activity of sulfate-reducing bacteria which produce 2 moles of bicarbonate for each mole of sulfate reduced to sulfide.

Initial assays for antibiotics and interesting natural products produced by halophiles are promising. Because actinomycetes have proven to be good sources of antimicrobial compounds, enrichments for these bacteria were tried but only spore-forming aerobic rods were isolated. However, these spore-formers are apparently novel halophilic bacteria and some show antimicrobial activity and the presence of potentially interesting compounds. Antibiotic activity associated with *A. salina* cysts was assayed because previous studies with marine shrimp and lobster eggs demonstrated that symbiotic surface-associated bacteria produced aromatic compounds that prevented the infection of the eggs by a fungus [18].

Artemia cysts and at least one bacterial isolate from unwashed cysts showed antibacterial activity. Yeasts and fungi, which as a group are osmotolerant, are rarely isolated from natural hypersaline environments [1]. Their lack of success could be due to antifungal compounds produced by the bacteria of the sediments and brines of these habitats.

ACKNOWLEDGMENTS

I thank the Hebrew University and the U. S. National Science Foundation for support. I thank Boaz Lazar, Russ Vetter and Bill Fenical for their collaboration and valuable discussions.

REFERENCES

[1]   B. Javor, "Hypersaline Environments: Microbiology and Biogeochemistry", Springer-Verlag, New York (1989)

[2]   J. J. Boon, Tracing the origin of chemical fossils in microbial mats: biogeochemical investigations of Solar Lake cyanobacterial mats using analytical pyrolysis methods *in* "Microbial Mats: Stromatolites", Y. Cohen, R. W. Castenholz and H. O. Halvorson, ed. MBL Lectures in Biology, Vol. 3, Alan R. Liss, New York (1984)

[3]   K. L. H. Edmunds and G. Eglinton, Microbial lipids and carotenoids and their early diagenesis in the Solar Lake laminated microbial mat sequence *in* "Microbial Mats: Stromatolites", Y. Cohen, R. W. Castenholz and H. O. Halvorson, ed. MBL Lectures in Biology, Vol. 3, Alan R. Liss, New York (1984)

[4]   P. Hirsch, Distribution and pure culture studies of morphologically distinct Solar Lake microorganisms *in* "Hypersaline Brines and Evaporitic Environments", A. Nissenbaum, ed. Developments in Sedimentology 28, Elsevier, New York (1980)

[5]   J. F. Stolz, Fine structure of the stratified microbial community at Laguna Figueroa, Baja California, Mexico: II. Transmission electron microscopy as a diagnostic tool in studying microbial communities in situ *in* "Microbial Mats: Stromatolites", Y. Cohen, R. W. Castenholz and H. O. Halvorson, ed. MBL Lectures in Biology, Vol. 3, Alan R. Liss, New York (1984)

[6]   E. D. D'Amelio, Y. Cohen and D. J. Des Marais, Association of a new type of gliding, filamentous, purple phototrophic bacterium inside bundles of

Microcoleus chthonoplastes in hypersaline cyanobacterial mats, *Arch. Microbiol.* 147: 213 (1987)

[7]   H. P. Dussault, Destruction of "red halophiles" by antagonistic bacteria, *Gaspé Fish. Exper. Stat. Note* 34: 3 (1954)

[8]   F. Rodriguez-Valera, G. Juez and D. J. Kushner, Halocins: salt-dependent bacteriocins produced by extremely halophilic rods, *Can. J. Microbiol.* 28: 151 (1982)

[9]   I. Meseguer and F. Rodriguez-Valera, Production and purification of halocin H4, *FEMS Microbiol. Lett.* 28: 177 (1985)

[10]   N. Gilboa-Garber, Direct spectrophotometric determination of inorganic sulfide in biological materials and other complex mixtures, *Anal. Biochem.* 43: 129 (1971)

[11]   R. D. Vetter, P. A. Matrai, B. J. Javor and J. O'Brien, Reduced sulfur compounds in marine environments: Analysis by high-performance liquid chromatography, *in* "Biogenic Sulfur in the Environment", E. S. Saltzman and W. J. Cooper, ed. ACS Symposium Series, v. 393, American Chemical Society, Washington, D.C. (1989)

[12]   B. Lazar, B. Javor and J. Erez, Total alkalinity in marine-derived surface brines and porewaters associated with microbial mats, *in* "Microbial Mats: Physiological Ecology of Benthic Microbial Communities", Y. Cohen and E. Rosenberg, ed. American Society for Microbiology, Washington, D.C. (1989)

[13]   J. D. Strickland and T. R. Parsons, "A Practical Handbook of Seawater Analysis", 2nd ed. Bull. Fish. Board Can. 167 (1972)

[14]   K. M. Johnson and J. McN. Sieburth, Dissolved carbohydrates in seawater. I. A precise spectrophotometric analysis for monosaccharides, *Mar. Chem* 5: 1 (1977)

[15]   K. M. Johnson, C. M. Burney and J. McN. Sieburth, Doubling the production and precision of the MBTH spectrophotometric assay for dissolved carbohydrates in seawater, *Mar. Chem.* 10: 467 (1981)

[16]   B. B. North, Primary amines in California coastal waters: utilization by phytoplankton, *Limnol. Oceanogr.* 20: 20 (1975)

[17]   M. Klug, P. Boston, R. François, R. Gyure, B. Javor, G. Tribble and A. Vairavamurthy, Sulfur reduction in sediments of marine and evaporite environments, *in* "The Global Sulfur Cycle", D. Sagen, ed. NASA Tech. Mem. 87570, Washington, D.C. (1985)

[18]   M. S. Gil-Turnes, Antimicrobial metabolites produced by epibiotic bacteria: their role in microbial competition and host defense, Ph.D. Thesis, Univ. of Calif. San Diego (1989)

TAXONOMY OF NEW SPECIES OF MODERATELY HALOPHILIC EUBACTERIA

Antonio Ventosa

Department of Microbiology
Faculty of Pharmacy
University of Sevilla
Sevilla
Spain

ABSTRACT

Moderately halophilic bacteria, i.e., those which can grow optimally in media containing 3 to 15% salts, are represented by archaeobacteria (methanogens) and eubacteria. Moderately halophilic eubacteria include a wide variety of Gram-positive and Gram-negative species; however, new groups are still to be taxonomically characterized. We present data on recent investigations that led us to propose three new taxa of moderately halophilic eubacteria. *Bacillus* sp. strain N23-2, a sporeforming Gram-positive rod with optimal growth at 15% salt and able to produce an extracellular nuclease, is included in a new species, *Bacillus halophilus*, on the basis of its peculiar phenotypic and chemotaxonomic features. Comparative studies of "*Chromobacterium marismortui*" ATCC 17056 and seven fresh isolates showed that they shared similar characteristics as well as high DNA-DNA homology ($\geq$66%), but differ from other related taxa, and we propose to place them in a new genus, *Chromohalobacter*, as *C. marismortui*. Finally, the new genus and species *Salinicoccus roseus* is proposed to accommodate a new non-motile Gram-positive moderately halophilic coccus that showed differences with previously described species with respect to its pigmentation, DNA base composition, murein type and other biochemical features.

INTRODUCTION

Moderately halophilic bacteria are defined by Kushner [1] as those that grow best in media containing 0.5-2.5 (ca. 3-15%) salt. This physiological group is represented by a wide variety of bacteria, both archaeobacteria and eubacteria [2]. The study of these microorganisms have been traditionally less attractive to the investigators than halobacteria or halococci, the extremely halophilic archaeobacteria that are also well adapted to live in hypersaline environments. Furthermore, the most interesting aspects of moderately halophilic bacteria studied by researchers were their physiological adaptations, that permitted them to live in habitats with a wide range of salt concentrations. However, very few studies were concerned with other aspects of moderately halophilic bacteria, as for example, their genetics or taxonomic distribution.

From a taxonomic point of view, moderately halophilic bacteria constitute a very heterogeneous group of microorganisms. Until recently there was a general belief that they were related to halobacteria, and speculation that they could be an intermediate evolutive group between extremely halophilic bacteria and marine or non-halophilic bacteria [3, 4]. Currently, and based on the studies of Woese and co-workers [5], it is recognized that phylogenetically the majority of moderately halophilic bacteria descend from a different branch than halobacteria. The latter belong to the archaeobacteria, together with methanogens and extreme thermophiles [5]. With the exception of three methanogenic moderately halophilic species recently described (*Methanohalophilus mahii* [6], *Methanohalophilus zhilinae* [7] and *Halomethanococcus doii* [8]) moderately halophilic bacteria are eubacteria. In the Approved Lists of Bacterial Names [9] published in 1980, only six moderately halophilic species were recognized; however, and on the basis of the many extensive taxonomic studies carried out during this decade, the number of species described increased drastically and at present there are rods, cocci and spirochetes, Gram-positive or Gram-negative, aerobic or anaerobic, and phototrophic or heterotrophic representatives. It should be pointed out that the majority of species are included in genera that group not only moderately halophilic bacteria, but also marine or non-halophilic species. the moderately halophilic eubacterial species validly published are: i) the phototrophic bacteria: *Rhodospirillum salexigens* [10], *Rhodospirillum salinarum* [11], *Ectothiorhodospira vacuolata* [12] and the recently described species *Chromatium salexigens* [13] (still to be validated by publication in the International Journal of Systematic Bacteriology), ii) the heterotrophic Gram-positive species: *Micrococcus halobius* [14], *Sporosarcina halophila* [15], *Marinococcus halophilus* [16] and *Marinococcus albus* [16]; and iii) the heterotrophic Gram-negative bacteria: *Vibrio costicola* [17], *Halomonas elongata* [18], *Halomonas subglaciescola* [19], *Halomonas halodurans* [20], *Halomonas halmophila* [21] (formerly designated *Flavobacterium halmephilum* [22]), *Deleya halophila* [23], *Haloanaerobium praevalens* [24], *Halobacteroides halobius* [25], *Halobacteroides acetoethylicus* [26], *Sporohalobacter lortetii* [27, 28], *Sporohalobacter marismortui* [28], *Spirochaeta halophila* [29], *Paracoccus halodenitrificans* [30], *Halovibrio variabilis* [31], and *Pseudomonas halophila* [31]. Until now two families have been proposed to accommodate moderately halophilic bacteria: the family Haloanaerobiaceae [32] (with the genera *Haloanaerobium, Halobacteroides* and *Sporohalobacter*) and the family Halomonadaceae [21] (with the genera *Halomonas* and *Deleya*). Besides these well characterized and validly published taxa, there are many other moderately halophilic bacteria that were isolated and studied from other standpoints but have not been studied taxonomically. Some typical examples are: "*Micrococcus varians* subsp. *halophilus*" [33], *Bacillus* sp. strain N23-2 [34], *Bacillus* sp. strain WN 13 [35],

"*Pseudomonas halosaccharolytica*" [36], *Acinetobacter* sp. strain 204-1 [37], and *Pseudomonas* sp. strain A-14 [38].

Recent studies suggest that moderately halophilic microorganisms not previously described may exist [39-43], and in fact, several new species will be proposed in the near future. We summarize here some of the more recent taxonomic studies that we have carried out in three groups of aerobic moderately halophilic eubacteria isolated from different saline environments.

NEW GROUPS OF MODERATELY HALOPHILIC EUBACTERIA

Onishi and co-workers [34] reported the isolation of a *Bacillus* sp. strain N23-2 that produced an extracellular nuclease; however, this organism has never been characterized taxonomically. We have studied this bacterium in detail and on the basis of its phenotypic and chemotaxonomic features we propose to accommodate it in a new species, *Bacillus halophilus* [44].

Bacillus halophilus is a Gram-positive rod, 0.5 to 1.0 by 2.5 to 9.0 μm, motile by peritrichous flagella and able to produce oval endospores centrally in non-swollen sporangia. It is a strict aerobe, catalase and oxidase positive and produces acid from D-glucose and other sugars. It grows between 3 and 30% salt and optimally at 15% salt. It does not reduce nitrate to nitrite nor hydrolyze gelatin, casein, starch or tyrosine. It hydrolyzes esculin and DNA. The cell wall peptidoglycan is of the directly cross-linked *meso*-diaminopimelic acid type. The main isoprenoid quinone is menaquinone with seven isoprene units. The mol% G+C content of the DNA is 51.5 [44]. To my knowledge this is the first moderately halophilic *Bacillus* species described; Weisser and Trüper [35] isolated a haloalkaliphilic *Bacillus* strain from the Wadi Natrun (Egypt) that differs from *B. halophilus* with respect to the spore shape and position, swollen sporangium, salt and pH range and optima, and utilization of glucose, maltose, ribose, sucrose and xylose. Two halophilic *Bacillus* strains recently isolated from Saudi Arabia [45] showed clear variations from *B. halophilus*, including morphological, physiological and biochemical characteristics as well as different G+C content (38 and 45 mol% for those strains and 51.5 mol% for *B. halophilus*).

On the basis of the pioneering studies that Elazari-Volcani [22] carried out in the Dead Sea, he isolated and characterized a pigmented moderately halophilic microorganism that he named as "*Chromobacterium marismortui*". However, this organism does not produce violacein nor does it have the typical flagellar arrangement of the genus *Chromobacterium*, and for those reasons was included in the 8th edition of Bergey's Manual of Determinative Bacteriology as an *incertae sedis* species [46]. Since it was not included in the Approved Lists of Bacterial Names [9], it is not currently accepted as a valid species.

In a recent comparative study of "*Chromobacterium marismortui*" ATCC 17056 and seven strains isolated from a saltern located in Alicante, Spain [47], we have found a very close relationship between all these microorganisms, and we have proposed to

47

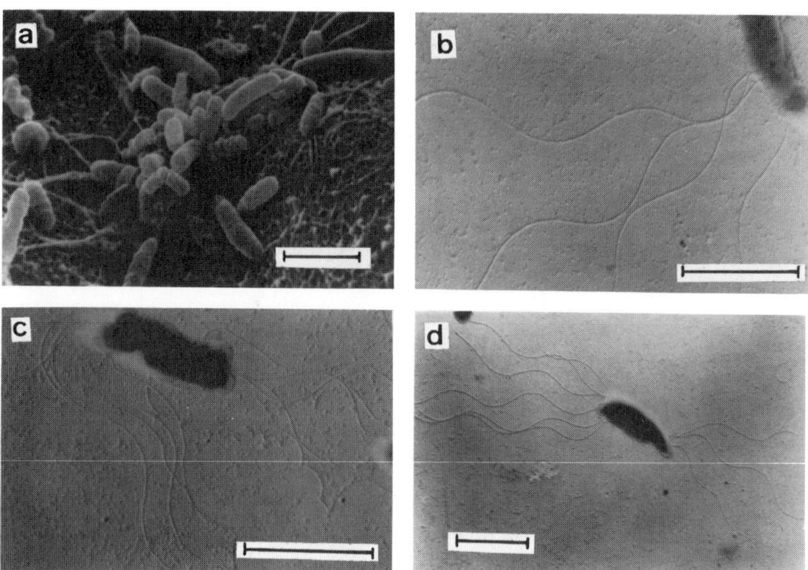

Figure 1. Microphotographs of *Chromohalobacter marismortui*. (a) Scanning electron micrograph of strain A-431; (b), (c) and (d) Transmission electron micrographs of strains ATCC 17056, A-65 and Elazari-Volcani strain, respectively. Bar = 3 μm.

include them in a new genus, *Chromohalobacter* , with the single species *C. marismortui* comb. nov., nom. rev. [48]. The phenotypic features that characterize this taxon are: they are Gram-negative rods, motile by peritrichous flagella (Figure 1). On solid media they produce colonies pigmented violet-blue to brown. They are chemoorganotrophs, strictly aerobic, showing optimal growth in media with 10% salt. Catalase positive and oxidase negative. They do not hydrolyze gelatin, casein, Tween 80, starch, esculin, DNA nor tyrosine. They produce acid from D-glucose and other sugars. DNA base composition ranges from 62.1 to 64.9 mol% G+C [48].

The DNA-DNA homology between *C. marismortui* ATCC 17056 and the seven isolates ranged between 66 and 100%, as determined by the competition procedure of the membrane filter method. However, very low levels of relatedness were obtained between this reference strain and other Gram-negative related species [48]. This new genus could be accommodated within the recently proposed family Halomonadaceae [21]. However, this should be confirmed by 16S ribosomal RNA cataloguing or other similar technique.

In a taxonomic study of Gram-positive cocci isolated from a saltern located in Alicante (Spain), we showed that the three predominant species were *Marinococcus halophilus*, *M. albus* and *Sporosarcina halophila* [49]. However, we observed other moderately halophilic Gram-positive cocci that were present in a minor proportion in this habitat. A representative of this group (strain no. 9) was studied taxonomically and the new name *Salinicoccus roseus* is proposed to name it (Ventosa et al., submitted for publication). This Gram-positive coccus grows optimally in media with 10% salt, it is

48

non-motile, non-sporeforming, pink to red pigmented, strictly aerobic and produces catalase and oxidase. It is able to hydrolyze gelatin, casein, starch, Tween 80, DNA and tyrosine. The murein is of the L-Lys-Gly$_5$ type. The DNA base composition is 51.2 mol% G+C. DNA-DNA relatedness experiments showed a very low homology with respect to other related species [14, 15, 15], and thus supported its inclusion in a new taxonomic entity.

ACKNOWLEDGEMENTS

I should like to thank Drs. M. Kocur, M. C. Gutiérrez, M. C. Márquez, M. T. García and M. Kamekura for permitting me to use some unpublished data and for stimulating discussion, and the Dirección General de Investigación Científica y Técnica and the Junta de Andalucía for financial support.

REFERENCES

[1]   D. J. Kushner, *in* "The Bacteria", Vol. VIII, C. R. Woese and R. S. Wolfe, eds., Academic Press, Orlando (1985)

[2]   A. Ventosa, *in* "Halophilic Bacteria", Vol. I, F. Rodriguez-Valera, ed., CRC Press, Boca Ratón (1988)

[3]   D. J. Kushner, *Adv. Appl. Microbiol.* 10:73 (1968)

[4]   D. J. Kushner, *in* "Microbial Life in Extreme Environments", D. J. Kushner, ed., Academis Press, London (1978)

[5]   C. R. Woese, *Microbiol. Rev.* 51:221 (1987)

[6]   J. R. Paterek and P. H. Smith, *Int. J. Syst. Bacteriol.* 38:122 (1988)

[7]   I. M. Mathrani, D. R. Boone, R. A. Mah, G. E. Fox and P. P. Lau, *Int. J. Syst. Bacteriol.* 38:139 (1988)

[8]   I. K. Yu and F. Kawamura, *J. Gen. Appl. Microbiol.* 33:303 (1987)

[9]   V. B. D. Skerman, V. McGowan and P. H. A. Sneath, *Int. J. Syst. Bacteriol.* 30:225 (1980)

[10]   G. Drews, *Arch. Microbiol.* 130:325 (1981)

[11]   H. Nissen and I. D. Dundas, *Arch. Microbiol.* 138:251 (1984)

[12]   J. F. Imhoff, B. J. Tindall, W. D. Grant and H. G. Trüper, *Arch. Microbiol.* 130:238 (1981)

[13]   P. Caumette, R. Baulaigue and R. Matheron, *Syst. Appl. Microbiol.* 10:284 (1988)

[14]   H. Onishi and M. Kamekura, *Int. J. Syst. Bacteriol.* 22:233 (1972)

[15]   D. Claus, F. Fahmy, H. J. Rolf and N. Tosonoglu, *Syst. Appl. Microbiol.* 4:496 (1983)

[16]   M. V. Hao, M. Kocur and K. Komagata, *J. Gen. Appl. Microbiol.* 30:449 (1984)

[17]   M. T. García, A. Ventosa, F. Ruiz-Berraquero and M. Kocur, *Int. J. Syst. Bacteriol.* 37:251 (1987)

[18]   R. H. Vreeland, C. D. Litchfield, E. L. Martin and E. Elliot, *Int. J. Syst. Bacteriol.* 30:485 (1980)

[19] P. D. Franzmann, H. R. Burton and T. A. McMeekin, *Int. J. Syst. Bacteriol.* 37:27 (1987)

[20] A. M. Hebert and R. H. Vreeland, *Int. J. Syst. Bacteriol.* 37:347 (1987)

[21] P. D. Franzmann, U. Weimeyer and E. Stackebrands, *Syst. Appl. Microbiol.* 11:16 (1988)

[22] O. B. Weeks, *in* "Bergey's Manual of Determinative Bacteriology", 8th ed., R. E. Buchanan and N. E. Gibbons, eds., Williams and Wilkins, Baltimore (1974)

[23] E. Quesada, A. Ventosa, F. Ruiz-Berraquero and A. Ramos-Cormenzana, *Int. J. Syst. Bacteriol.* 34:287 (1984)

[24] J. G. Zeikus, P. W. Hegge, T. E. Thompson, T. J. Phelps and T. A. Langworthy, *Curr. Microbiol.* 9:225 (1983)

[25] A. Oren, W. G. Weisburg, M. Kessel and C. R. Woese, *Syst. Appl. Microbiol.* 5:58 (1984)

[26] S. Renjipat, T. A. Langworthy and J. G. Zeikus, *Syst. Appl. Microbiol.* 11:16 (1988)

[27] A. Oren, *Arch. Microbiol.* 136:42 (1983)

[28] A. Oren, H. Pohla and E. Stackebrandt, *Syst. Appl. Microbiol.* 9:239 (1987)

[29] E. P. Greenberg and E. Canale-Parola, *Arch. Microbiol.* 110:185 (1976)

[30] M. Kocur, *in* "Bergey's Manual of Systematic Bacteriology", Vol I. N. R. Krieg, ed., Williams and Wilkins, Baltimore (1984)

[31] C. Fendrich, *Syst. Appl. Microbiol.* 11:36 (1988)

[32] A. Oren, B. J. Paster and C. R. Woese, *Syst. Appl. Microbiol.* 5:71 (1984)

[33] M. Kamekura, T. Hamakawa and H. Onishi, *Appl. Environ. Microbiol.* 44:994 (1982)

[34] H. Onishi, T. Mori, S. Takeuchi, K. Tani, T. Kobayashi and M. Kamekura, *Appl. Environ. Microbiol.* 45:24 (1983)

[35] J. Weisser and H. G. Trüper, *Syst. Appl. Microbiol.* 6:7 (1983)

[36] T. Hiramatsu, Y. Ohno, H. Hara, I. Yano and M. Masui *in* "Saline Environments", H. Morishita and M. Masui, eds., Business Centre for Academic Societies Japan, Tokyo (1980)

[37] H. Onishi and O. Idaka, *Can. J. Microbiol.* 24:1017 (1978)

[38] D. Van Qua, U. Simidu and N. Taga, *Can. J. Microbiol.* 27:505 (1981)

[39] T. N. Zhilina, *Syst. Appl. Microbiol.* 7:216 (1986)

[40] M. C. Marquez, A. Ventosa and F. Ruiz-Berraquero, *J. Gen. Microbiol.* 133:45 (1987)

[41] E. Quesada, M. J. Valderrama, V. Bejar, A. Ventosa, F. Ruiz-Berraquero and A. Ramos-Cormenzana, *Syst. Appl. Microbiol.* 9:132 (1987)

[42] H. Shiba, H. Yamamoto and K. Horikoshi, *FEMS Microbiol. Lett.* 57:191 (1989)

[43] D. Giani, D. Jannsen, V. Schostak and W. E. Krumbein, *FEMS Microbiol. Ecol.* 62. 143 (1989)

[44] A. Ventosa, M. T. García, M. Kamekura, H. Onishi and F. Ruiz-Berraquero, *Syst. Appl. Microbiol.* in press.

[45] A. A. Salamah, *J. Univ. Kuwait (Sci.).* 15:313 (1988)

[46] P. H. A. Sneath, in "Bergey's Manual of Determinative Bacteriology", 8th ed., R. E. Buchanan and N. E. Gibbons, Williams and Wilkins, Baltimore (1974)

[47]   A. Ventosa, E. Quesada, F. Rodriguez-Valera, F. Ruiz-Berraquero and A. Ramos-Cormenzana, *J. Gen. Microbiol.* 128:1959 (1982)

[48]   A. Ventosa, M. C. Gutiérrez, M. T. García and F. Ruiz-Berraquero, *Int. J. Syst. Bacteriol.* 39:382 (1989)

[49]   A. Ventosa, A. Ramos-Cormenzana and M. Kocur, *Syst. Appl. Microbiol.* 4:564 (1983)

PHENOTYPIC CHARACTERIZATION OF HALOPHILIC BACTERIA

FROM GROUND WATER SOURCES IN THE UNITED STATES

Russell H. Vreeland and J. H. Huval

Department of Biology
West Chester University
West Chester, Pennsylvania 19383
United States of America

ABSTRACT

Many areas of the U. S. mainland contain extensive underground salt formations. These formations are large bedded evaporite deposits which were formed when prehistoric seas became landlocked and dried. Several of these ancient salt beds have now been penetrated by meteoric waters which are slowly solubilizing the salts. Microbiological sampling of the brines being produced revealed that these artesian brines contain a wide variety of halophilic bacteria.

The isolates included moderate to extremely halophilic and euryhaline bacteria. Numerical taxonomic characterization showed that several of the microorganisms isolated from a portion of the Permian basin belonged to the family *Halomonadaceae*. Isolates obtained from the oldest portion of this basin, the so called Hennessey formation, never grouped together with bacteria obtained from one of the younger parts of the basin.

In addition to the various eubacteria, five extremely halophilic bacteria were isolated from these subterranean brines. The taxonomic analysis of these bacteria showed that they had a surprisingly low level of similarity to the *Halobacterium*. The levels of similarity were in fact so low (less than 70% S_j) that there is reason to believe that these bacteria represent a new taxonomic group.

More intensive taxonomic characterization of all of these organisms is now underway. The results of the numerical examinations will be discussed in terms of the evolution of these bacteria and significance of their presence in subsurface salt formations.

General and Applied Aspects of Halophilic Microorganisms
Edited by F. Rodriguez-Valera, Plenum Press, New York, 1991

53

Over its many eons of existence the North American land mass has frequently been inundated by ocean waters. This is particularly true of the land areas located between the two great mountain ranges (the eastern Appalachians and western Rockies). As the land rose to its present elevation these ancient seas became land locked and slowly dried out depositing huge salt beds which were often several hundred meters thick and which today outline the final positions of these seas (A in Figure 1). Over time these large evaporite deposits were covered by sedimentary and consolidated rock layers. Some, such as those in the southern U.S. (formed 100 million years before present B.P.) (Figure 1) were exposed to severe geological forces and were converted into salt domes. Others such as the Permian basin in the midwestern U.S. (B in Figure 1) have either been penetrated by meteoric waters or have remained in a relatively pristine condition. Microbiological samples obtained from the southern salt domes and from the Permian basin contained a wide variety of halophilic bacteria. In some cases these subterranean brines contained extremely halophilic bacteria which were probably not present in the meteoric fresh water that became brine as it moved through the formations.

The existence of living microorganisms in underground salt formations was suggested prior to the onset of this study [1, 2]. Unfortunately the combination of sampling errors and the fact that the researchers concentrated on non-halophilic bacteria stimulated numerous questions regarding the validity of their work. This issue has not however died completely and the possible existence of bacteria locked into salt crystals has been recently suggested [3, 4].

The bacteria described in this report were isolated in 1985 from brine present in a salt dome in Louisiana and from brines flowing out of the Permian basin in Oklahoma. The salt dome brine sample was obtained from the now abandoned salt

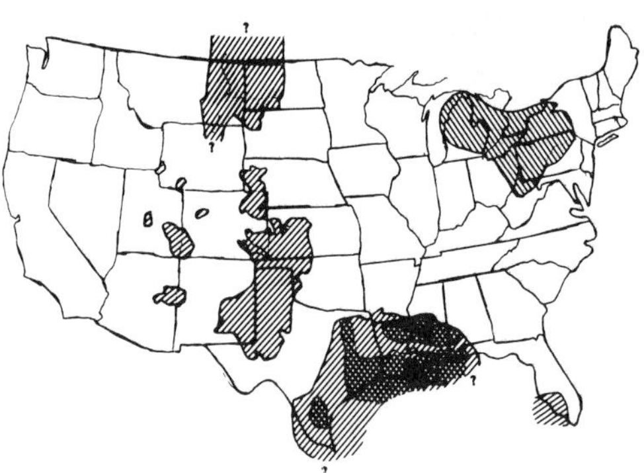

Figure 1.   Map of the U.S. mainland showing the location of the major underground salt deposits. Striped areas are bedded deposits, crossed hatched regions indicate salt domes.

mine located under Jefferson Island Louisiana. This mine was flooded in 1980 when the drill bit from an oil drilling operation accidently punched through the side of an upper chamber of the mine [5]. This accident allowed fresh water from Lake Peignur to flood the mine. During this catastrophe approximately 1.3×10^7 M^3 of water (NaCl <1%) filled the mine cavity. Since this catastrophe the mine has been closed and all salt recovery has been abandoned. The Permian samples arose from two different regions of this basin. One sample obtained from the Acme salt brine company in southern Oklahoma contained salt from the 100 million year old Flowerpot formation and the second which was obtained in Great Salt Plains National Park in northern Oklahoma contained 250 million year old salt from the Hennesey formation [6]. The water that forms these particular brines originates as meteoric rain water which percolates into the ground and enters the formation in eastern Colorado. The water that forms the brine is significantly younger than the solubilized salt [6].

The samples obtained from the brines containing Permian salt invariably contained numerous halophilic bacteria. The bacteria in these brines ranged from moderate to extreme halophiles and euryhaline bacteria [7]. The moderate and euryhaline forms were present in all the brines while the extreme halophiles were only present in the very concentrated brines. This paper describes the results of the first phenetic characterizations of the various halophiles isolated from the brines. The data show that while many of the euryhaline isolates resemble the salt tolerant bacteria isolated from traditional surface samples the extremely halophilic forms show some distinct and intriguing differences when compared to some other extreme halophiles.

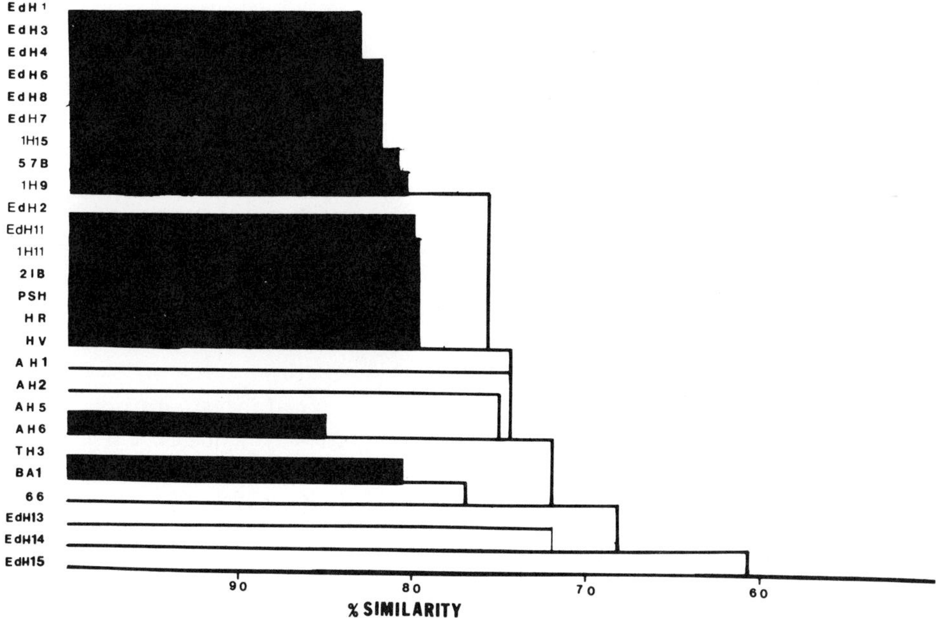

Figure 2.   Phenogram (Sj) showing the similarity between the euryhaline bacteria isolated from groundwater and those isolated from more conventional environments.

RESULTS AND DISCUSSION

Figure 2 shows the results of the computer analysis of the taxonomic data developed for the euryhaline bacteria isolated from the Permian salt brines. The computer comparison showed that these cultures could be distributed into four phenons distinct at the 79% similarity level (S_j).

This would indicate that these bacteria probably represent separate species of the Halomonadaceae family. All of these bacteria were rod shaped with rounded ends which occurred singly, in pairs and in chains. All of these bacteria are Gram negative, aerobic, catalase positive, grew optimally at pH's from 7-9 at 30°C. These euryhaline organisms produced white to cream coloured opaque colonies on complex media. None of the bacteria produced acid or gas from mannose, sucrose, lactose or gluconic acid. These bacteria did not produce hydrogen sulfide, and none gave positive reactions for the Voges-Proskauer or methyl red tests. Antibacterials such as valinomycin, aphidicolin and monensin did not inhibit the growth of any of these organisms.

The largest of the phenons (Phenon A, Figure 2) produced in this study contained the reference strain *Halomonas elongata* and its biotype [8]. In addition the group included 6 isolates from Enid Oklahoma and one organism obtained from the Pacific Ocean. These bacteria produced umbonate, smooth, glistening colonies on solid complex media containing 8% salt. They do not grow on any media below pH 6, will grow in a candle jar, or in thioglycolate and are not sensitive to penicillin, bacitracin, tetracycline or streptomycin. The bacteria in this group utilized aesculin, starch, gluconic acid, mannose, lactose, sucrose and glycerol as their sole source of carbon and energy in defined media.

The second bacterial phenon (Phenon B, Figure 2) included the reference strain *H. halodurans* and its biotype [9] as well as five other strains, two each from Enid and Great Salt Lake and one isolate from the Pacific Ocean. The properties of this phenon are similar to those of Phenon A except that these organisms were sensitive to penicillin and ampicillin and did not reduce nitrate or utilize cellobiose as readily as did the bacteria in Phenon A. In addition these bacteria produced larger colonies on solid complex media and most produced a golden pigment. The two phenons A & B were similar to each other at the 75% S_j level.

The third phenon produced by this data contained two strains isolated from the Acme Salt Company and did not include any reference organisms. These bacteria were sensitive to ampicillin, were oxidase positive, and decarboxylated lysine. These organisms did not utilize mannose or oxidize cellobiose. They were similar to the other two phenons at the 72% S_j level.

The last euryhaline phenon (Phenon D, Figure 2) contained three cultures one of which was the Dead Sea isolate known as Ba_1, one isolate which was obtained from a Permian salt brine in Texas and one Pacific Ocean isolate. These strains were considerably different from the other phenons described here. The organisms produce golden colonies with dull, rough surfaces both will grow at pH 5 and in a candle jar.

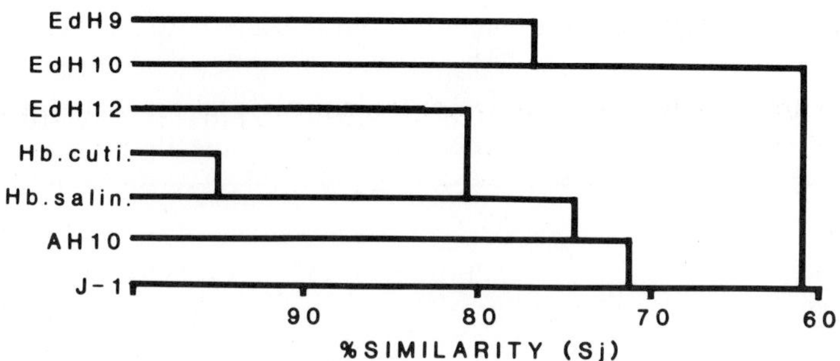

Figure 3. Phenogram (Sj) showing the similarity between extreme halophiles isolated from U.S. groundwaters.

The cultures grow only at the surface of thioglycolate broth, reduce nitrate to nitrite, produce indole from tryptophan, decarboxylate ornithine and utilize mannose, lactose, sucrose, glycerol, aesculin, cellobiose, starch and gluconic acid. The members of this phenon were resistant to streptomycin, tetracycline and bacitracin. They gave negative responses for urease and lysine decarboxylase. This phenon joined the other three phenons at a similarity level of 72% S_j.

The presence of euryhaline bacteria in these subsurface brines is an interesting but not totally unexpected occurrence. Basically bacteria of this type would be able to enter the meteoric fresh waters as these waters percolate into the formations. Then because of their halotolerance they could adapt to the increasing salt concentrations and travel through the formations. The same argument could not be used to explain the presence of extremely halophilic archaeobacteria in the concentrated brines.

During the course of this work several concentrated brine sources were sampled. These particular brines contained anywhere from 20 to 32% (w/v) NaCl and all contained at least some extreme halophiles. Figure 3 shows the results of the numerical taxonomic comparison of these bacteria with two well known *Halobacteria*. While this comparison is not by any means complete at the time of this writing the results are very interesting. First Figure 3 shows that all of these isolates are distinctly different from the classical "*Halobacterium cutirubrum*" and *Halobacterium salinarium*.

This in itself may not be exciting, what is exciting is the fact that while these halophiles are also very different from each other these differences correspond to the ages of the formations from which the brines arose. The cultures labeled with the EdH prefix arose from the brines containing 250 million year B.P. Hennesey salts, while the AH culture came from the 150 million year B.P. Flower Pot formation and J-1 came from the brines in the flooded salt dome (ca. 100 million years B.P.).

The other truly exciting aspect of these results is that all of these organisms are classic extreme halophiles which lyse in salt concentrations below 15 to 20%, grow optimally in NaCl concentrations above 20% (w/v) and produce bright red to orange pigments (Table 1). All of these bacteria also have typical halobacterial ether linked

Table 1. Characteristics of the extremely halophilic bacteria isolated from salt brines.

Characteristics	EdH10	EdH9	EdH12	AH10	JI-1
Gram Rxn.	-	-	-	-	-
Optimum T°C	37-45	37-45	37-45	37-35	37-45
Optimum NaCl	25	25	25	25	30
Minimum NaCl	20	20	20	20	20
Lysis in Hypotonic solutions +	+	+	+	+	
Growth at pH 4.0	-	-	-	-	+
Growth at pH 9.0	+	+	-	+	+
Red/Orange Pigment	+	+	+	+	+
Oxidase	-	-	-	-	+
Catalase	+	+	+	+	+
Urea Hydrolysis	-	-	-	-	-
Starch Utilized	+	+	+	-	+
Glycerol Utilized	+	+	-	-	-
Gluconic acid Utilized	-	+	-	+	+
Acid from Sugars	-	-	-	-	+
Nitrate Reduction	-	-	-	-	-
Bacitracin Sensitive	+	+	-	-	-
Ether Lipids Present	+	+	+	+	+

lipids although the lipid patterns of EdH9 and EdH10 are quite different from those of other known halophiles (W. D. Grant, personal communication).

As stated earlier the exact taxonomic position of these halophilic archaeobacteria has yet to be determined and cannot be answered in this manuscript. It is however apparent that these organisms are true extreme halophiles. What is not aparent from the available data is how these bacteria got into the brines in the first place.

The obvious mode of entry would be the original waters that entered the mine and formations. However tests conducted during this taxonomic work show that all of these extreme halophiles lyse in either fresh water or brines containing less than 15% NaCl. In fact the field samplings showed that these bacteria were only present in the concentrated brines. This was especially evident in the Enid Ok. samples where we had access to two sites that were less than 1/4 mile apart and were each discharging brine from the same stratigraphic layer. The only differences between these two samples was the fact that one brine contained 25% NaCl and had extreme halophiles and halotolerant bacteria while the second brine contained 8% NaCl and yielded only halotolerant isolates. In the case of the flooded salt mine (isolate J-1, Figure 3) the lake that flooded the mine contained less than 1% salt which could not support this particular bacterium (Table 1).

A second mode of entry would be from surface contamination of the shafts and pipes containing the brines sampled. Since the areas sampled are open to the atmosphere this source of entry cannot be ruled out at present. What can be said in this regard is that none of the environments studied to date have any hypersaline lakes which contain extreme halophiles and which could serve as a contamination source. In fact in the case of the Louisiana sample the amount of rainfall, the geology of the region, and its topography actually preclude the development of hypersaline systems.

A third possibility is that these organisms arose through a series of mutational events which caused a eubacterium to become an extreme halophile. Given the rather considerable molecular differences existing between the archaeobacteria and eubacteria it is difficult to see how any series of mutational events would convert one into the other. Further all of these extreme halophiles arose from samples that would have been dark yet all of them have bright red carotenoid, light absorbing pigments. Finally, in the case of isolate J-1 these mutational events would have had to occur within a five year period (the elapsed time between the flooding and sampling events) which would mean that this organism's mutation rate far exceeded anything known to date.

The only other obvious source for these bacteria would be the salt crystals, or perhaps the brine inclusions in the crystals [3]. This would of course imply that these bacteria were either locked into the crystals at the time of formation or were somehow trapped in the connate brines located between the crystals. This would be an exciting possibility since it would mean that these bacteria were present at the time these salts were deposited and had somehow survived the ensuing millenia.

At this time our lab is engaged in a project to examine the microbiology of a part of the Permian Basin known as the Salado formation. This particular part of the Permian has been studied extensively by geologists and geochemists [10, 11, 12]. The formation is 250 million years old, has never been penetrated by meteoric waters and has never been disrupted by geologic forces. This study site is basically pristine and exists today in the same way it was deposited. It is our hope that these on-going studies will help answer some of the questions raised during the preliminary studies described in this report.

REFERENCES

[1]   H. J. Drombrowski. *Second Symp. on Salt* 1:215 (1961)
[2]   P. Tasch, *Univ. of Wichita Bull.* 39:3 (1963)
[3]   C. F. Norton and W. D. Grant, *J. Gen. Microbiol.* 134:1365 (1988)
[4]   E. Roedder. *Amer. Mineralogist* 69:413 (1984)
[5]   M. Gold. *Science 81*, Nov:57 (1981)
[6]   L. Jordan and D. L. Vosburg. *Okla. Geol. Survey Bull.* 102:76 (1963)
[7]   R. H. Vreeland, *Critical Rev. Microbiol.* 14:311 (1987)
[8]   R. H. Vreeland, C. D. Litchfield, E. L. Martin and E. Elliot, *Int. J. System. Bacteriol.* 30:485 (1980)

[9]   A. M. Hebert and R. H. Vreeland, *Int. J. System. Bacteriol.* 37:347 (1987)
[10]   D. J. Borns and J. C. Stormount, *Sandia Report* 87-1375:1 (1988)
[11]   C. L. Stein, *Sandia Report* 85-0321:1 (1985)
[12]   C. L. Stein, *Sandia Report* 83-0451:3 (1985)

PART II

PHYSIOLOGY

HALOPHILES OF ALL KINDS: WHAT ARE THEY

UP TO NOW AND WHERE DO THEY COME FROM?

Donn J. Kushner
Department of Microbiology
University of Toronto
Toronto, ON M5S 1A8
Canada

ABSTRACT

The author presents some personal history of his involvement with halophilic bacteria, including his current work on moderate halophiles, and much older accounts and speculations on the history and origin of certain extreme halophiles.

INTRODUCTION

This is a combined account of an opening plenary lecture on historical aspects of studies of halophiles, and a talk given in the physiology section of the meetings. Both dealt with past work that includes major contributions by those attending this meeting (reviewed, when not specifically cited, in References 1-3).

MORE OR LESS RECENT HISTORY

My own acquaintance with the world of halophiles began in 1961, when I joined the group of Dr. N. E. Gibbons at the National Research Council of Canada in Ottawa. Dr. R. M. Baxter was still in this laboratory. His and Dr. Gibbons' suggestion [4] was probably the first that the very high salt requirement of most enzymes of extreme halophiles was caused by their acidic nature; that is, that the high concentration of cations was needed to prevent mutual repulsion between negative charges on the protein molecules.

The acidic nature of halophilic proteins has been very well documented [5, 6], and indeed, the ability of malic dehydrogenase and other dehydrogenases to bind very large amounts of Na^+ and K^+ ions is probably due to strategically located negative groups on the outside of these proteins [5]. Dr. J. Lanyi, however, has pointed out that the action of salts is also due to their maintaining hydrophobic bonds in proteins, whose proportion of non-polar amino acids is much lower than the proteins of non-halophiles [6].

For some years, it seemed as if red halophiles had limited metabolic capabilities: though rods and spheres had been found and named all isolates of each form seemed much alike [7]. We know now that isolation of bacteria that form red colonies at high salt concentrations and rich media almost guarantees that a limited number of species will be found, likely only two, *Halobacterium salinarium* and the closely related or identical *H. halobium* and *H. cutirubrum* and *Halococcus morrhuae*, all of which require a number of amino acids for growth.

We now know, through the work of Dr. L. Hochstein and his colleagues and of the Spanish hosts of this conference, that many halophilic archaeobacteria can grow using such simple carbon sources as glucose or succinate, and NH_4^+ as nitrogen source [8, 9]. As will appear in this conference, such organisms are well suited for genetic studies.

Another memory is of the question I posed to Dr. S. T. Bayley, some 28 years ago, about the nature of the ribosomes of halobacteria: whether the ribosomal proteins and nucleic acids could really stay together in the high salt concentrations they were exposed to. It turned out that the ribosomes <u>required</u> high salt concentrations, specifically high KCl concentrations, to remain intact, and that in low salt concentrations the ribosomal particles fell apart, losing most of their proteins [10]. This observation was the start of other studies on *in vitro* protein synthesis by extracts of halobacteria, work that is being continued today in the laboratory of Dr. Amils and others.

My long association with Dr. M. Kates was partly responsible for first bringing me to Ottawa. It has been gratifying to see that the ether linked lipids, which he first described in the halobacteria, have now proved, with some variations, to be the basic lipids for all the archaeobacteria [11]. Indeed, the halobacteria are probably being studied now more in virtue of their unusual biochemical properties, including the possession of bacteriorhodopsin and related pigments, and for their position among the archaeobacteria, than for their halophilic nature.

The halophilic eubacteria have attracted less attention than the halophilic archaeobacteria, but they do pose some unusual problems. One of these is the <u>range</u> of salt concentration over which they can grow, say 0.5 - 3.5 M NaCl for *Vibrio costicola* compared to 3 - 5 M NaCl for *Halobacterium salinarium*. This has raised the question, what is the mechanism of adaptation to such different salt concentrations? Will, for example, different proteins be made at the lowest and highest salt concentration in the growth range? Thus far there is little evidence that this happens. To some extent, indeed, aerobic halophilic eubacteria behave as if they have high salt concentrations on the outside and low salt concentrations on the inside.

This statement is based on the general observation that intracellular enzymes of several aerobic halophilic eubacteria are inhibited by the levels of salt (NaCl or KCl) concentration that measurements of the cell-bound cations suggest are present inside the cell. One partial explanation for this is that, in fact, these bacteria exclude Cl⁻ ions, so that their concentration inside is 1/10 or less the concentration outside. It now seems that Cl⁻ ions are toxic for a number of enzymatic reactions and other biological processes, in aerobic halophilic eubacteria and, likely, in non-halophilic eubacteria, which also exclude these ions.

In contrast, halophilic anaerobic eubacteria, as well as halophilic archaeobacteria, have about as high internal as external Cl⁻ concentrations [1, 12].

We have studied the site of toxicity of Cl⁻ ions in the protein-synthesizing system, poly-uridylic acid (polyU)-directed incorporation of phenylalanine of the moderate halophile, *V. costicola*. Together with Dr. M. Kogut, who introduced us to studies of *in vitro* protein synthesis, we had shown earlier that this system functioned only in <u>low</u> salt concentrations [13]. The reason for this puzzling result was that we were using the wrong anion. In this system, protein synthesis was completely inhibited by 0.6 M NaCl but was slightly stimulated by 0.6 M Na-glutamate [14]. This was also true for a similar *in vitro* protein synthesizing system of an unidentified moderate halophile, NRCC 41227, (M. Kamekura and D. J. Kushner, unpublished). Density-gradient experiments with the *V. costicola* system showed that Cl⁻ ions prevented the attachment of ribosomes to the artificial messenger, polyU. Glycine betaine (betaine) and glutamate, both of which are present inside cells of *V. costicola*, partly protect *in vitro* protein synthesis, and also the attachment of ribosomes to polyU, from the toxic action of Cl⁻ ions [15].

Studies of the action of Cl⁻ and other ions might be more meaningful if carried out on a more natural *in vitro* protein synthesizing system. Although no natural mRNAs of *V. costicola* (or of any halophilic eubacteria) have been isolated, we found that the RNA of the R_{17} bacteriophage of *Escherichia coli* can serve as a messenger for *in vitro* protein synthesis by *V. costicola*, as measured by the incorporation of ^{14}C-valine. This incorporation requires the phage RNA, a cellular supernatant remaining after centrifuging at 150,000 x G, and an energy source. Furthermore, it was stimulated by fMet-tRNA, which is used in the initiation of protein synthesis. During incubation, a protein was formed with the same electrophoretic mobility as the R_{17} coat protein (which is rich in valine), the major gene product of this RNA. The endogenous mRNA system of *V. costicola*, whose sensitivity to salts has been previously described [13], was also studied. The endogenous system uses the natural RNAs present in the cytoplasm and is measured by the incorporation of a mixture of amino acids.

The antibiotics, kasugamycin and aurin tricarboxylic acid, which inhibit initiation, had a much stronger effect on the R_{17} system than on the endogenous one, while chloramphenicol and neomycin which affect polypeptide elongation, inhibited both systems equally. Thus, the R_{17} system seemed to require initiation, while in the endogenous system many of the ribosomes are apparently already involved in protein synthesis.

Both systems were strongly stimulated by NH_4^+ ions and inhibited by Cl^- ions. The R_{17} system was especially sensitive to Cl^- ions, and functioned over a narrower range of salt concentration than the endogenous system; both showed optimal activity in 200-300 mM ammonium glutamate. Betaine stimulated both systems and partly counteracted the toxic effects of Cl^- ions.

The R_{17} system has permitted us to study initiation factors and their response to salts. Initiation factors can be removed from *E. coli* ribosomes by the action of 1 M NH_4Cl, but this NH_4Cl concentration has no effect on the ribosomes of *V. costicola*; however, initiation factors are removed from the latter ribosomes by 3.0 - 3.5 NH_4Cl. Higher concentrations of this salt will inactivate the initiation factors, through an effect on the factors themselves, rather than on the ribosomes. In the R_{17} system, high Cl^- ions prevent the attachment of initiation factors to fmet-tRNA, but not to the R_{17} RNA; the former effect is partly reversed by betaine [16].

In addition to stimulating and protecting protein synthesis from the action of Cl^- ions, betaine has previously been shown to stimulate active transport by *V. costicola* at high NaCl concentrations [17]. We have now studied the pathways of its synthesis in *V. costicola*, in relation to the external NaCl concentration. We observed, as have others [18], that increasing the NaCl concentration in a complex medium (in our case, 1% proteose peptone + 1% tryptone) also increased intracellular betaine concentration. For cells grown in the presence of 0.5 M NaCl, incubation in the presence of 1.0 M NaCl, in a complex medium, caused an increase in intracellular betaine (from 0.07 M to 0.62 M). This increase was inhibited about 90% by chloramphenicol, implying that a protein(s) must be synthesized in response to the higher NaCl concentration, likely one involved in transport of betaine from the complex medium into the cell. Since chloramphenicol has little effect on the development of salt-tolerant transport when *V. costicola* is exposed to higher salt concentrations [17], these experiments suggest that betaine itself is not the sole agent responsible for such salt-tolerant transport, though it may be involved.

When cells were grown in a synthetic medium, which contained choline, we found that choline was transformed to betaine through the action of choline dehydrogenase, located on the cells' membranes followed by the action of betaine aldehyde dehydrogenase, a cytoplasmic enzyme Both of these enzymes could function in fairly high concentrations of NaCl; the first enzyme was stimulated by increasing NaCl concentrations. Formation of the enzymes seemed dependent on the presence of choline in the medium and was stimulated by increased NaCl concentrations.

STILL LESS RECENT HISTORY

I will close by returning to the more glamorous halophilic archaeobacteria. Many serious studies are being carried out on the evolution of these creatures, on their relation to other bacteria, including other archaeobacteria. Some will be discussed in this meeting. But some tribute should be paid to the way in which the halophilic archaeobacteria have stimulated the imagination of those who have studied and loved them.

Halophilologists are proud of the long history of their favourite microorganisms. Baas-Becking's classic article refers to a Chinese treatise, possibly as much as 5000 years old, that describes the red colour of solar salterns, a colour that we now know is due mainly to halophilic archaeobacteria [19]. Dr. B. Volcani's description of these organisms in the Dead Sea [20], and the active studies of other scientists in Israel have given an almost Biblical air to some descriptions of red halophilic archaeobacteria in their natural environments. And from somewhere or other, the following simple tale has sprung:

The First Halophile

Everyone knows how God destroyed Sodom and Gommorah, and all the Cities of the Plain; how Lot, Abraham's nephew, with his wife and two daughters, fled from the city before it was destroyed; how all the fugitives obeyed the angels' orders not to look back on the burning city, except for Lot's wife, who was turned into a pillar of salt.

But no one knows, until now, of a conversation that took place in Heaven on that same day.

"What is that dreadful noise?" The speaker was the Archangel Basil, who had been put in charge of biological classification and who had, thus far, found it a heavy job. He looked over the parapets of Heaven from his rapidly growing list of earthly creatures.

"You can't expect me to leave here without saying goodbye to _anyone_!" Mrs. Lot's voice rose through the ether, clearer than it had ever sounded on earth.

"Mmh," said her husband and mumbled something.

"So, they're sinners," Mrs. Lot said, apparently in reply to his statement. "No one's perfect!"

"Ecch!" her husband replied.

"And where will I find salt? Don't you realize that most places don't have any?"

The Archangel Basil heard a chuckle behind him. "_That_ one's going to cause trouble," a mild, polite voice remarked. It was Little Vishnik, one of Vishnu's many manifestations, a visiting god from a neighbouring heaven.

The Archangel looked up from his list of good beasts. "Humans always do," he said.

"And what about my clothes?" Mrs. Lot cried. "My red beehive coat: where will I find one like that?"

"Yichh!" her husband said.

"A beehive coat?" The archangel Basil rolled his eyes at Little Vishnik, then glanced uneasily at the Lord who was still sleeping behind his great beard. When God awoke, His wrath was usually terrible.

"She means a honeycomb pattern," Little Vishnik explained. "A regular hexagon, with each spot surrounded by six others."

The Archangel closed his eyes. "Don't bother me with mathematics too. My computer handles all my calculations."

"You say I can always get another!" the angry voice cried from below. "Do you think those Samarian peasants know how to sew? And what would they do with the purple patches? They'd make me look like a flag-pole!"

"Such vanity." The Archangel Basil shook his head primly.

"Oh no," Little Vishnik said, "It is a very subtle arrangement of purple that appears best on dull, airless days; very pretty indeed."

The Archangel glanced at the Lord. "<u>He</u> won't like it."

"No, He won't," Little Vishnik agreed cheerfully, "I sometimes wonder what your boss <u>does</u> like."

No answer came to this question. Instead, from below, there was a piercing but soft and mysterious song.

Little Vishnik smiled. "Does she do that often?"

"Not often enough," the Archangel sighed. Then he corrected himself: "That is, we still haven't figured out the meaning of her song."

"Of course not," Little Vishnik agreed, happily. "Perhaps you never will."

The singing stopped. "So your Uncle Abraham made a deal that if there were five righteous ones the city would be spared!" Mrs. Lot cried. "Big deal! What about all the rest of us, who were too busy cooking and cleaning to be sinners? And what about the children?"

At last her husband's voice could be heard. "Blasphemy!" he cried.

Up in Heaven, Little Vishnik glanced at the sleeping Lord God Jehovah. "It's lucky He can't hear her."

The Archangel nodded. "But He'll wake up soon. Then it's bye-bye time for Sodom." He frowned sadly. Even minor gods and Archangels are not comfortable with humans who pose difficult questions.

"What will happen to her?" Little Vishnik asked.

The Archangel Basil pushed the keys of his computer. "It's all arranged," he announced. "She'll run with her family, but of course she'll turn back to see what's happening to her city. Her body will become a pillar of salt."

"And her spirit?" asked Little Vishnik.

The Archangel frowned. "What would _you_ do with it?"

"In _our_ heaven, the spirit often migrates to a lower animal."

"A camel, perhaps? the Archangel suggested.

"I don't advise it. Have you ever seen a truly angry camel?"

"A jackal? A rat? A spider? A gadfly?"

Little Vishnik, who had shaken his head at each of his companion's suggestions, paused for an instant at the gadfly. "No," he said regretfully, nodding at the sleeping Lord. "He'd swat it in a second."

"I wouldn't want that either," the Archangel agreed.

Mrs. Lot cried "And we won't have any salt there either, I'm sure of it!"

"Is there no other creature on your list?" Little Vishnik asked.

The Archangel looked at his list again, rolling it up from the clouds below. "None of them seems suitable."

"You'll have to make a new one, then," said Little Vishnik firmly.

The Archangel cheered up, "Why not?"

"It should ask questions we can't answer," said Little Vishnik."

"Absolutely!" The Archangel keyed this into his computer. "And it should like salt."

"Very much," Little Vishnik agreed. "But it should be very small."

"Quite." The Archangel pushed another key. "So small as to escape His notice."

"It should have a coat, with a honeycomb pattern and purple patches."

"Why not?" said the Archangel, pushing another key. "Should it move?"

"It should certainly swim."

"Right," the Archangel said, "Anything else?"

"It should like salt."

"I've already said that." But the Archangel pushed another key. "Now it will like salt <u>very much</u>. Should it sing?"

"One of these days, perhaps." Little Vishnik winked.

The Archangel pushed more keys. "Now it's all arranged."

The two heavenly beings beamed at each other, well pleased with the new creature they had designed. They were especially pleased that it had been programmed into the celestial computer, whose circuits cannot be altered Now they waited with pleasant anticipation for the Lord to awake and turn His bright, burning eyes on the doomed Cities of the Plain.

REFERENCES

[1]   D. J. Kushner, Life in high salt and solute concentrations: halophilic bacteria, *in* D. J. Kushner, ed, "Microbial Life in Extreme Environments", Academic Press, London (1978)

[2]   D. J. Kushner, The Halobacteriaceae, *in* C. R. Woese and R. S. Wolfe, eds, "The Bacteria", Vol. 8, Academic Press, Orlando (1985).

[3]   F. Rodriguez-Valera, ed. "Halophilic Bacteria" in 2 volumes, CRC Press, Boca Raton (1988)

[4]   R. M. Baxter and N. E. Gibbons, Effects of sodium and potassium chloride on certain enzymes of *Micrococcus halodenitrificans* and *Pseudomonas salinaria, Can. J. Microbiol.* 2: 599 (1956)

[5]   H. Eisenberg and E. J. Wachtel, Structural studies of halophilic proteins, ribosomes and organelles of bacteria adapted to extreme salt concentrations, *Ann. Rev. Biophys. Biophys.. Chem.* 16: 69 (1987)

[6]   J. K. Lanyi, Salt dependent properties of proteins from extremely halophilic bacteria, *Bacteriol. Rev.* 38: 272 (1974)

[7]   R. R. Colwell, C. D. Litchfield, R. F. Vreeland, L. A. Kiefer and N. E. Gibbons. Taxonomic studies of red halophilic bacteria, *Int. J. Syst. Bacteriol.* 29:379 (1979)

[8]   L. I. Hochstein, The physiology and metabolism of the extremely halophilic bacteria, *in* "Halophilic Bacteria", Vol. 2, F. Rodriguez-Valera, ed, CRC Press, Boca Raton (1988)

[9]   G. Juez, Taxonomy of extremely halophilic archaebacteria, *in* "Halophilic Bacteria", Vol. 2, F. Rodriguez-Valera, ed, CRC Press, Boca Raton (1988)

[10]   S. T. Bayley and D. J. Kushner, The ribosomes of the extremely halophilic bacterium *Halobacterium cutirubrum*, *J. Mol. Biol.* 9: 654 (1964)

[11]   M. Kamekura and M. Kates, Lipids of halophilic archaeobacteria *in* "Halophilic Bacteria", Vol. 2, F. Rodriguez-Valera, ed, CRC Press, Boca Raton (1988)

[12]   D. J. Kushner and M. Kamekura, Physiology of halophilic eubacteria *in* "Halophilic Bacteria", Vol. 1, F. Rodriguez-Valera, ed, CRC Press, Boca Raton (1988)

[13]   R. M. Wydro, M. Madira, T. Hiramatsu, M. Kogut and D. J. Kushner, Salt-sensitive *in vitro* protein synthesis by a moderately halophilic bacterium, *Nature (Lond).* 169: 824 (1977)

[14]   M. Kamekura and D. J. Kushner, Effect of chloride and glutamate ions on *in vitro* protein synthesis by the moderate halophile *Vibrio costicola. J. Bacteriol.* 160:385 (1984)

[15]   C. G. Choquet, M. Kamekura and D. J. Kushner, *In vitro* protein synthesis by the moderate halophile *Vibrio costicola*: site of action of Cl⁻ ions, *J. Bacteriol.* 171: 880 (1989)

[16]   C. G. Choquet and D. J. Kushner, Use of natural mRNAs in the cell-free protein-synthesizing system of the moderate halophile *Vibrio costicola*, *J. Bacteriol.* 172: 3462 (1990)

[17]   D. J. Kushner, F. Hamaide and R. A. MacLeod, Development of salt-resistant active transport in a moderately halophilic bacterium. *J. Bacteriol.* 153: 1163 (1983)

[18]   J. F. Imhoff and F. Rodriguez-Valera, Betaine is the main compatible solute of halophilic eubacteria. *J. Bacteriol.* 160: 478 (1984)

[19]   L. G. M. Baas-Becking, Historical notes on salt and salt manufacture. *Sci. Month.* 32: 434 (1931)

[20]   B. E. Volcani, Studies of the microflora of the Dead Sea, Ph.D. thesis, Hebrew University, Jerusalem (1940)

MECHANISM OF CHLORIDE TRANSPORT IN
HALOPHILIC ARCHAEBACTERIA

Janos K. Lanyi
Department of Physiology and Biophysics
University of California
Irvine CA 92717
U.S.A.

INTRODUCTION

The halophilic archaebacteria, i.e. the various *Halobacterium* and *Natronobacterium* species (and possibly others), accumulate KCl rather than organic compounds to balance the high osmotic pressure of the nearly saturated NaCl solution which constitutes their environment. Because of this, chloride ions present a special bioenergetic problem to them, which is not normally of great importance in other organisms. The problem is that while the uptake and retention of the positively charged K ions are consistent with the inside negative membrane potential across the cytoplasmic membrane, the negatively charged chloride ions will be ejected rather than accumulated by any passive permeability pathway. Thus, to reach the several molar intracellular concentrations of KCl required, chloride has to be transported into the cells by an active mechanism, i.e. one which is capable of "uphill" transport, against the prevailing direction of the electrochemical potential of chloride.

Two mechanisms of active chloride uptake have been described so far in the halophiles: a) a light-driven electrogenic chloride pump, halorhodopsin [1, 2, 3], which is similar in many respects to the better known proton pump, bacteriorhodopsin [3, 4, 5], and b) a protonmotive force-driven chloride pump [6], which couples the transport of chloride to the proton gradient created by any of the proton transport systems available to these organisms, either during illumination via bacteriorhodopsin, or also in the dark, via the electron transport system. Little is known about mechanism b), other than its existence, and this article will focus, instead, on halorhodopsin as a chloride pump [7, 8]: what can be said about its structure, its similarities to and differences from bacteriorhodopsin, and the molecular mechanism of the transport. It is fortunate that two examples of halorhodopsins are now available for structural and functional comparisons: one from halobacteria, named simply halorhodopsin, and another from natronobacteria, named pharaonis halorhodopsin [9].

General and Applied Aspects of Halophilic Microorganisms
Edited by F. Rodriguez-Valera, Plenum Press, New York, 1991

Although made in much smaller amounts than bacteriorhodopsin by most strains, halorhodopsin can be prepared in very pure form in tens of mg quantities for biochemical and biophysical studies [10, 11, 12, 13], and will form proteoliposomes upon reconstitution with lipids, active in light-driven chloride transport experiments [7, 10, 14]. Moreover, as with the other retinal pigment, single-turnover measurements with flash-excitation allow following spectroscopic changes, e.g. in the visible region, which reveal the cyclic reaction accompanying chloride transport [15, 16, 17, 18, 19, 20, 21, 22]. Thus, halorhodopsin is a unique system for the study of anion transport. This article is not intended to give an exhaustive overview of the halorhodopsin literature, but to provide a progress report on the author's ten year old quest in search of answers hidden in halorhodopsin.

STRUCTURE

The bacterial rhodopsins are typically of 26 kDa size, i.e. they contain a polypeptide chain just long enough for seven trans-membrane helical segments, and short interhelical connections. So it is with the halorhodopsins also. The primary structures of the two halorhodopsins are known from DNA sequencing, and confirmed by partial amino acid sequences [9, 23]. They share a number of residues of obvious or seeming importance with bacteriorhodopsin: a lysine in the middle of helix G, to which retinal is bound via a Schiff base, two aspartates, on helices D and G, the latter probably forming an ion-pair with the Schiff base, two arginines, on helices C and F, and a number of tryptophans which probably form the retinal-binding pocket. The common residues undoubtedly fulfill structural functions, and ensure specific interaction with the retinal which shifts its absorption spectrum from about 440 nm in a model protonated Schiff base to 570 - 580 nm in the retinal-proteins. Equally significantly, other residues in bacteriorhodopsin are not found in the halorhodopsins. Two aspartates, both on helix C, recently implicated in accepting and donating a proton from and to the Schiff base in bacteriorhodopsin [24, 25, 26, 27, 28, 29, 30], are missing in the halorhodopsins, in keeping with the fact that the latter proteins do not transport protons.

The orientation of halorhodopsin in the membrane is in the same sense as bacteriorhodopsin, i.e. with the C terminus facing the cytoplasm [31, 32], in spite of the fact that transport in the two systems proceeds in opposite directions: chloride is translocated into the cells as mentioned above, while the protons are extruded into the medium.

Unlike bacteriorhodopsin, which forms an extended 2-dimensional crystalline array in the cytoplasmic membrane of the halophiles, halorhodopsin is not found in such "patches". However, CD spectra [10, 13, 33], and hydrodynamic data [34] obtained with detergent-solubilized halorhodopsin, suggest that like the other pigment, this protein also forms trimers of well-defined geometry.

SPECTROSCOPIC PROPERTIES

After a period of incubation in the dark, the retinal in halorhodopsin assumes a composition of about 60% 13-*cis* and 40% all-*trans* isomers [35]. Continuous illumination with blue light reverses this ratio in favour of the all-*trans* configuration, but a second illumination with red light shifts the equilibrium back towards 13-*cis* [35, 36]. Greatly prolonged illumination with red light produces a mixture containing large amounts of 9-*cis* retinal as well, an effect fully reversible by re-illumination with blue light [37]. Each of these halorhodopsin states has a characteristic absorption maximum; that of all-*trans* halorhodopsin is at 578 nm. These large-scale light-dependent spectroscopic shifts proceed with low quantum efficiencies, and the turn-over of the pigment is too slow to cause chloride transport. There is evidence, however, which suggests that absorption of a photon by the all-*trans* pigment only will produce transport [35], and thus the previous illumination history of halorhodopsin can be expected to have effects on its physiological function.

Chloride transport is associated with a distinct reaction sequence, "photocycle", initiated with high quantum efficiency (about 0.3 [18, 38]) by light absorption, which leads back within a few tens of msec to the original pigment. In the presence of chloride salts, photoexcitation of all-*trans* halorhodopsin produces first a species which absorbs at 600 nm [39, 40, 41], which can be called HR-K on the basis of analogies with the similar bacteriorhodopsin photointermediate. At room temperature, in less than 60 nsec HR-K is transformed into a species which absorbs at 580 nm (HR-KL), and the latter, in turn, is transformed in about 1 usec into a blue-shifted species absorbing at 526 nm, named HR-L [42]. The decay of HR-L produces two species: a small amount of a transient red-shifted species (absorption maximum 638 nm), called HR-O, accompanies the regeneration of halorhodopsin (HR) with a half-life of about 8 msec [19, 42]. The HR-L to HR-O reaction is chloride-dependent: investigations of this chloride dependency [17, 19, 22] confirmed an earlier suggested model for this part of the photocycle [16, 18], according to which HR-O is produced from HR-L by chloride release (and *vice versa*), while HR is produced directly from HR-O. The retinal in HR-K [40], HR-L and HR-O [17, 43] is 13-*cis*; presumably it is the HR-O ---→ HR reaction which includes reisomerization to all-*trans*. At physiological chloride concentrations the rate constants of these reactions are such that first an apparent equilibration of HR-L and HR-O takes place within 1 msec, followed by decay of both to HR. To account for the chloride ion lost in the HR-L ---→ HR-O step, a chloride uptake step is included after the recovery of HR from HR-O, which would then reproduce the original pigment [16, 18]. Although not seen after photoexcitation, the chloride uptake should result in a small red-shift in the absorption maximum, corresponding to an earlier described chloride-dependent equilibrium [11, 44] observable independently of illumination (cf. below). However, this step is probably much too rapid to be detected after the relatively slow reaction which precedes it. As expected on the basis of the intimate involvement of chloride in the photocycle, when chloride is replaced with nitrate the pigment undergoes a different photocycle, which is not accompanied, presumably, by anion transport. This reaction sequence resembles that seen in chloride, but HR-O is produced directly from a precursor named HR-KO, and no intermediate like HR-L is observed [22].

According to the scheme described above, the cyclic photoreaction is accompanied by cyclic chloride release and uptake. These could constitute the transport cycle, provided that the release and uptake were to take place at the inside and outside surfaces of the membrane, respectively. Thus, the interconversions of the photointermediates described above may represent steps along the chloride translocation pathway. As expected from this, nitrate, an anion which is transported poorly by halorhodopsin but as well as chloride by pharaonis halorhodopsin, permits the appearance of a chloride-type photocycle only partially and only at high concentration in halorhodopsin [22], but fully and under all conditions in pharaonis halorhodopsin [45].

The photointermediates of halorhodopsin bear strong resemblance, in their spectra, order of appearance after the photoexcitation, and the kinetics of their rise and decay, to the photointermediates of bacteriorhodopsin [3, 42]. What is missing, however, is a species which would correspond to the M intermediate of bacteriorhodopsin (which absorbs at 410 nm and arises between L and O), where the Schiff base is deprotonated. Rather, such a deprotonated species in halorhodopsin is produced as a side-reaction, with low efficiency, probably from HR-O [16]. The HR-M produced reprotonates very slowly (tens of mins), unless a weak base, such as azide, is added which then acts as a mobile conductor of the proton to and from the Schiff base [16, 46]. Thus, at pH > 7 sustained illumination normally results in the trapping of a fraction of the pigment in the M form. Lack of the Schiff base deprotonation in the photocycle sequence is consistent with the lack of a proton acceptor residue in halorhodopsin corresponding to asp-85 [23]. The acceleration of the de- and reprotonation of the Schiff base by azide suggests that when the Schiff base does deprotonate in this pigment, the proton is lost to the medium by traversing across a significant impermeable barrier in the protein [46]. A similar acceleration of the reprotonation of the Schiff base by azide was recently described for bacteriorhodopsin, in which asp-96 was replaced with asparagine [30].

ANION BINDING SITES

In the presence of various anions, such as chloride and nitrate, the absorption maximum of halorhodopsin is shifted slightly but in characteristic directions, and from this effect a model containing two binding sites was proposed [44]. Recently, a number of observations necessitated the revision of the details of this model. In the following, the data and a description of the revised mode are given.

Unlike the case with bacteriorhodopsin, the Schiff base of halorhodopsin reversibly deprotonates in the dark with an apparent pK of 7.4 in the presence of sulfate, an anion which interacts poorly if at all with the pigment [44, 47]. When chloride, nitrate, or some other anions are added the pK is shifted upwards, by as much as 1.5 pH units depending on the hydrated size of the anion [44, 47], suggesting an electrostatic effect of the negative charge of the anions on the Schiff base. Resonance Raman spectra of high resolution indicated that the C=N stretching vibration of the protonated Schiff base is changed under these conditions: with chloride or bromide a single frequency at 1633 1/cm is obtained but with nitrate two frequencies, at 1633 1/cm and at 1645 1/cm, are seen at saturating anion concentrations [48]. Perchlorate gave only the higher frequency. Thus, the data suggested the existence of two binding sites, one specific for polyatomic anions (site I), while the other (site II) occupied by chloride (or bromide), as well as sone polyatomic anions, such as nitrate.

This two-site model is consistent with the anion specificity and concentration dependency of transport, and the finding of two photocycles described above. Transport of anions, as assayed with passive proton flux into envelope vesicles which follows the anion in the presence of protonophores [8], recently confirmed the transport of nitrate, albeit at higher concentrations and with lower maximal velocity than chloride [45]. Consistently with this, nitrate, as well as chloride, elicited the photocycle containing HR-L associated with transport [22, 45], as mentioned above, but at higher concentration and only partially. Transport and the appearance of HR-L after photoexcitation are therefore attributed to anion binding to site II.

If chloride and nitrate will both bind to site II, they should produce the same maximal effects, at their (different) saturating concentrations. The observed partial effects of nitrate are thus not explained by this model, but they are consistent with the recent finding of mutual inhibition between the two binding sites [49]. The evidence is as follows. Chloride will, through binding to site II as stated above, cause site II type effects (transport and the photoproduction of HR-L). Nitrate will do the same, but when added in addition to chloride, it will partially reverse these effects, presumably through binding to site I. Thus, the results suggest that the two sites are mutually exclusive: either the effects of anion binding at the two sites counteract one another, or more likely, anion binding at either site prevents binding at the other site.

The location of sites I and II in the halorhodopsin molecule is problematical. It is interesting therefore, that pharaonis halorhodopsin seems to lack site I: it transports nitrate as well as chloride, and exhibits only an HR-L containing photocycle, in which an O-like intermediate is not detectable [45]. Comparison of the primary structures of the two halorhodopsins reveals [9] that there are two differences which might account for the missing site I in the pharaonis pigment. These are: a) arg-103, located in halorhodopsin between helices B and C (on the extracellular side) is replaced by valine in pharaonis halorhodopsin, and b) the arginine-rich AB interhelical loop in halorhodopsin (net charge + 4, located on the cytoplasmic side) is replaced by a relatively neutral segment (net charge + 1) in pharaonis halorhodopsin. Site II is almost certainly on the extracellular side, since this site has the same anion specificity as the transport, and the latter is initiated from the external side of the cells. The finding of two mutually exclusive sites suggests therefore two alternatives: either direct competition between bound anions through steric and/or electrostatic effects in a region accessible to the extra-cellular side, which contains sites I and II in close proximity (in case of a), or interaction between the two sites located on opposite sides of the membrane, through conformational effects which are communicated from one site to the other across the width of the protein in the membrane (in case of b).

SPECULATIONS ON THE ANION TRANSPORT MECHANISM

The concept of an ion pump requires the delivery of the ions to a switch which consists of an irreversible translocation mechanism, and the conduction of the ions past the barrier to the other side. In bacteriorhodopsin the switch for proton transport probably consists of the movement of the Schiff base during the isomerization of the retinal around its 13-14 double-bond, which occurs between the loss and the regain of its proton, although the additional requirement of a conformational change of the protein has also been suggested [50].

Analogously, the crucial step in chloride transport is the movement of the chloride ion across a barrier, from a site which communicates with the cell exterior, to a site which communicates with the cell interior. It has been suggested [3] that this is accomplished by the chloride ion becoming, transiently, a counter-ion to the positively charged Schiff base, and is thereby moved the required distance when the retinal chain undergoes its configurational changes. The translocation step is not yet placed in the halorhodopsin photocycle, but is likely to be either at or before the HR-L $\longrightarrow$ HR-O reaction, in which the chloride is released. That this step might include the translocation, as well as the release, is suggested by the fact that the reaction is suspiciously slow (in the msec range) for the simple release of an ion from a site which already faces the right aqueous phase. It is tempting to assign the roles of uptake and release to sites II and I, respectively, even though the location of site I, as discussed above, is not established. Furthermore, the concept of these sites is derived mainly from equilibrium phenomena, and the sites (particularly site I) might not remain unchanged during whatever conformational alterations might occur during the photocycle of the pigment. However, the observations that pharaonis halorhodopsin does not exhibit the effects of anion binding to site I, and in its photocycle the L $\longleftrightarrow$ O equilibrium is considerably perturbed [45], are important clues which link site I and anion release to one another.

Thus, a general picture of how chloride translocation might be accomplished in halorhodopsin is beginning to emerge, but crucial data which would lead to a specific model are still missing. What is needed is to identify amino acid residues directly or indirectly involved in the anion binding, determine their locations relative to the membrane surface and the retinal Schiff base and assign their participation in the transport to the already known reaction steps in the photocycle. In the short run, site-specific mutagenesis of the protein will provide some of these answers. In the long run, a high-resolution crystal structure for halorhodopsin will also be necessary.

REFERENCES

[1]   P. Hegemann, J. Tittor, A. Blanck and D. Oesterhelt, Progress in halorhodopsin, *in* "Retinal Proteins", Y. A. Ovchinnikov, ed., pp. 333-352, VNU Science Press, Utrecht (1987)

[2]   J. K. Lanyi *Ann. Rev. Biophys. Biophys. Chem.* 15: 11 (1986)

[3]   D. Oesterhelt and J. Tittor, *TIBS* 14: 57 (1989)

[4]   J. K. Lanyi, Bacteriorhodopsin and related light-energy converters, *in* "Comparative biochemistry: bioenergetics", L. Ernster, ed, pp. 315-350, Elsevier, Amsterdam (1984)

[5] W. Stoeckenius and R. A. Bogomolni, *Ann. Rev. Biochem.* 51: 587 (1982)

[6] A. Duschl and G. Wagner. *J. Bacteriol.* 168: 548 (1986)

[7] E. Bamberg, P. Hegemann and D. Oesterhelt. *Biochemistry* 23: 6216 (1984)

[8] B. Schobert and J. K. Lanyi, *J. Biol. Chem.* 257: 10306 (1982)

[9] J. K. Lanyi, A. Duschl, G. W. Hatfield K. May and D. Oesterhelt, *J. Biol. Chem.* 265: 1253–1260 (1990)

[10] A. Duschl, M. A. McCloskey and J. K. Lanyi, *J. Biol. Chem.* 263: 17016 (1988)

[11] T. Ogurusu, A. Maeda and T. Yoshizawa, *J. Biochem.* 95: 1073 (1984)

[12] M. Steiner and D. Oesterhelt, *EMBO J.* 2: 1379 (1983)

[13] Y. Sugiyama and Y. Mukohata, *J. Biochem.* 96: 413 (1984)

[14] R. A. Bogomolni, M. E. Taylor and W. Stoeckenius. *Proc. Natl. Acad. Sci. USA* 81: 5408 (1984)

[15] N. Hazemoto, N. Kamo, Y. Terayama, Y. Kobatake and M. Tsuda, *Biophys. J.* 44: 59 (1983)

[16] P. Hegemann, D. Oesterhelt and M. Steiner, *EMBO J.* 4: 2347 (1985)

[17] J. K. Lanyi and V. Vodyanoy. *Biochemistry* 25: 1465 (1986)

[18] D. Oesterhelt, P. Hegemann and J. Tittor, *EMBO J.* 4: 2351 (1985)

[19] J. Tittor, D. Oesterhelt, R. Maurer, H. Desel and R. Uhl, *Biophys. J.* 52: 999 (1987)

[21] H. J. Weber and R. A. Bogomolni, *Photochem. Photobiol.* 33: 601 (1981)

[22] L. Zimányi and J. K. Lanyi, *Biochemistry* 28: 5172 (1989)

[23] A. Blanck and D. Oesterhelt, *EMBO. J.* 6: 265 (1987)

[24] M. S. Braiman, T. Mogi, T. Marti, L. J. Stern, H. G. Khorana and K. J. Rothschild, *Biochemistry* 27: 8516 (1988)

[25] H. J. Butt, K. Fendler, E. Bamberg, J. Tittor and D Oesterhelt. *EMBO J.* 8:1657 (1989)

[26] M. Holz, L. A. Drachev, T. Mogi, H. Otto, A. D. Kaulen, M. P. Heyn, V. P. Skulachev and H. G. Khorana, *Proc. Natl. Acad. Sci. USA* 86: 2167 (1989)

[27] T. Marinetti, S. Subramaniam, T. Mogi, T. Marti and H. G. Khorana, *Proc. Natl. Acad. Sci. USA* 86: 529 (1989)

[28] T. Mogi, L. J. Stern, T. Marti, B. H. Chao and H. G. Khorana, *Proc. Natl. Acad. Sci. USA* 85: 4148 (1988)

[29] J. Soppa, J. Otomo, J. Straub, J. Tittor, S. Meessen and D. Oesterhelt, *J. Biol. Chem.* 264: 13049–13056 (1989)

[30] J. T. Hor, C. Soell, D. Oesterhelt, H. -J. Butt and E. Bamberg, *EMBO J.* 8: 3477–3482 (1989)

[31] K. M. May, F. A. Jay and D. Oesterhelt, *J. Biol. chem.* 263: 13623 (1988)

[32] B. Schobert, J. K. Lanyi and D. Oesterhelt, *EMBO J.* 7: 905 (1988)

[33] C. A. Hasselbacher, J. L. Spudich and T. G. Dewey, *Biochemistry* 27: 2540 (1988)

[34] E. S. Schegk and D. Oesterhelt, *EMBO J.* 7: 2925 (1988)

[35] J. K. Lanyi *J. Biol. Chem.* 261: 14025 (1986)

[36] S. O. Smith, M. J. Marvin, R. A. Bogomolni and R. A. Mathies, *J. Biol. Chem.* 259: 12326 (1984)

[37] L. Zimányi and J. K. Lanyi, *Biophys. J.* 52: 1007 (1987)

[38] J. K. Lanyi, *Biochem. Biophys. Res. Comm.* 122: 91 (1984)

[39] H. J. Polland, M. A. Franz, W. Zinth, W. Kaiser, P. Hegemann and D. Oesterhelt. *Biophys. J.* 47: 55 (1985)

[40] K. J. Rothschild, O. Bousche, M. S. Braiman, C. A. Hasselbacher and J. L. Spudich. *Biochemistry* 27: 2420 (1988)

[41] L. Zimányi and J. K. Lanyi, *Biochemistry* 28: 1662 (1989)

[42] L. Zimányi, L. Keszthelyi and J. K. Lanyi, *Biochemistry* 28: 5165 (1989)

[43] J. K. Lanyi, *FEBS Lett.* 175: 337 (1984)

[44] B. Schobert, J. K. Lanyi and D. Oesterhelt, *J. Biol. Chem.* 261: 2690 (1986)

[45] A. Duschl, J. K. Lanyi and L. Zimányi. *J. Biol. Chem.* 265: 1261-1267 (1990)

[46] J. K. Lanyi, *Biochemistry* 25: 6706 (1986)

[47] B. Schobert and J. K. Lanyi, *Biochemistry* 25: 4163 (1986)

[48] C. Pande, J. K. Lanyi and R. H. Callender, *Biophys. J.* 55: 425 (1989)

[49] J. K. Lanyi, A. Duschl, Gy. Váró and L. Zimányi, *FEBS Lett.* (in press).

[50] S. P. Fodor, J. B. Ames, R. Gebhard, E. M. van der Berg, W. Stoeckenius, J. Lugtenburg and R. A. Mathies, *Biochemistry* 27: 7097 (1988)

PROCESSES OF ADAPTATION OF DIFFERENT CELL-LINES OF

DUNALIELLA TO WIDELY DIFFERING SALT CONCENTRATIONS

Margaret Ginzburg

Plant Biophysical Laboratory
Botany Department
The Hebrew University
Jerusalem, Israel

ABSTRACT

Experiments on the green unicellular alga *Dunaliella parva* show that cultures growing in different salt concentrations vary in tolerance to changes in salt concentration. Cultures grown in 0.5M NaCl tolerated abrupt increases to 2M but viability fell to 1% in 3M NaCl. Viability was higher after downward changes. Slow changes in concentration were better tolerated. This *Dunaliella* cell-line consists of several physiological phenotypes with the same genetic composition but different environmental requirements for growth. Experiments on salt effects on carbonic anhydrase are also reported.

INTRODUCTION

Members of the genus *Dunaliella*, a green unicellular alga, are able to grow in saline media. Isolates have been obtained from waters containing salt varying in concentration from 20 mM to approaching saturation. The genus is the most extreme known among green plants in its adaptation to salt. Isolates from marine and from hypersaline environments have been referred to as halotolerant and halophilic respectively. There is no precise physiological or genetic definition of these terms.

When *Dunaliella* cells are subjected to a change in salt concentration, the cells become modified in several ways depending on the direction and size of the change in salt concentration. An immediate observable change in cell-volume occurs as the cells

General and Applied Aspects of Halophilic Microorganisms
Edited by F. Rodriguez-Valera, Plenum Press, New York, 1991

shrink or swell in accordance with the laws of osmotic pressure [1]. Such alterations in cell-volume are made possible by the elastic cell-envelope which surrounds the cell. With passage of time the cells regain their former volume; there are also temporary changes in the extent of membrane material and in the activity of ion-pumps so Na^+ and Cl^- which may have entered the cell may be extruded. There are long-term changes in the rates of respiration, photosynthesis, and in concentration of cell glycerol [2]. If the cell-culture is growing at the time of the change in salt concentration, growth stops. It is presumably only when the various physiological processes have returned to a steady coordinated state that cell-division is resumed. It therefore seemed to us that the ability to divide was an essential criterion for the full adaptation of *Dunaliella* cells to a new NaCl concentration and could be determined by performing viability tests immediately after changing the NaCl concentration of a culture.

The purpose of this work has been to define the limits of tolerance of *Dunaliella* cells to changes in salt concentration and to throw light on the mechanism of coordination of the reactions involved in adaptation to a change in salt concentration. An attempt will be made to define the difference between halotolerant and halophilic cell-lines. In addition, experiments are reported on the enzyme carbonic anhydrase (CA), to demonstrate its role in the adaptation of *Dunaliella* cells to new NaCl concentrations.

METHODS

Cell strains and methods of culture. Most of the work described was performed with *D. parva* 19/9 Lerche, a species with green cells (cell-volume: $\sim100.10\text{-}18$ m^3). Certain experiments were repeated with *D. tertiolecta* and cell-line A12, both with small green cells. The cell-lines D13 and C9AA have larger cells ($\sim800.10^{-18}$ m^3) usually reddish in colour, and were also used in some experiments. Methods of culture and the basic medium used have already been described [3].

Viability tests. A standard method was used [4]. Three sets of 10 tubes each containing 10ml medium were sterilised and inoculated with 16.4, 1.64 or 0.16 cells per type. The number of tybes per set to develop green cultures was recorded; the ratio gave the Most Probable BNumber of viable organisms per inoculum in the first dilution, according to the statistical Tables of Halvorson and Ziegler (1933) quoted in Meynell [4]. This number was compared with the total number of cells initially introduced into the tubes, as determined by microscopic counting. It took 12 and 40 days for cultures to develop in 0.5M NaCl and 3.5M NaCl respectively.

Cell counts. Counts were made by means of a Neuberger Improved hemacytometer. Not less than 600 cells per sample were counted.

Adjustment of NaCl concentration by drip. The NaCl concentration was increased by introducing medium with 5M NaCl into cell suspensions through fine tubing. Flow was regulated by means of a peristaltic pump; the flow rate varied from 3 to 100 ml hr^{-1}. In order to bring about a decrease in NaCl concentration, medium without NaCl was pumped into cell suspension in medium with 3.5M NaCl.

Measurement of activity of carbonic anhydrase. Note was taken of the time needed for the pH of an ice-cold solution to fall from 8.0 to 7.0 in presence and absence of a cell extract containing carbonic anhydrase [5]. The extract consisted of the supernatant obtained from lysed cells after centrifugation at 50,000g for 30 min at 4oC.

RESULTS

The range of adaptability of cells of a given *Dunaliella* cell-line was determined from the proportion of cells to survive in samples in which salt concentrations were abruptly changed in an incremental series. Upward and downward changes were tested. Survival was measured by viability tests of the sort described under Methods and by microscopic counting of intact cells. (Tables 1 and 2). Table 1 is the result of viability tests performed on *Dunaliella parva* 19/9 cells transferred from 0.5M NaCl to higher salt concentrations. On transfer to 1.5M and 2M the % survival was >27. After transfer to 3M NaCl, the % viable cells was low (~1) and dropped to 0.01 in 3.5M NaCl.

Downward transfer from 3.5M to 0.5M NaCl was studied in 4 *Dunaliella* strains by counting the proportion of intact cells under the microscope (Table 2). Cells with an intact membrane can be distinguished from those in which the inner contents are escaping because of tears in the cell-membrane. Such cells are no longer viable. In *D. parva* 19/9 only 20% of the cells remained intact 60 min after the transfer. In the 3

Table 1.   Survival of *Dunaliella parva* 19/9 cells transferred from low to higher NaCl concentration Survival was measured by the ability of individual cells to divide and produce suspensions in liquid culture. Results of 2 experiments, I and II are given.

Range of transfer, M.	% survival	
	I	II
0.5 → 0.5	100	100
0.5 → 1.5	57	100
0.5 → 2	27	100
0.5 → 3	1	0.01
0.5 → 3.5	0.01	

Table 2. Survival of *Dunaliella* cells transferred from 3.5M to 0.5M NaCl. Live cells counted as those with an intact membrane.

Cell-line	Time after transfer min.	Total cells counted	Intact cells	% survival
D. parva 19/9	15	1133	555	49
"	15	289	122	42
"	60	289	60	21
"	120	289	55	19
	24 hours	289	58	20
D. tertiolecta	15	554	282	51
A12	15	571	410	72
"	15	1349	900	67
D13	15	151	71	47
C9AA	15	97	52	67

other strains tried, measurements were made 15 min after transfer and % survivval ranged from 47 to 70. The experiments demonstrate that there are limits to the tolerance of cells to abrupt changes in salt concentration. Also, survival after downward transfer is higher than after upward transfer (3.5M → 0.5M: 20%; 0.5M → 3.5M: 0.01%).

Large changes in concentration were tolerated better when they took place over a period of several hours (figures 1 and 2). Cells adapted to 3.5M NaCl resumed growth in 0.5M NaCl if the change took 1.5 h or more: the number of cells to develop into cultures equalled the total number of cells plated out, within 95% probability limits. For cells starting in 0.5M NaCl, the increase to 3.5M NaCl had to take 24-48 hours for survival to approach 100%. The experiments demonstrate once more that the processes of downward and upward adaptation are different.

Experiments were performed to determine whether cells growing in 3.5M NaCl retained the ability to grow at this concentration after transfer to 0.5M NaCl (Table 3). When such cells were transferred to 0.5M, 23% remained intact after 150 min (see also Table 2). Of these, only 1.5% were intact after transfer back to 3.5M NaCl (Table 3). It was concluded that 98% of *D. parva* 19/9 cells adapted to 3.5M NaCl lost the ability to grow at this concentration within 150 min of being transferred to 0.5M NaCl. It follows that ability to grow in high salt concentrations is not a genetic property, but rather requires a period for physiological adaptation.

Table 3. Effect of short time in 0.5M NaCl on *D. parva* 19/9 cells grown in 3.5M NaCl. Cell counts were made by means of an Improved Neubauer hemacytometer.

	Total number	Intact	% dead
Cells transferred from 3.5M NaCl to 0.5M NaCl:-			
after 15 min (group A)	185	170	8
after 150 min (group B)	185	143	23
After re-transfer to 3.5M NaCl:-			
group A	65	56	14
group B	65	1	98.5

The activity of carbonic anhydrase (CA), measured per cell, increases x2.5 with the NaCl concentration of the growth medium (figure 3). The results are similar to those obtained for *D. tertiolecta* [5]. It should be noted that the CO_2 content of the medium falls as the salt concentration rises and the increase in enzyme activity may e a response to the fall in CO^2. There is an inhibitory action of salt on the enzyme: 50% inhibition occurs with 800 mM NaCl, though considerable activity is retained even in 4M NaCl. It is concluded that the large increase in CA per cell that occurs with increasing [NaCl] in the medium is correlated with the lowered CO_2 content of the medium and also with the fall in activity of the enzyme itself.

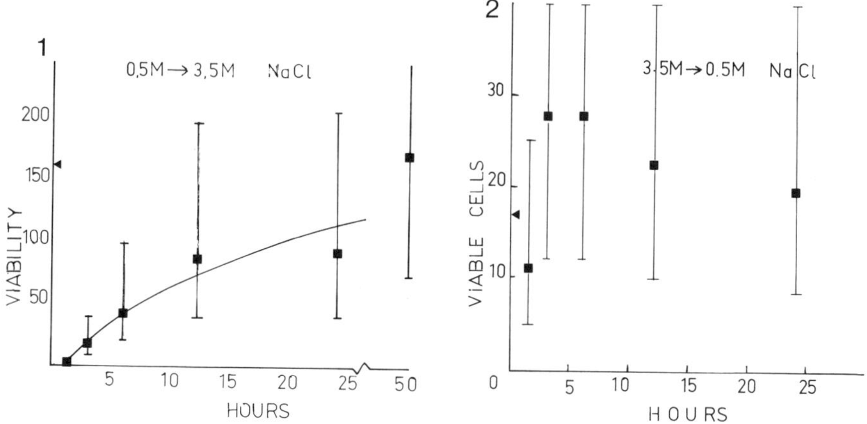

Figures 1 and 2. Effect of slow transition from 0.5M to 3.5M NaCl (Fig. 1) and from 3.5M to 0.5M NaCl (Fig. 2) on viability of *D. parva* 19/9. Time taken for the transition on the horizontal axis, in hours. ◄ :total number of cells tested for viability. Vertical bars indicate the 95% confidence limits.

Table 4. Effect of inhibitors on carbonic anhdrase. Data for erythrocytes (RBC) and *Chlamydomonas* from Bundy and Cote [6].

Inhibitor	RBC	*Chlamydomonas*	*Dunaliella*
	concentration causing 50% inhibition (M)		
acetazolamide (diamox)	5.10^{-7}	$7.8.10^{-9}$	$3.8.10^{-9}$
sodium azide		$3.9.10^{-5}$	$3.7.10^{-1}$
sodium chloride		$5.6.10^{-2}$	$7-8.10^{-1}$

There are similarities between the CAs of *Dunaliella Chlamydomonas* and erythrocytes (Table 4). In contract, the CA from higher plants is less sensitive to acetazolamide [6]. The *Dunaliella* CA is characterised by its relative insensitivity to CL^{-} and to azide, both members of the Hofmeister series of anions.

When the salt concentration of the medium of *D. parva* 19/9 cells in 1.5M NaCl is increased to 3M, an increase in CA activity can be detected within 15 min (Figure 4).

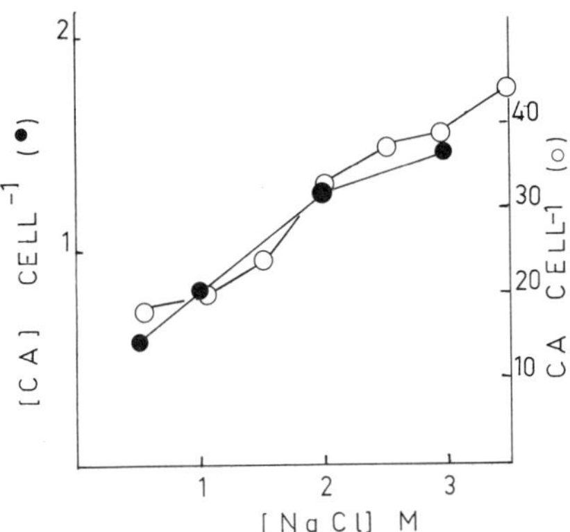

Figure 3.   Variation of CA activity [CA] with NaCl concentration in growth medium. (●): *D. parva* 19/9 (o) *D. tertiolecta* [7].

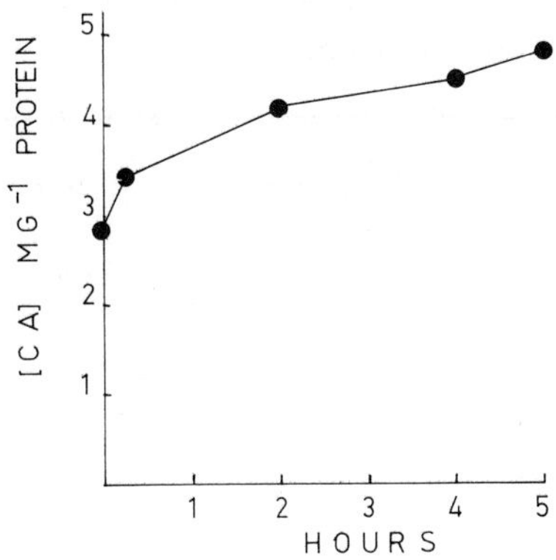

Figure 4.   Increase in [CA] in *D. parva* 19/9 after raising the NaCl concentration from 1.5M to 3M at time 0.

DISCUSSION

It appears that the vast majority (98%) of *D. parva* 19/9 cells were unable to withstand an abrupt rise in salt concentration from 0.5M to 3.5M NaCl (Table 1), presumably due to the swamping of the cell-interior by salt. 80% of the cells of this strain were unable to withstand an abrupt fall in salt concentration due to tearing of the cell membrane following the rapid osmotic intake of water. However, if abrupt changes are avoided, there seems no limit to the cells' ability to withstand slow, continuous adjustments in either direction.

No evidence has been found of sub-strains adapted to one concentration-range or another: every cell is totipotent with respect to salt concentration but only potentially so since the change-over from one state to another takes several hours.The results have demonstrated that there exist physiological phenotypes (i.e. states) within the same cell-line of *Dunaliella*, with the same genes but growing at different salt concentrations. It is worthwhile asking whether the division of the genus into halophilic and halotolerant types may not be an artifact arising from the difference in salt concentration of the original habitats from which the different types were taken. At any rate, before labelling a strain as "halophilic" it would be well to see whether it cannot grow equally well or better at 0.5M NaCl by using the methods outlined in this paper for transforming one physiological phenotype into another.

The present work has presented evidence for difference in CA activity of the different physiological phenotypes of *D. parva* 19/9. To our knowledge, this is the first time that an enzyme has been found to be involved in the adaptation of a cell-line from one salt concentration to another.

There is little or no information on the coordination of the reactions characterising the different physiological phenotypes in *Dunaliella*. It is the task of future research to discover whether this genus contains a global regulatory network of the sort known in Escherichia coli [8] to coordinate all the reactions affected by the salt concentration of the medium.

ACKNOWLEDGEMENTS

Much of the technical work described in this paper was performed by Mrs. Mira Cohen. The help of Professor Ben-Zion Ginzburg is also gratefully acknowledged.

REFERENCES

[1]   H. Gimmler, R. Schirling and U. Tobler. Cation permeability of the plasmalemma of the halotolerant alga *Dunaliella parva. Zeitschr. f. Pflanzenphysiologie* 83:145 (1977).

[2]   M. Ginzburg. *Dunaliella*: a green alga adapted to salt, in: "Adv. Bot. Res." vol. 14. J. Callow, ed. Academic Press, London (1987).

[3]   M. Ginzburg and B. Z. Ginzburg. Influence of age of culture and light intensity on solute concentrations in two *Dunaliella* strains. *J. exp. Bot.* 36:701 (1985).

[4]   G. G. Meynell and E. Meynell. "Theory and Practice in experimental Bacteriology", Cambridge University Press (1965).

[5]   K. M. Wilbur and N. G. Anderson. Electrometric and colorimetric determination of carbonic anhydrase. *J. Biol. Chem.* 176:147 (1948).

[6]   H. F. Bundy and S. Cote. Purification and properties of carbonic anhydrase from *Chlamydomonas reinhardtii. Phytochemistry* 19:2531 (1980).

[7]   A. H. Latorella and R. L. Vadas. Salinity adaptation by *Dunaliella tertiolecta. J. Phycol.* 9:273 (1973).

[8]   S. Gottesman, Bacterial regulation: global regulatory networks. *Ann. Rev. Genetics* 18:415 (1984).

OSMOREGULATION IN *RHIZOBIUM MELILOTI*: CONTROL OF GLYCINE BETAINE BIOSYNTHESIS AND CATABOLISM

Daniel Le Rudulier and Jean-Alain Pocard

Laboratoire de Biologie Végétale et Microbiologie
URA CNRS 79
Université de Nice-Sophia Antipolis
Parc Valrose
F-06034 Nice Cédex, France

ABSTRACT

In the root nodule bacterium *Rhizobium meliloti*, glycine betaine and choline can support growth, and they are also effective osmoprotectants. A kinetic approach has been used to investigate the possibility of multiple transport systems. Three separate transport activities for choline have been distinguished: two high-affinity transport activities, one induced by choline itself, and one low-affinity transport activity. Only the induced high- and low-affinity activities are important for osmoregulation. In the case of glycine betaine a high-affinity transport activity strongly stimulated by high osmolarity has been characterized. A periplasmic glycine betaine-binding protein has been found. Cells never accumulate choline which is rapidly converted into glycine betaine. Very high intracellular levels of glycine betaine have been observed in cells maintained under salt stress whereas a strong catabolism exists in low-salt-grown cells. High osmolarity in the medium decreased the activities of enzymes involved in the degradation of glycine betaine but not those of enzymes that lead to its biosynthesis from choline.

INTRODUCTION

Many bacteria respond to inhibitory levels of osmolarity in the environment by accumulating small organic compounds in the cytosol. These compounds, termed osmoprotectants, function by increasing the internal osmotic strength, thus allowing the cell to avoid cytoplasmic dehydration [1, 2]. Glycine betaine is an effective osmoprotectant in the enteric bacteria *Escherichia coli*, *Salmonella typhimurium*, and *Klebsiella pneumoniae* [3, 4]. In *E. coli*, glycine betaine cannot be used as a building block for cellular components; it functions only as an osmoprotectant [5]. Glycine betaine also plays a similar role in *Rhizobium meliloti*, the root nodule symbiont of alfalfa. It has been shown to protect the growth of the free-living bacterium [6] and to enhance symbiotic nitrogen fixation of young nodulated alfalfa seedlings under osmotic stress [7]. Furthermore, it was demonstrated that labeled glycine betaine accumulates in the cytosol of *R. meliloti* cells when cultures are grown at inhibitory osmolarity. However, this species catabolizes glycine betaine at low osmolarity, salvaging both carbon and nitrogen [6].

A great variety of aerobic and anaerobic bacteria have the ability to use choline as a carbon, nitrogen, and/or energy source [8 - 10]. In *R. meliloti*, choline can support growth in media of low osmolarity. This molecule is the precursor of glycine betaine; therefore, choline is indirectly an effective osmoprotectant in media of high osmotic strength [11].

In the present contribution, we wish to summarize the osmotic control of glycine betaine biosynthesis and degradation in *R. meliloti*. The focus will be on three areas in which recent advances have been made. These are: (1) characterization of three choline transport activities, (2) evidence for the presence of a periplasmic glycine betaine-binding protein, (3) enzymological data concerning the pathway of glycine betaine biosynthesis and catabolism.

CHOLINE TRANSPORT ACTIVITIES

A kinetic approach was used to investigate the possibility of multiple transport systems in *R. meliloti* 102F34. Figure 1 shows the double-reciprocal plots of choline transport activities. When cells were grown in minimal medium with D, L-sodium lactate (0.1%) and sodium aspartate (0.13%), two separate transport activities were distinguished by their relative affinities for choline: a high-affinity and a low-affinity transport activity with K_ms of 6.7 and 100 μM, respectively (Figure 1, a and b). Both activities showed significant levels of substrate inhibition, so that the observed V_{max}s were only 4.3 and 40.5 nmol/min per mg of protein for the high- and low-affinity activities, respectively, whereas the calculated maximum velocity for both activities was 50 nmol/min per mg of protein. When cells were grown with 0.3 M NaCl, the K_m and W_{max} for the high-affinity activity decreased 13- and 56-fold, respectively (Table 1). Thus, an increase in substrate affinity was correlated with a decrease in velocity. The substrate inhibition previously observed had disappeared (Figure 1a). On the contrary, the kinetic parameters determined for the low-affinity activity were not significantly modified by high osmotic strength in the medium (Table 1). In this case also, the

90

substrate inhibition completely disappeared when the culture was salt stressed (Figure 1b).

Kinetic parameters of choline uptake were also determined from cultures grown in medium with choline added. Under this condition, a third kinetically distinct choline transport activity was observed (Figure 1c; Table 1). In low-salt medium, the K_m and V_{max} were 0.4 μM and 26 nmol/min per mg of protein, respectively. Furthermore, this new high-affinity activity was also inhibited by choline but only at very high concentrations, so that the observed maximum velocity was nearly the same as the extrapolated value. The effect of NaCl in the medium on the choline activated high-affinity uptake system activity also served to distinguish this activity from the

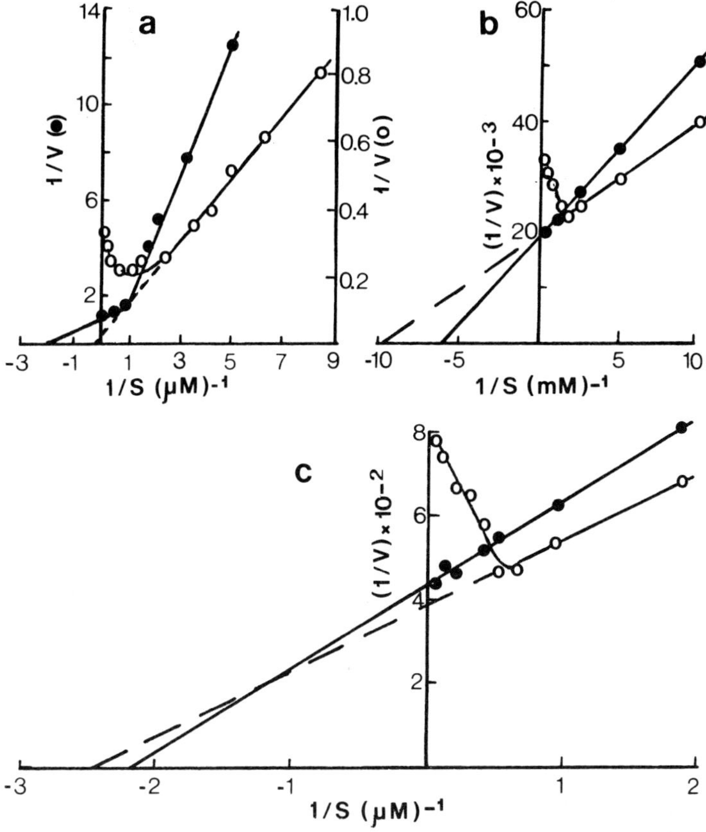

Figure 1.   Double reciprocal plots of choline transport activities in *R. meliloti* 102F34. (a) High-affinity transport activity; (b) Low-affinity transport activity; (c) High-affinity transport activity from cells grown in the presence of 7 mM choline. Symbols: O, cells grown in minimal medium; ●, cells grown in minimal medium plus 0.3 NaCl.

previous high-affinity activity. With salt-stressed cells (0.3 M NaCl), the substrate inhibition completely disappeared (Figure 1c), whereas the K_m and V_{max} values remained about the same. Thus, the value for V_{max} was 25-fold higher than that from cells grown at high osmolarity without choline. This increase in transport rate may be helpful for osmoprotection [12].

Finally, all the results in Figure 1 and Table 1 demonstrate that *R. meliloti* 102F34 possesses at least three separate transport activities, each one has its own unique set of kinetic parameters when cells are grown in the absence or the presence of NaCl added to the medium. It is very common for bacteria to have multiple transport systems for uptake of important substrates. Such multiple transport systems are usually distinguished genetically, since kinetic data alone can never prove multiple uptake systems. Further work is needed to determine whether or not these uptake activities share any (or all) structural or regulatory elements. We are currently investigating this question.

Table 1. Kinetic parameters of choline transport in *R. meliloti* 102F34[1]

Activity	Medium	Km (μM)	V_{max}[2] calculated	observed
High affinity	No salt	6.7	50	4.3
	0.3 M NaCl	0.5	0.9	
Low affinity	No salt	100	50	40.5
	0.3 M NaCl	150	50	
Choline activated	No salt	0.40	26	21.8
high affinity	0.3 M NaCl	0.45	22.7	

1. For choline-activated cultures, 7 mM choline was added to the medium.
2. V_{max} are expressed as nmol/min per mg of protein. Due to the substrate inhibition which occurred with cultures grown without NaCl, the observed maximum rates were lower than the V_{max} values obtained from double reciprocal plots.

GLYCINE BETAINE TRANSPORT ACTIVITY

Using a kinetic approach with glycine betaine concentration from 3.5 to 400 μM, we have characterized only one high-affinity uptake activity. When cells were grown in minimal medium, in the presence (0.3 M NaCl) or in the absence of added salt, transport was a saturable function of substrate concentration. Double-reciprocal plots gave a straight line, indicating that the uptake follows typical Michaelis–Menten kinetics. The apparent K_m was 5.6 μM with low-salt grown-cells and 9 μM with high-salt-grown cells. The small difference in the apparent K_m values is insignificant and cannot be related to different transport systems. However, maximal velocities were quite different in both growing conditions: V_{max} varied from 5.5 to 20.5 nmol/min per mg of protein with low- or high-salt-grown cells, respectively.

The time course of glycine betaine uptake by cells grown at increased salt concentration was followed. Cells grown in minimal medium exhibited a relatively small

92

glycine betaine uptake activity. The uptake was linear during only 7 to 10 min, and then a plateau was reached. The addition of NaCl in the growth medium strongly stimulated glycine betaine transport: the initial rate was increased and constant during, at least, 20, 30 or 60 min in the presence of 0.15, 0.3 or 0.5 M NaCl, respectively. Consequently, after 1 hour the transported quantity of glycine betaine was 75, 300, 710 and 1150 nmol per mg of protein when cells were grown in medium with 0, 0.15, 0.3 or 0.5 M NaCl, respectively. Calculations of intracellular water space determined by the method of Stock et al. [13], revealed an intracellular concentration of 18 mM in minimal medium-grown cells, 308 and 640 mM in 0.3 or 0.5 M NaCl-grown cells, respectively.

In *E. coli* and *S. typhimurium*, glycine betaine transport has been studied extensively, and two transport systems have been identified [14, 15, 16]. One system encoded by the *proU* locus, is binding protein-dependent [17, 18]. In *R. meliloti* betaine transport is much less understood. However, several lines of evidence support the conclusion that a periplasmic binding protein is present in cells grown in medium of high osmolarity (0.3 M NaCl); but, so far, it is not known whether this protein in involved in transport. First, the cold osmotic shock [19] was not followed by a significant reduction of glycine betaine transport. Second, glycine betaine-binding

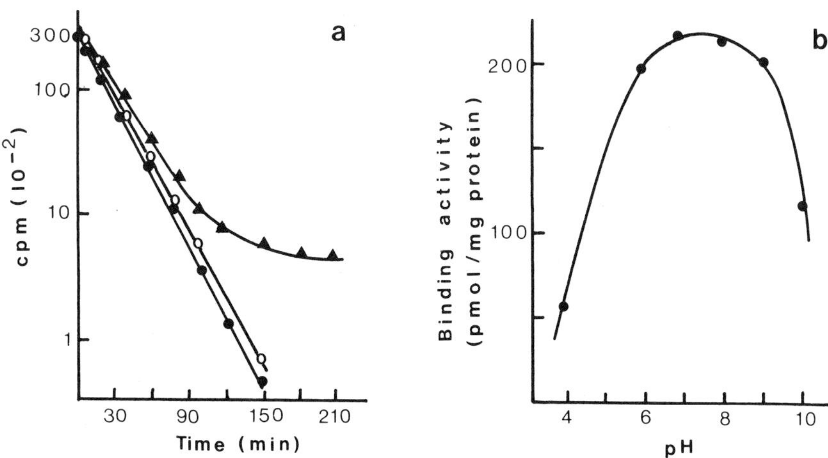

Figure 2. Evidence for a periplasmic glycine betaine binding protein in high-salt-grown cells of *R. meliloti*. (a) Dyalysis technique; binding of ^{14}C-glycine betaine was assayed in 10 mM Tris-HCl, pH 7.5. Symbols: ●, control experiment without shock fluid; ○, shock fluid from no-salt grown cells; ▲, shock fluid from 0.3 M NaCl-grown cells. (b) Effect of pH on glycine betaine binding activity. Proteins were from 0.3 M NaCl-grown cells.

activity was present in the crude periplasmic osmotic shock fluid. We have measured the binding of ^{14}C-glycine betaine to periplasmic proteins using a dialysis technique based on the retention phenomenon of ligands by binding proteins. Figure 2a shows a rapid linear exit of free glycine betaine from a dialysis bag that contained only buffer or shock fluids from cells grown in the absence of NaCl. In contrast, when shock fluids from high-salt-grown cells are used, there was an initial rapid efflux of excess of glycine betaine followed by a slow dissociation of bound substrate. As an alternative, binding assays were also done by the filter-binding procedure in which periplasmic proteins were incubated with ^{14}C-glycine betaine, precipitated by ice-cold saturated ammonium sulphate, and collected by filtration onto nitrocellulose filter. Betaine-binding activity at different substrate concentrations was assayed in periplasmic fractions isolated from cells grown at high osmolarity. The K_D for binding was found to be approximately 3 μM. This is in good agreement with the K_t for glycine betaine transport (9 μM). The pH dependence of the binding activity was assayed (Figure 2b). Optimum pH was between 7.0 and 8.0. Third, using a non-denaturing polyacrylamide gel electrophoresis with ^{14}C-glycine betaine added to the periplasmic shock fluids, we were able to show that the complex is sufficiently stable to be detected by autoradiography after 14 hours migration. Similar results have been obtained previously with ^{14}C-γ butyrobetaine, using a minigel system [20]. With this procedure only one radioactive band was found with extract from high-salt-grown cells, showing the presence of a binding protein in this condition. As observed in *E. coli* and *S. typhimurium* this protein is certainly osmotically induced. An apparent molecular weight of 32,000 was determined by SDS-PAGE.

GLYCINE BETAINE BIOSYNTHESIS AND CATABOLISM

The proposed pathway of glycine betaine biosynthesis and degradation was determined previously [6, 11] by two methods: we have compiled data on the degradation of both 1,2-^{14}C-choline and 1,2-^{14}C-glycine betaine and also tested the ability of the suspected intermediates to act as sole carbon sources.

$$\text{Choline} \xrightarrow{1} \text{Betainal} \xrightarrow{2} \text{Glycine betaine} \xrightarrow{3} \text{Dimethylglycine}$$

$$\xrightarrow{4} \text{Sarcosine} \xrightarrow{5} \text{Glycine} \xrightarrow{6} \text{Serine} \xrightarrow{7} \text{Pyruvate}$$

When cells were grown in low-osmolarity medium, choline and glycine betaine were rapidly degraded. In contrast, when cultures were grown at high osmolarity, glycine betaine was accumulated. It is clear that once choline is transported into the cells it can quickly be converted to glycine betaine.

Enzyme activity assays were used as another method of investigating this pathway. Table 2 lists all of the enzymes in the proposed pathway. All of the activities were low, in the absence of choline in the growth medium. They increased from 6-fold to more than 250-fold when the culture was supplemented with choline. When cultures were grown in the presence of NaCl, the specific activities of the enzymes that catalyze

Table 2. Effect of choline and high osmolarity in the medium on the enzymatic activities of the glycine betaine metabolic pathway in *R. meliloti* 102F34

Enzyme	Specific activity [1] of enzyme from culture with :		
	No addition	Choline 7 mM	Choline-NaCl 7 mM 0.5M
1. Choline oxidase	<1.7	71	72
2. Betaine aldehyde dehydrogenase	<1	250	410
3. Glycine betaine transmethylase	<0.03	0.18	0.13
4. Dimethylglycine dehydrogenase	3.6	54	33
5. Monomethylglycine dehydrogenase	5.7	102	68
6. Serine transhydroxymethylase	1.9	29	5.4
7. Serine dehydratase	10	450	120

1. Specific activities are expressed as nmol/min per mg of protein.

glycine betaine biosynthesis either remained constant (choline oxidase) or increased (betainal dehydrogenase), while the enzyme activities involved in glycine betaine degradation decreased, compared with those from cultures grown at low osmolarity. High osmolarity inhibited mainly the latter part of the pathway that leads to glycine betaine degradation. The activity of serine dehydratase, the last enzyme involved in glycine betaine degradation, decreased steadily as NaCl in the medium was increased. The maximum decrease occurred at 0.5 M NaCl, the highest concentration tested. In contrast, 0.5 M NaCl or KCl added to the assay mixture *in vitro* did not change the serine dehydratase activity in either cell extracts or toluene-treated cells.Hence, salt regulates serine dehydratase activity but does not act directly on the enzyme by altering the protein structure.

The main conclusion of this report is that glycine betaine functions in both nutrition and osmoprotection in *R. meliloti*. However, other osmoprotectants such as proline betaine may be critical for osmotic tolerance. Since alfalfa itself accumulates this betaine in the nodules when osmotically stressed (Gloux and Le Rudulier, manuscript in preparation), it may be that proline betaine can play an important role in the protection of osmotically stressed bacteroids. We are investigating this possibility.

REFERENCES

[1]   A. D. Brown, Microbial water stress, *Bacteriol. Rev.* 40: 803 (1976)

[2]   P. H. Yancey, M. E. Clark, S. C. Hand, R. D. Bowlus and G. N. Somero, Living with water stress: evolution of osmolyte systems, *Science* 217: 1214 (1982)

[3]   D. Le Rudulier and L. Bouillard, Glycine betaine, an osmotic effector in *Klebsiella pneumoniae* and other members of the *Enterobacteriaceae*, *Appl. Environ. Microbiol.* 46: 152 (1983)

[4]   D. Le Rudulier, A. R. Strom, A. M. Dandekar, L. T. Smith and R. C. Valentine, Molecular biology of osmoregulation, *Science* 224: 1064 (1984)

[5]   B. Perroud and D. Le Rudulier, Glycine betaine transport in *Escherichia coli* osmotic modulation, *J. Bacteriol.* 161: 393 (1985)

[6]   T. Bernard, J. A. Pocard, B. Perroud and D. Le Rudulier, Variations in the response of salt-stressed *Rhizobium* strains to betaines, *Arch. Microbiol.* 143: 359 (1986)

[7]   J. A. Pocard, T. Bernard, G. Goas and D. Le Rudulier, Restauration partielle, par la glycine betaïne, de l'activité fixatrice d'azote de jeunes plantes de *Medicago sativa* L. soumises à un stress hydrique, *C. R. Acad. Sci.* 298: 477 (1984)

[8]   H. R. Hayward and T. C. Stadtman, Anaerobic degradation of choline. I. Fermentation of choline by an aerobic cytochrome-producing bacterion, *Vibrio cholinophagum* N. sp. *J. Bacteriol.* 78: 557 (1959)

[9]   G. J. J. Kortstee, The aerobic decomposition of choline by microorganisms, *Arch. Microbiol.* 71: 235 (1970)

[10]   J. H. F. G. Heijthuijsen and T. A. Hansen, Betaine fermentation and oxydation by marine *Desulfuromonas* strains, *Appl. Environ. Microbiol.* 55: 965 (1989)

[11]   L. T. Smith, J. A. Pocard, T. Bernard and D. Le Rudulier, Osmotic control of glycine betaine biosynthesis and degradation in *R. meliloti*, *J. Bacteriol.* 170: 3142 (1988)

[12]   J. A. Pocard, T. Bernard, L. T. Smith and D. Le Rudulier, Characterization of three choline transport activities in *R. meliloti*: modulation by choline and osmotic stress, *J. Bacteriol.* 171: 531 (1989)

[13]   J. B. Stock, B. Rauch and S. Roseman, Periplasmic space in *Salmonella typhimurium* and *Escherichia coli*, *J. Biol. Chem.* 252: 7850 (1977)

[14]   G. May, E. Faatz, M. Villarejo and E. Bremer, Binding protein dependent transport of glycine betaine and its osmotic regulation in *Escherichia coli* K 12, *Mol. Gen. Genet.* 205: 225 (1986)

[15]   J. Cairney, I. R. Booth and C. F. Higgins, *Salmonella typhimurium proP* gene encodes a transport system for the osmoprotectant betaine, *J. Bacteriol.* 164: 1218 (1985a)

[16]   J. Cairney, I. R. Booth and C. F. Higgins, Osmoregulation of gene expression in *Salmonella typhimurium*: *proU* encodes an osmotically induced betaine transport system, *J. Bacteriol.* 164: 1224 (1985b)

[17]   A. Barron, J. U. Jung and M. Villarejo, Purification and characterization of a glycine betaine binding protein from *Escherichia coli*, *J. Biol. Chem.* 262: 11841 (1987)

[18]   C. F. Higgins, L. Sutherland, J. Cairney and I. R. Booth, The osmotically regulated *proU* locus of *Salmonella typhimurium* encodes a periplasmic betaine-binding protein, *J. Gen. Microbiol.* 133: 305 (1987)

[19]   H. C. Neu and L. A. Heppel, The release of enzymes from *Escherichia coli* by osmotic shock and during the formation of spheroplasts, *J. Biol. Chem.* 240: 3685 (1965)

[20]   S. Nobile and J. Deshusses, Detection of ligand-protein binding by direct electrophoresis of the complex, *J. Chromatogr.* 449: 331 (1988)

IS THE NA⁺-ACTIVATED NADH-QUINONE-ACCEPTOR OXIDOREDUCTASE IN MARINE BACTERIA AND MODERATE HALOPHILES A PRIMARY ELECTROGENIC NA⁺ PUMP?

Robert A. MacLeod

Department of Microbiology
Macdonald College
McGill University
21,111 Lakeshore Rd.
Ste Anne de Bellevue, Québec
Canada H9X 1C0

ABSTRACT

Tokuda and Unemoto observed that a concentration of the protonophore carbonyl cyanide m-chlorophenylhydrazone (CCCP) which completely collapsed the membrane potential of the marine bacterium *Vibrio alginolyticus* 138-2 and the moderate halophile *Vibrio costicola* at pH 7 only partially and transiently collapsed it at pH 8.5. They also found that a concentration of CCCP which inhibited Na^+ extrusion from and Na^+ dependent amino acid uptake by *V. alginolyticus* 138-2 at pH 7.0 did not do so at pH 8.5. To explain these observations they proposed the existence of a primary electrogenic Na^+ extrusion system which was uncoupler resistant and functioned best at alkaline pH. The organisms examined have been shown to have a Na^+-activated NADH:quinone-acceptor oxidoreductase with a pH optimum at 8. Since mutants of *V. alginolyticus* 138-2 lacking this enzyme were sensitive to CCCP at pH 8.5, it was concluded that the enzyme functions as the proposed electrogenic Na^+ pump. Hamaide et al., however, demonstrated Na^+/H^+ antiport activity at pH 8.5 in *V. costicola* which was sensitive to the protonophores CCCP and 3,3',4',4-tetrachlorosalicylanilide (TCS). MacLeod et al. found that CCCP inhibited Na^+-dependent amino acid uptake into a number of marine bacteria and *V. costicola* at pH 8.5 if the concentration was increased sufficiently while TCS was almost as inhibitory in this capacity at pH 8.5 as at 7. Relatively low concentrations of TCS collapsed the membrane potential of *V. alginolyticus* 118 at pH 8.5 and prevented Na^+ extrusion from the cells. These findings suggested that NADH oxidation at pH 8.5 in this organism and *V. costicola* leads to the extrusion of protons which in turn cause Na^+ to be pumped out of the cells via a Na^+/H^+ antiporter. Reasons for the differences in the conclusions reached are discussed.

REVIEW AND ASSESSMENT

All the Gram-negative marine bacteria that have been examined in detail using chemically-defined media, some 700 strains and species, have been found to require Na^+ specifically for growth [1]. In the limited number of these which have been examined further, the bacteria also have a requirement for Na^+ for membrane transport [2, 3, 4]. One organism which has been examined in detail in our laboratory, *Alteromonas haloplanktis* 214 has been found to require Na^+ for the uptake of every metabolite tested. This included all the amino acids taken up by the cells, tricarboxylic acid cycle intermediates and the inorganic ions, K^+ and $PO_4\equiv$ [3, 5].

In 1977 Unemoto et al [6] discovered another function of Na^+ in the metabolism of the marine bacterium *V. alginolyticus* 138-2 and the moderate halophile *V. costicola*. They found that the NADH oxidase activities of membrane preparations of these organisms were strongly and specifically stimulated by Na^+. Those of *Escherichia coli* were not. Because Na^+ dependent activation of NADH oxidase

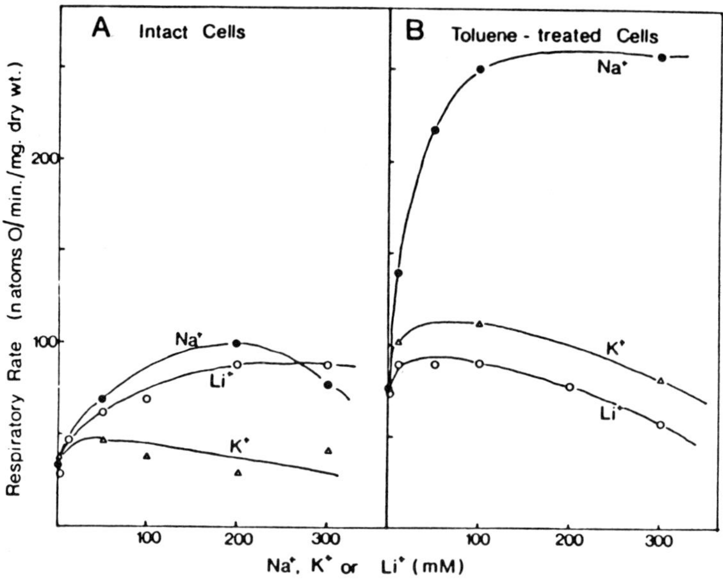

Figure 1. Effects of Na^+, K^+ and Li^+ on the rate of oxidation of NADH by intact cells (A) and by toluene-treated cells (B) of *A. haloplanktis* 214. Figure from [7] (reprinted with permission).

occurred in the halophilic organisms but not in *E. coli* they suggested that this property might prove to be a characteristic feature of halophilic bacteria.

Studies in our laboratory confirmed that Na^+ strongly stimulated NADH oxidation in a number of marine bacteria [7]. This is shown with *A. haloplanktis* 214 in Figure 1 where we compared the effect of ions on NADH oxidation by intact and toluene-treated cells. Intact cells can be seen to oxidize NADH at a low rate by a process which is stimulated by Na^+ or Li^+ but not by K^+. When the cells were made permeable to NADH by treatment with toluene, NADH was oxidized in the absence of Na^+ but at a much faster rate in its presence. The requirement for Na^+ for maximum rate of oxidation of NADH is 100 to 200 mM while the internal concentration of Na^+ in actively metabolizing cells is so low that no flux of $^{22}Na^+$ into the cells with AIB could be demonstrated [8]. Internal K^+, however, is of the order of 400 mM [9]. At 400 mM, K^+ strongly spared the requirement for Na^+ for maximum rate of NADH oxidation by toluene treated cells of *A. haloplanktis* [7] and by membrane preparations of *V. alginolyticus* [6] thus explaining how NADH can be oxidized at a high rate inside the cells, even when the intracellular Na^+ concentration is very low.

Unemoto and Hayashi [10] subsequently showed that the site of Na^+ dependent activation in the respiratory chain of *V. alginolyticus* 138-2 was the NADH: quinone acceptor oxidoreductase.

In 1981 Tokuda and Unemoto [11] observed that the membrane potential of K^+-depleted and Na^+-loaded *V. alginolyticus* 138-2 was completely collapsed at pH 6 to 7 by the protonophore carbonyl cyanide m-chlorophenylhydrazone (CCCP) but was only partially and transiently collapsed by the same uncoupler at the same concentration at pH 8.5. Since the concentration of protonophore used caused the intracellular pH to decrease, it was concluded that the uncoupler had made the membrane permeable to protons. They also found that respiration-dependent Na^+ extrusion from the cells as well as α-aminoisobutyric acid (AIB) transport into the cells was inhibited by CCCP at pH 6.5 but not at pH 8.5. To explain these observations the authors proposed the existence of an uncoupler insensitive primary electrogenic Na^+ extrusion system (a Na^+ pump) which operates at alkaline pH.

Tokuda [12] and Tokuda and Unemoto [13] reported the isolation of mutants of *V. alginolyticus* which lacked the Na^+-activated NADH oxidase activity and which had a Na^+ extrusion system which at pH 8.5 was sensitive to CCCP. A spontaneous revertant of one of these mutants had these properties restored to those characteristic of the wild-type strain. The authors thus concluded that the Na^+-activated NADH-quinone acceptor oxidoreductase served as the primary electrogenic Na^+ pump in *V. alginolyticus* 138-2. This enzyme was found to have a pH optimum about 8. Since growth of *V. alginolyticus* and *V. costicola* in complex medium was not inhibited at pH 8.5 by a concentration of CCCP which was inhibitory at pH 6.5 the authors concluded that the organisms exhibited two different types of energetics depending on the external pH. At acidic pH, only the proton motive force is generated as the immediate result of respiration. At alkaline pH an electrochemical potential of Na^+ is generated by a primary electrogenic Na^+ pump [14].

Since then, Tsuchiya and Shinoda [15] observed respiration-driven Na^+ extrusion from *Vibrio parahaemolyticus* which was insensitive to CCCP at pH 8.5. Ken-Dror et al. [16] found that the uncoupler, carbonyl cyanide p-trifluoromethoxyphenylhydrazone (FCCP), at the concentration tested actually stimulated slightly the rate of Na^+ extrusion from the halotolerant bacterium Ba_1 at pH 8.5. Dibrov and co-workers [17] concluded that results they obtained with inverted membrane vesicles of *V. alginolyticus* 138-2 were in agreement with the suggestion of Tokuda and Unemoto that a primary electrogenic Na^+ pump associated with NADH oxidation was present in this organism. More recently, Dimroth and Thomer [18] observed that inverted membrane vesicles of *Klebsiella pneumoniae* accumulated Na^+ ions upon NADH oxidation. Na^+ uptake at pH 8.5 was slightly stimulated by the concentration of FCCP tested and they concluded from this and other observations that the organism had a respiratory Na^+ pump similar to that in *V. alginolyticus* 138-2. Müller et al. [19] have observed Na^+-extrusion from the methanogen *Methanosarcina barkeri* which is insensitive to the protonophore 3,3',4',5-tetrachlorosalicylanilide (TCS) at pH 6.8. They have concluded that this organism has a primary electron transport-driven Na^+ pump.

In view of all the work that has been conducted by Tokuda and Unemoto and co-workers on Na^+-extrusion by *V. alginolyticus* and *V. costicola* and the apparent confirmation of this work in these and other organisms by other workers, why is there a question as to whether the Na^+-activated NADH-quinone acceptor oxidoreductase operates as a primary electrogenic pump at alkaline pH in marine bacteria and moderate halophiles?

Hamaide, Kushner and Sprott [20] working with a strain of the moderate halophile *V. costicola* found that when anaerobic cells of the organism suspended in 1 M KCl buffered at pH 8.5 were pulsed either with O_2 or with NaCl, acidification of the medium occurred. This was followed in each case by the slower return of the medium to its original pH. The protonophores CCCP and TCS prevented the acidification. Acidification on pulsing with O_2 indicates that efflux of protons from *V. costicola* occurs at pH 8.5 during respiration. Acidification on pulsing with NaCl suggests that a Na^+/H^+ antiporter is functioning in *V. costicola* at pH 8.5. Prevention of net proton movement by the protonophores suggests that these reagents speed the resorption of protons by the cells.

When anaerobic cells of *V. costicola* were suspended in 1 M NaCl at pH 8.5 and pulsed with O_2 an alkalinization of the medium resulted. This alkalinization was prevented by CCCP and TCS. These results are in accord with the prediction that

Na^+/H^+ antiport activity should be most rapid at alkaline pH to maintain a pH gradient of orientation acidic inside. This could be achieved if the stoichiometry of proton efflux during respiration to proton influx and Na^+ efflux through the antiporter was 2:3:2 as predicted by Krulwich [21] for alkalophilic bacteria, Figure 2. According to this model the primary pump is a respiration driven proton pump. The proton gradient is converted to an inwardly directed Na^+ gradient by a Na^+/H^+ antiporter. The stoichiometry of Na^+ and H^+ movements accounts for the development of a $\Delta\Psi$, inside negative, a cytoplasm acidic relative to the outside and a Na^+ motive force suitable for driving Na^+-dependent transport processes. Hamaide et al. [22] have shown that *V. costicola* transports α-aminoisobutyric acid (AIB) at pH 8.5 by a Na^+-dependent process which is inhibited by both the protonophores CCCP and TCS.

Kakinuma and Unemoto [4] observed that Na^+-dependent sucrose uptake by *V. alginolyticus* 138-2 was completely inhibited at pH 7 by the protonophore CCCP but only partially at pH 8.5. A mutant of the organism which lacked the Na^+-activated NADH oxidase system had a Na^+-dependent sucrose uptake system which was sensitive to CCCP at pH 8.5. From this, these workers concluded that sucrose uptake by *V. alginolyticus* 138-2 at pH 8.5 was driven by the Na^+ electrochemical gradient generated by a protonophore-resistant primary electrogenic Na^+ pump.

We examined the effect of protonophore concentration on amino acid uptake by 3 marine bacteria, the moderate halophile *V. costicola* and *E. coli* [23], Figure 3. All three protonophores tested, CCCP, FCCP and TCS inhibited transport into the three

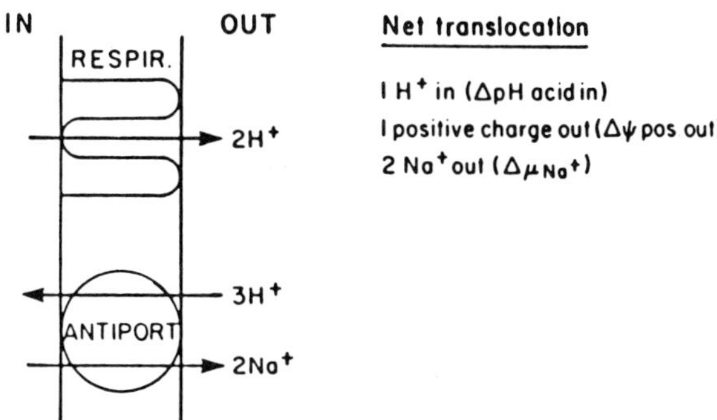

Figure 2. Example of how an electrogenic Na^+/H^+ antiporter together with a primary proton pump might produce a $\Delta\Psi$ inside negative, a Na^+ motive force and net acidification of the cell interior. Figure from [21] (reprinted with permission).

marine bacteria and *V. costicola* completely at pH 7.0 when added at 1 to 5 μM (10-50 nmoles/mg cell dry weight). For *E. coli* 5 to 10 times as much of the protonophores was required for complete inhibition of uptake. At pH 8.5 the extent of the inhibitory effect varied depending on the protonophore and the organism. CCCP was less effective at pH 8.5 than at 7 with all of the organisms but was still completely effective at concentrations ranging from 10 to 50 μM depending on the organism. For two of the marine bacteria and *V. costicola* the protonophore TCS was almost as effective at pH 8.5 as at 7 at inhibiting transport. It was more effective than CCCP at pH 8.5 for *A.*

Table 1.   Effect of TCS concentration on the $\Delta\Psi$ of *V. alginolyticus* 118 at pHs 7.0 and 8.5. Table from [23] (reprinted with permission).

TCS[a]	$\Delta\Psi$[b] at pH:	
	7.0	8.5
0	−166 (±3.6%)	−189 (±6%)
5 (1.65)	0	−128 (±20%)
10 (3.30)	0	−120 (±17%)
20 (6.60)	0	0

[a] Values represent the micromolar concentration (nanomoles of TCS per milligram of cell dry weight).

[b] Values represent the average (average deviation) of triplicate determinations.

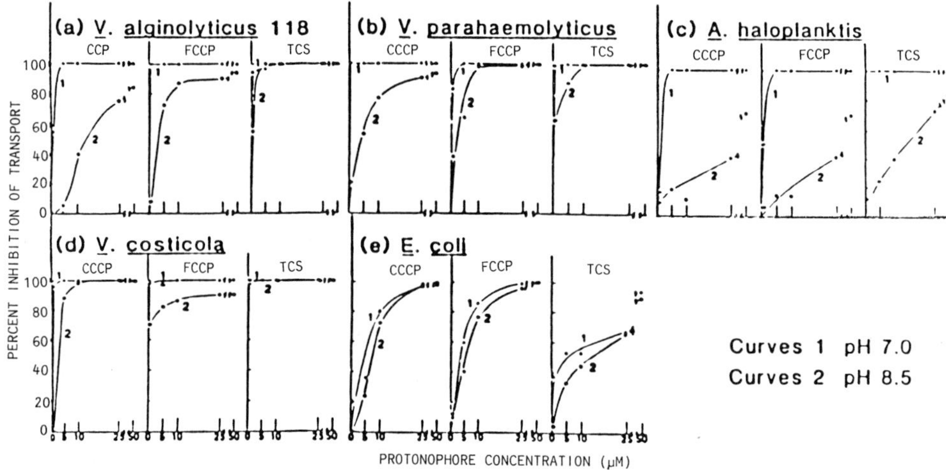

Figure 3. Capacity of the protonophores CCCP, FCCP and TCS to inhibit the initial rate of uptake of α-aminoisobutyric acid (Figures a, b, d) and D-alanine (Figures c, e) into the various organisms indicated, at pH 7.0 and 8.5. Figure from [23] (reprinted with permission).

102

haloplanktis. Only for *E. coli* was TCS less effective at pH 8.5 than the other protonophores. For all of the organisms the effect of FCCP was intermediate between that of CCCP and TCS.

Thus for three organisms which have been shown to have a Na^+ activated NADH oxidase, the protonophore TCS at low concentration and CCCP and FCCP at higher concentrations can inhibit transport at pH 8.5. When the effect of TCS on the membrane potential was tested using one of these organisms, *V. alginolyticus* 118, it can be seen (Table 1) that TCS could collapse $\Delta\Psi$ completely at pH 8.5 though more was required at pH 8.5 than at pH 7. These results indicate that TCS is capable of acting as a protonophore in this organism at pH 8.5.

The fact that TCS acts as a protonophore at pH 8.5 and therefore can collapse the membrane potential does not rule out the possibility that the membrane potential was generated by a protonophore-insensitive primary electrogenic Na^+ pump. To determine if such a pump was operative we examined the effects of inhibitors on Na^+ extrusion from sodium-loaded cells of *V. alginolyticus* 118 using techniques similar to those applied by Tokuda [24]. The results (Figure 4) show that TCS did indeed, like the respiratory inhibitors KCN and HOQNO, prevent Na^+ extrusion by the cells at pH 8.5 while CCCP was less effective.

The results of Hamaide et al. with *V. costicola* and of ourselves with *alginolyticus* 118 support the conclusion that Na^+ extrusion from these organisms at pH 8.5 occurs by way of a Na^+/H^+ antiporter not by means of a primary electrogenic Na^+ pump. How then could so many workers seemingly confirm the findings of Tokuda and co-workers using other organisms? This is because the distinction between the operation of a primary Na^+ pump and a secondary Na^+/H^+ antiporter has rested primarily upon lack of inhibition of Na^+ extrusion at pH 8.5 by CCCP.

CCCP is known to function poorly as a protonophore at alkaline pH [21]. That it is less effective than TCS at alkaline pH can be attributed to its lower pKa value, Figure 5. Protonophores are weak acids and both dissociated and undissociated species are believed to be required in the membrane to produce a cycle which transports protons across the membrane [25]. As the pH rises the concentration of the undissociated form of the protonophore could be expected to become too low to maintain the cycle.

Although CCCP is not as effective as TCS as a protonophore at pH 8.5, our data suggests that if its concentration is raised sufficiently at this pH it can be inhibitory. Most of the workers using CCCP have used it at pH 8.5 at concentrations which our data suggests could not be expected to be inhibitory [23]. Some concentrations may permit the entry of protons as Tokuda and Unemoto have shown [26], but not at a rate sufficient to balance the rate of extrusion of protons via the primary proton pump.

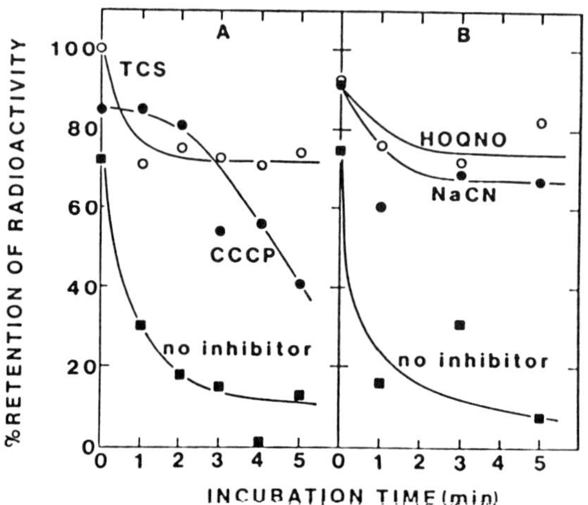

Figure 4. Effect of inhibitors on the extrusion of Na^+ from Na^+-loaded cells of *V. alginolyticus* 118 at pH 8.5. Figure from [23] (reprinted with permission).

Figure 5. Structures and pKa values of the protonophores studied (see [23] see [23] for sources of data).

There are obviously other factors besides pH which determine the sensitivity of different micro-organisms to protonophores. *E. coli* was one of the least sensitive of the organisms tested to all three protonophores, *V. costicola* was one of the most. Studies by Hopfer and co-workers [27] have shown that differences in phospholipid composition of artificial bilayer membranes can affect their penetrability by protonophores. The insensitivity of *E. coli* may possibly be related also to a relative lack of penetrability of its outer membrane by protonophores.

Although the results of Hamaide et al. with *V. costicola* and our observations with *V. alginolyticus* 118 do not support the conclusion that a protonophore-resistant primary electrogenic Na^+ pump operates in these organisms at alkaline pH, Tokuda has isolated two mutants of *V. alginolyticus* 138-2 which at pH 8.5 were more sensitive than the parent cells to CCCP during growth and transport, extruded Na^+ by a CCCP-sensitive mechanism and lacked Na^+-activated NADH oxidase activity. In a spontaneous revertant of one of these mutants CCCP insensitivity and Na^+-activated NADH oxidase activity were both restored. These mutants, however, need to be examined further. The CCCP sensitivity of the mutants at pH 8.5 suggests that in the mutants the Na^+ extrusion system is a Na^+/H^+ antiporter, a system not detected in the parent cells. Although the obtaining of a spontaneous revertant by Tokuda suggests that a single mutation was involved, the possibility that the mutation exhibits pleiotropic properties must be considered. Mutations to non-alkalophily in *Bacillus alcalophilus* showed pleiotropic characteristics even though evidence that the changes could result from a single mutation was presented [28]. Thus one of the characteristics of the mutation in *V. alginolyticus* 138-2 could be an increased penetrability of the cytoplasmic membrane by CCCP.

Although the evidence that Hamaide et al. and ourselves have obtained does not support the conclusion that the Na^+-activated NADH-quinone acceptor oxidoreductase is a primary electrogenic Na^+ pump in the marine bacteria and moderate halophile examined, Müller et al. [19] have provided evidence for electron-transport-driven sodium extrusion during methanogenesis from formaldehyde and molecular hydrogen by *Methanosarcina barkeri*. They found that a concentration of TCS which at pH 7 stopped methanogenesis, collapsed the proton motive force and caused a decline of intracellular ATP concentration, failed at the same pH to prevent Na^+ extrusion from the cells when formaldehyde but not when methanol was the substrate.

Part of the appeal of considering that a primary electrogenic Na^+ pump operates at alkaline pH in marine bacteria and moderate halophiles is the difficulty of accounting for the high phosphorylation potential in the cells when the pH gradient is in the reverse direction and the proton motive force is apparently low. This is a special problem for alkalophilic bacteria whose optimal pH for growth on a nonfermentable carbon source is pH 10.5. Guffanti and Krulwich [29] have shown that ATP synthesis in these cells at this pH is driven by an imposed pH but not by an imposed pNa^+. They conclude that when natural proton pumps energize ATP synthesis in these organisms they may deliver at least some of the necessary protons to the ATP synthase by a pathway that avoids complete equilibration of the protons with the highly alkaline medium. This concept of localized proton pathways in cell membranes is not new. It was proposed by R. J. P. Williams a number of years ago [30].

Since the range of pH in sea water is 7.5 to 8.5, marine bacteria would be expected to have somewhat less of a problem than alkalophiles generating a proton motive force. In *V. costicola* at pH 8.5 most of the proton motive force is derived from the membrane potential [31]. The possibility that local proton pathways can account for the high membrane potentials in *V. costicola* and marine bacteria at alkaline pH needs to be considered.

REFERENCES

[1]   J. L. Reichelt and P. Baumann, *Arch. Microbiol.* 97: 329 (1974)

[2]   G. R. Drapeau and R. A. MacLeod, *Biochem. Biophys. Res. Comm.* 12: 111 (1963)

[3]   R. Droniuk, P. T. S. Wong, G. Wisse and R. A. MacLeod, *Appl. Environ. Microbiol.* 53: 1487 (1987)

[4]   Y. Kakinuma and T. Unemoto, *J. Bacteriol.* 163: 1293 (1985)

[5]   H. M. Hassan and R. A. MacLeod, *J. Bacteriol.* 121: 160 (1975)

[6]   T. Unemoto, M. Hayashi and M. Hayashi, *J. Biochem. (Tokyo)* 82: 1389 (1977)

[7]   G. Khanna, L. DeVoe, L. Brown, D. F. Niven and R. A. MacLeod, *J. Bacteriol.* 157: 59 (1984)

[8]   G. D. Sprott, J. P. Drozdowski, E. L. Martin and R. A. MacLeod, *Can. J. Microbiol.* 21: 43 (1975)

[9]   J. Thompson and R. A. MacLeod *J. Biol. Chem.* 248: 7106 (1973)

[10]   T. Unemoto and M. Hayashi, *J. Biochem. (Tokyo)* 85: 1461 (1979)

[11]   H. Tokuda and T. Unemoto, *Biochem. Biophys. Res. Comm.* 102: 265 (1981)

[12]   H. Tokuda, *Biochem. Biophys. Res. Comm.* 114: 113 (1983)

[13]   H. Tokuda and T. Unemoto, *J. Biol. Chem.* 259: 7785 (1984)

[14]   H. Tokuda and T. Unemoto, *J. Bacteriol.* 156: 636 (1983)

[15]   T. Tsuchiya and S. Shinoda *J. Bacteriol.* 162: 794 (1985)

[16]   S. Ken-Dror, R. Preger and Y. Avi-Dor, *Arch. Biochim. Biophys.* 244: 122 (1986)

[17]   P. A. Dibrov, V. S. Kostyrko, R. L. Lazarova, V. P. Skulachev and I. A. Smirnova, *Biochim. Biophys. Acta* 850: 449 (1986)

[18]   P. Dimroth and A. Thomer, *Arch. Microbiol.* 151: 439 (1989)

[19]   V. Müller, C. Winner and G. Gottschalk, *Eur. J. Biochem.* 178: 519 (1988)

[20]   F. Hamaide, D. J. Kushner and G. D. Sprott, *J. Bacteriol.* 161: 681 (1985)

[21]   T. A. Krulwich, *Biochim. Biophys. Acta* 726: 245 (1983)

[22]   F. Hamaide, G. D. Sprott, and D. J. Kushner, *Biochim. Biophys. Acta* 766: 77 (1984)

[23]   R. A. MacLeod, G. A. Wisse and F. L. Stejskal, *J. Bacteriol.* 170: 4330 (1988)

[24]   H. Tokuda, *Methods Enzymol.* 125: 520 (1986)

[25]   S. G. A. McLaughlin and J. P. Dilger, *Physiol. Rev.* 60: 825 (1980)

[26]   H. Tokuda and T. Unemoto, *J. Biol. Chem.* 257: 10007 (1982)

[27]   U. Hopfer, A. L. Lehninger and W. J. Lennarz, *J. Membr. Biol.* 3: 142 (1970)

[28]   R. J. Lewis, E. Kaback and T. A. Krulwich, *J. Gen. Microbiol.* 128: 427 (1982)

[29]   A. A. Guffanti and T. A. Krulwich, *J. Biol. Chem.* 263: 14748 (1988)

[30]   R. J. P. Williams, *FEBS Lett.* 85: 9 (1978)

[31]   F. Hamaide, D. J. Kushner and G. D. Sprott, *J. Bacteriol.* 156: 537 (1983)

IONIC RELATIONS AND POLYOL METABOLISM OF MARINE FUNGI IN RELATION TO THEIR ENVIRONMENT

D. H. Jennings

Department of Genetics & Microbiology
The University
P. O. Box 147
Liverpool L69 3BX
U.K.

ABSTRACT

The filamentous Hyphomycete *Dendryphiella salina* is the most extensively studied marine fungus. It has been shown by radiotracer flux analysis and X-ray microanalysis that the cytoplasmic concentrations of K, Na and Cl in mycelium gowing in 500 mM NaCl is of the order of 51-88, 74-139 and 160 mM respectively. There appears to be no accumulation of salt in vacuoles. *In vivo* studies of the effect of K and NaCl on enzymes are in keeping with the above values. Polyols make a major contribution to the osmotic ballast. The total concentration at any one external water potential is relatively constant. However the proportions of the four individual polyols, glycerol, erythritol, arabitol and mannitol may differ according to the solute generating the water potential. Growth of the marine yeast *Debaryomyces hansenii* in continuous culture suggests that polyols as well as producing compatible osmotic ballast are also, through their metabolism, involved in energy dissipation via futile cycles under conditions when growth becomes limited by conditions other than carbon supply. Primary and secondary transport in marine fungi is proton- and not sodium-based but a large volume cell wall may be involved in maintaining the appropriate proton electrochemical potential gradient.

INTRODUCTION

I wish to discuss what we know about the ability of higher marine fungi to generate the appropriate internal osmotic potential when growing in their natural

General and Applied Aspects of Halophilic Microorganisms
Edited by F. Rodriguez-Valera, Plenum Press, New York, 1991

environment. By higher marine fungi I mean those members of the Ascomycotina and related members of the Fungi Impertecti [1]. No significant physiological or biochemical studies have yet been made on marine members of the Basidiomycotina or the Oomycetes [2]. We need to remember also that there is a group of lower fungi, the Thraustochytrids and their allies, which have a macro-requirement for sodium in the growth medium [2]. There is debate however about whether or not these organisms are indeed fungi [3].

There are several reasons for interest in how higher marine fungi grow in their marine environment. First, we need to know whether this group of walled-organisms use similar mechanisms for generating turgor as other such organisms which grow in the sea. Second, there is a particular mycological interest. It is believed that certain species have evolved from marine algal ancestors while other marine fungi are secondarily marine, having evolved from terrestrial counterparts which became adapted to their marine environment [4]. Here I shall be considering the two best studied marine fungi the filamentous Hyphomycete *Dendryphiella salina* for which we have most information about physiology and biochemistry and the yeast *Debaryomyces hansenii* for which there is less but nevertheless substantial information. Both fungi are likely to be secondarily marine.

GENERATION OF THE OSMOTIC POTENTIAL IN *DENDRYPHIELLA SALINA*

Four polyols are routinely found in growing mycelium of *D. salina* namely arabitol, erythritiol, glycerol and mannitol [5]. In non-growing mycelium only arabitol and mannitol are present in any amount. Glycerol is rapidly broken down when mycelium is transferred to medium containing only glucose. When such mycelium is exposed to increased concentrations of sodium chloride, both arabitol and mannitol increase in concentration within the mycelium [6]. In growing mycelium it is glycerol which becomes the most abundant polyol within mycelium when there are high concentrations of sodium chloride outside [5]. But, when other solutes bring about an equivalent reduction in the medium water potential, glycerol may make a much less significant contribution to the mycelial osmotic potential. Indeed, when inositol is used to generate the low medium water potential, glycerol is absent in the mycelium [5]. The striking feature of polyol metabolism in mycelium growing in the media of low water potential is that the total soluble carbohydrate content of the mycelium is relatively constant at any one medium water potential irrespective of the solute used to generate that potential. In most instances, i.e. when salts are used, soluble carbohydrate equates with polyol. With inositol, it is the cyclitol accumulated inside as well as the polyol which must be included in the calculation. Thus, within the fungus there are mechanisms regulating the contribution of soluble carbohydrates to the osmotic potential. One point of regulation is the flux of glucose to glycogen, such that soluble carbohydrate is removed to insoluble form [7]. Other points of regulation are probably in the balance of $NADP^+$, $NADPH$, NAD^+ and $NADH$ in the hyphae, the balance

Table 1. Concentrations of K, Na and Cl in the wall, cytoplasm and vacuoles of hyphae of *Dendryphiella salina* grown in the presence of 500 mM NaCl as determined by radiotracer flux analysis (R) and X-ray microanalysis (X). Results are given in terms of analysed volume, those in brackets have been expressed on a tissue water basis.

	Concentration (mM)				
	K		Na		Cl
	R	X	R	X	X
Wall	16.3	74	357	172	210
Cytoplasm	50.8(64)	88(110)	73.9(92.6)	139(173)	159(19)
Vacuole		69(86)		159(199)	138(173)

changing according to the solute absorbed to contribute to the non-polyol part of the internal osmotic potential [8].

The exact contribution of polyols to the internal osmotic potential depends on the assessment of the concentration of inorganic ions within the protoplasm of the mycelium. Two methods have been used to make an assessment of the necessary values – radiotracer flux analysis [9] and X-ray microanalysis [10]. The first procedure gives values for the wall and the protoplasm – the efflux curve does not allow discrimination between cytoplasm and vacuole –while the latter gives values for wall, cytoplasm and vacuoles. Table 1 shows that the cytoplasmic concentrations (on the basis of analysed volume) of K, Na and Cl are 51-88, 74-139 and 160 mM respectively.

For K and Na there is very reasonable agreement for the results for cytoplasmic concentrations obtained by the two methods; the lack of agreement for values for the wall is discussed later. Even when concentrations are calculated on a tissue water basis (and there is argument as to whether this necessarily gives a better assessment of the actual concentrations present), the values are not much higher: 64-110, 93-173 and 199 mM for K, Na and Cl respectively.

Stereological analysis of transmission electron micrographs has allowed an assessment of the volume fractions of wall cytoplasm and vacuole [11]. From such analysis, it is possible to more properly appraise the contribution of vacuoles to the sequestration of ions within the protoplasm (Table 2). Vacuoles are only a small fraction (17%) of the protoplasm of mycelium growing in the presence of 500 mM NaCl. It is clear that only a small concentration (no more than c39 mM) is found in them of potassium and sodium chlorides. Thus unlike higher plants, vacuoles do not have a role in the generation of turgor through the accumulation of a substantial volume of a high concentration of sodium chloride.

Other measurements [5, 9] have shown that the osmotic potential of the cytoplasm of *D. salina* is made up of c29-39% polyols and c40% (at a maximum of 188 mM) potassium and sodium chlorides. Even if the concentration of these salts were to be, as given in Table 2 in terms of tissue water content, around 283 mM, this concentration would not necessarily interfere unduly with metabolic processes. It has been shown [12] that the presence of 200 mM NaCl at pH 7.4 has virtually no effect on the *in vitro* activity of glucose 6-phosphate dehydrogenase and, while malate dehydrogenase is inhibited, there is still significant (86% of control) activity.

Some comment is necessary concerning the discrepancy between the results for radiotracer flux analysis and X-ray microanalysis for K and Na in the wall (more strictly in the case of flux analysis one should be speaking of the free space). This

Table 2. The contribution of K, Na and Cl within either cytoplasm or vacuole to the total protoplasmic content of that ion in hyphae of *Dendryphiella salina* grown in the presence of 500 mM NaCl. Also given are the volume fractions (V_f) and the K:Na ratios within the two compartments.

	V_f	Concentration[1]			K:Na
		K	Na	Cl	
Cytoplasm	0.83	73(86)	115.4(81)	132(85)	0.63
Vacuole	0.17	11.7(14)	27(19)	23.5(15)	0.43

[1]Values are expressed as μ moles g^{-1} analysed volume as calculated from X-ray microanalysis data and volume fractions of the compartments. Bracketed values are the percentage contribution of a compartment to the total protoplasmic content of an individual ion species.

component has been shown stereologically to comprise 40% of the hyphal volume. At this stage it seems most probable that the discrepancy is due to different handling procedures for the mycelium prior to analysis.

PRIMARY ACTIVE TRANSPORT AND SELECTION FOR POTASSIUM AGAINST SODIUM

It has been demonstrated unambiguously that primary transport in *D. salina* is an electrogenic proton pump whose physiological characteristics *in situ* have much in common with that of *Neurospora crassa* [13]. Enzymatic evidence is compatible with an ATPase with a pH optimum of 6.5 - 7.0 [14], in spite of previous evidence to the contrary [15]. Glucose is transported into *D. salina* by a proton symport [13]. Thus it appears that in this fungus transport is proton-based; there is no evidence to date of a sodium extrusion pump. At present it is not clear how K : Na selectivity is maintained, though there is no doubt about the presence of a highly specific potassium transport system which is dependent on the presence of calcium [16, 17]. With respect to the generation of the necessary proton-motive force for transport of other solutes, it is possible that the large volume hyphal wall is a necessary adjunct [10]. In the alkaline medium of the sea, the pH component of the force is much lower than in terrestrial fungi, if protons were extruded across the plasmalemma directly into sea water. The large volume of the wall might reduce the flow of protons into the external medium, thus increasing the pH gradient. Certainly there is clear evidence that at low temperature (4°) loss of potassium from the wall has a very high half-time (2-4h) [18] compared with what has been found for *N. crassa* at 20° (1.2 min) [19].

There is enzymic evidence for a proton extrusion pump in *Debaryomyces hansenii* similar to that in *N. crassa* and *D. salina* [20]. There are indications also that, consequent upon growth in the presence of macro-amounts of sodium, there is produced within the cell an ATPase with an alkaline pH (8.5) optimum. This suggests for this marine yeast that there is accumulation of sodium in the vacuole but this supposition needs to be investigated in detail.

POLYOLS AS PHYSIOLOGICAL BUFFERING AGENTS INVOLVED IN ENERGY SPILLAGE

Chemostat studies on *D. hansenii* have shown that, under conditions of carbon limitation, the yield of cells becomes increasingly less than expected from standard growth kinetics as the growth rate moves towards the maximum [21]. Under saline conditions, the divergence between actual and expected becomes greater as the concentration of NaCl in the medium is increased. This might suggest that the maintenance requirements become greater under these conditions. In fact, it seems more likely that growth is limited by factors other than availability of carbon, particularly so under saline conditions. A probable factor is the inability of transport systems at high growth rates to cope with the demands of the decreased medium water potential such that the necessary internal osmotic potential is maintained in the cells to generate the turgor. At any rate, the consequences are that as the specific growth rate is increased by increasing the concentration of glucose in the medium, growth becomes no longer limited by carbon *per se*. The observations for *D. hansenii* mirror what has been observed for a range of microorganisms where growth ceases to be limited by availability of specific nutrients [22, 23]. Indeed because of lack of such limitation there becomes an imbalance between the rate of ATP production by catabolism and the rate at which ATP can be consumed by anabolism. In this respect, it appears as if there is inadequate regulation of cellular processes such that ATP has to be constantly spoiled. One way in which this can be achieved is via futile or substrate cycles [24].

Glycerol and mannitol as well as contributing to the internal osmotic potential of higher marine fungi are also involved in substrate cycles; the evidence that this is so is to be presented elsewuere [25]. Essentially, not only can there be a net flux of carbon into these two polyols which may or may not involve utilisation of ATP but there is also turnover without net synthesis which does involve the breakdown of ATP. Since these two polyols are also compatible solutes, stores of reducing power and sinks for and sources of protons, they have been described as physiological buffering agents [8]. It should be noted that there are a range of such agents relating to particular parts of metabolism within a fungus. But in higher marine fungi it is clear that of the physiological buffering agents the polyols, glycerol and mannitol, have a particularly significant role for the growth of these fungi in the sea.

REFERENCES

[1]   J. Kohlmeyer and E. Kohlmeyer, "Marine Mycology", Academic Press, New York (1979)

[2]   D. H. Jennings, Some aspects of the physiology and biochemistry of marine fungi. *Biol. Rev.* 58: 423 (1983)

[3]   A. H. L. Chamberlain and S. T. Moss, The Thraustochytrids: a protist group with mixed affinities. *Bio Systems*. 21: 341 (1988)

[4]   J. Kohlmeyer, Taxonomic studies of the marine Ascomycotina, *in* "The biology of marine fungi", S. T. Moss, ed. Cambridge University Press, Cambridge (1986)

[5]   J. M. Wethered, E. C. Metcalf and D. H. Jennings, Carbohydrate metabolism in the fungus *Dendryphiella salina*. VIII. The contribution of polyols and ions to the mycelial solute potential in relation to the external osmoticum. *New Phytol.* 101: 631 (1985)

[6]   D. H. Jennings, Cations and filamentous fungi: invasion of the sea and hyphal functioning, *in* "Ion transport in plants", W. P. Anderson, ed. Academic Press, London (1973)

[7]   J. C. B. MacDermott and D. H. Jennings, The relationship between the uptake of glucose and 3-*O*-methyl glucose and soluble carbohydrate and polysaccharide in the fungus *Dendryphiella salina*. *J. Gen. Microbiol.* 97: 193 (1976)

[8]   D. H. Jennings, Polyol metabolism in fungi, *Adv. Microbial Physiol.*, 25: 150 (1984)

[9]   N. J. W. Clipson and D. H. Jennings, The role of sodium and potassium in the generation of the osmotic potential of the marine fungus *Dendryphiella salina*. *Mycol. Res.* (in press) (1990)

[10]   N. J. W. Clipson, M. A. Hajibagheri and D. H. Jennings, Ion compartmentation in the marine fungus *Dendryphiella salina* in response to salinity: X-ray microanalysis. *J. Exp. Bot.* 41: 199 (1990)

[11]   N. J. W. Clipson, D. H. Jennings and J. L. Smith, The response to salinity at the microscopic level of the marine fungus *Dendryphiella salina* Nicot and Pugh as investigated stereologically. *New Phytol.* 113: 21 (1989)

[12]   F. M. Paton and D. H. Jennings, Effect of sodium and potassium chloride and polyols on malate and glucose 6-phosphate dehydrogenases from the marine fungus *Dendryphiella salina*. *Mycol. Res.* 92: 470 (1988)

[13]   J. M. Davies, C. Brownlee and D. H. Jennings, Electrophysiological evidence for an electrogenic proton pump and proton symport of glucose in the marine fungus *Dendryphiella salina*. *J. Exp. Bot.* 41: 449 (1990)

[14]   A. Garrill and D. H. Jennings, unpublished data (1989)

[15]   M. F. J. Galpin and D. H. Jennings, A plasma-membrane ATPase from *Dendryphiella salina*: cation specificity and interaction with fusicoccin and cyclicAMP. *Trans. Br. Mycol. Soc.* 75: 37 (1980)

[16]   E. B. G. Jones and D. H. Jennings, The effect of cations on the growth of fungi. *New Phytol.* 64: 85 (1965)

[17]   J. M. Davies, Ion transport in the marine fungus *Dendryphiella salina*, Ph.D. thesis, University of Liverpool (1989)

[18]   D. H. Jennings and J. S. Aynsley, Compartmentation and low temperature fluxes of potassium in mycelium of *Dendryphiella salina*. *New Phytol.* 70: 713 (1971)

[19]   C. W. Slayman and C. L. Slayman, Potassium transport in *Neurospora*. Evidence for a multisite carrier at high pH. *J. Gen. Physiol.* 55:758 (1970)

[20]   J. G. Commerford, P. T. N. Spencer-Phillips and D. H. Jennings, Membrane-bound ATPase activity, the properties of which are altered by growth in saline conditions isolated from the marine yeast *Debaryomyces hansenii*. *Trans. Br. Mycol. Soc.* 85: 431 (1985)

[21]   R. M. Burke and D. H. Jennings, The effect of sodium chloride on the growth characteristics of the marine yeast *Debaryomyces hansenii* in batch and continuous culture under carbon limitation and under potassium limitation. *Mycol. Res.* (in press)(1990)

[22]   D. W. Tempest and O. M. Neijssel, Growth yield values in relation to respiration, *in* "Diversity of bacterial respiratory systems, Vol. 1." C. J. Knowles, ed. C.R.C. Press, Boca Raton, (1980)

[23]   D. W. Tempest and O. M. Neijssel, The status of Y_{ATP} and maintenance energy as biologically interpretable phenomena. *Ann. Rev. Microbiol.* 38: 459 (1984)

[24]   E. A. Newsholme and B. Crabtree, Substrate cycles in metabolic regulation and in heat generation. *Biochem. Soc. Symp.* 41: 61 (1976)

[25]   D. H. Jennings and R. M. Burke, Compatible solutes – the mycological dimension and their role as physiological buffering agents. *New Phytol.* submitted. (1990)

SALT ADAPTATION OF *ECTOTHIORHODOSPIRA*

Johannes F. Imhoff, Toni Ditandy and Bernhard Thiemann

Institut für Mikrobiologie und Biotechnologie
Rheinische Friedrich-Wilhelms-Universität
Meckenheimer Allee 168
D-5300, F.R.G.

Ectothiorhodospira species are anoxygenic phototrophic bacteria which taxonomically belong to the purple sulfur bacteria. The great majority of the known anoxygenic phototrophic bacteria live in fresh water and marine habitats. Among the few species that require elevated salt concentrations are *Chromatium salexigens, Rhodospirillum salexigens, Rhodospirillum salinarum*, and *Ectothiorhodospira* species [1, 2].

THE NATURAL ENVIRONMENT

In nature *Ectothiorhodospira* species have been found in marine locations, in hypersaline environments and in concentrated brines of alkaline soda lakes. This presentation will be concerned with isolates from alkaline soda lakes of the Wadi Natrun in Egypt. These lakes have a negligible content of the divalent cations magnesium and calcium. The predominant cation is sodium (potassium concentrations are a few millimolar), which is accompanied in most of the lakes by balanced concentrations of sulfate, chloride and carbonate [3]. The unusual ionic composition and the strong alkalinity distinguishes these lakes from other well known hypersaline habitats like Great Salt Lake, Dead Sea and marine salterns (see [4]) and offers highly selective conditions for the mass development of *Ectothiorhodospira* species. Other halophilic phototrohpic purple bacteria, *Rhodospirillum* species and *Chromatium salexigens*, are dominant in marine salterns and require media of near neutral pH and an ion content typical of marine waters, including significant concentrations of divalent cations [5, 6, 7].

SALT RESPONSES OF *ECTOTHIORHODOSPIRA*

Halophily and halotolerance are two different aspects of life at high salt concentrations, that have to be distinguished clearly. Halophily is demonstrated by the requirement of high salt concentrations for optimum growth. Halotolerance qualitatively and quantitatively describes the ability to grow at salt concentrations higher than optimum. According to the salt concentrations required for optimum growth, halophilic bacteria were classified as non-halophilic (below 0.2 M NaCl), slightly halophilic (0.2 to about 1.0-1.2 M NaCl), moderately halophilic (about 1.0-1.2 to 2.0-2.5 M NaCl), and extremely halophilic (more than 2.0-2.5 M NaCl)[8]. The same ranges of salt concentrations were recommended to describe the degree of halotolerance, i.e. slightly, moderately and extremely halotolerant.

Salt Optima of *Ectothiorhodospira*

Salt optima of various *Ectothiorhodospira* species range from about 5% to 25% salinity. Salt optima of *E. mobilis* strains isolated from the Wadi Natrun were at about 5% salinity. Much higher salt concentrations were required by strains of *E. halochloris, E. abdelmalekii* and *E. halophila* [9]. *E. halophila* strains formed two groups. Strains of the first group had optima close to 15% salinity, strains of the second group, all isolated from the Wadi Natrun, had extremely high salt optima around 25% salinity. Strains of the second group even grow in saturated salt solutions and represent the most halophilic eubacteria known. They require higher salt concentrations for optimum growth than most of the extremely halophilic archaeobacteria [10].

Ionic Growth Requirements

Growth responses clearly indicate that a distinct but species or strain specific salinity is required by all *Ectothiorhodospira* isolates. In order to determine whether this salt dependence reflects a requirement for high osmolality, for a high but nonspecific ionic environment, or for a specific ion at high concentrations, a series of growth experiments were made. Three strains were selected for our studies, *E. mobilis* strain 9903 with optimum salinity at 5%, and two *E. halophila* strains (9630 and 9624) with salt optima at 15% and 25% salinity, respectively.

Inorganic anions. Although chloride is by far the dominant anion of the environment, growth of *E. mobilis* in the absence of added chloride was possible. Sodium chloride could be replaced by sodium sulfate and sodium carbonate [11]. Also sulfate is not essential for growth and may be omitted completely from the growth media. Concentrations of carbonate (200 mM in our standard medium) which is used to buffer the media and serves as a carbon source for *Ectothiorhodospira*, can be reduced considerably (Imhoff, unpublished results). It is concluded that *E. mobilis* does not require high concentrations of any specific anion.

Potassium. The content of potassium in the medium can be manipulated over a wide range of concentrations without significant changes in the growth response. A specific requirement for increased concentrations of potassium was not found in any of the halophilic *Ectothiorhodospira* species. In cultures of extremely halophilic *E. halophila* strains grown at 15% and 25% salinity, respectively, in the absence of added potassium (the actual potassium content of these media was less than 0.35 mM and 0.58 mM, respectively), no significant decrease of the cell yield during a number of successive transfers could be found compared to cultures grown with addition of 0.3 to 300 mM KCl.

Sodium. All results obtained so far point to a requirement for high concentrations of sodium ions. During cultivation of the cells at different osmolalities by isoosmotic variation of the concentrations of NaCl and KCl, growth responses of the three strains were different.

Growth optima of *E. mobilis* strain 9903 showed a clear correlation to the sodium concentration. Among all conditions tested the maximum growth rate was at 3% NaCl without addition of KCl, i.e. in standard medium with 5% salinity. Irrespective of the osmotic conditions and the concentration of KCl, the highest growth rates were always found in media containing 3% NaCl (equivalent to a total of 780 mM sodium). This indicates that *E. mobilis* has a specific requirement for the sodium ion. Evidently the standard medium at 5% salinity offers both, the optimum sodium requirement (780 mM) and the optimum osmotic requirement (1.29 osmolal).

Best growth of *E. halophila* strain 9624 was at 23% NaCl in the absence of added KCl (standard medium of 25% salinity, 12.05 osmolal). At and above a minimum NaCl concentration of 13%, both KCl and NaCl could replace each other isoosmotically. Growth optima correlated strikingly well with the osmotic conditions of 12.05 osmolal. In addition a distinct sodium optimum was found at 18% NaCl (20% standard medium). The results revealed discrepancies between osmotic and sodium optimum.

Optimum growth of *E. halophila* strain 9630 was found at 13% NaCl in the absence of added KCl (15% standard medium, osmolality of 5.20). Presumably this medium offers both, optimum osmotic conditions and optimum NaCl concentration. At higher NaCl concentrations, KCl stimulated growth up to concentrations of about 0.9 M, despite the fact that due to its addition osmolality and ion content of the media were increased. This stimulation could well be due to positive effects upon the osmotic adaptation by KCl. Growth stimulation by KCl under similar conditions was not found in the two other *Ectothiorhodospira* strains.

Influence of glycine betaine and glycerol. The isoosmolar replacement of NaCl by glycine betaine, added in concentrations of 1 M and 2M to the media, was not possible. In no case did betaine at such high concentrations stimulate growth; in most strains it showed strong growth inhibition. Nor could glycerol replace sodium in the growth media; it inhibited growth at concentrations exceeding 500 mM, and in the most halophilic strain from 100 mM upwards. It is concluded that inorganic salts cannot be replaced by betaine or glycerol to adjust the external osmolality, but that high concentrations of inorganic ions are required.

CYTOPLASMIC SOLUTE CONCENTRATIONS

Inorganic Ions

A specific requirement for sodium exists in all *Ectothiorhodospira* species, although the concentrations required for optimum growth and the maximum concentrations tolerated vary between the species. Sodium apparently is actively extruded from the cytoplasm of these bacteria. Although actual sodium concentrations of the cytoplasm have not been measured, several facts indicate that cytoplasmic sodium concentrations under active growth conditions have to be kept quite low. Low cytoplasmic sodium concentrations are indicated by enzyme inhibition at low salt concentrations, accumulation of molar concentrations of organic compatible solutes, and active transport of sodium out of the cells under illumination. Already low concentrations (less than 100 mM) of sodium salts significantly inhibit various cytoplasmic enzyme activities, such as isocitrate dehydrogenase, glutamate dehydrogenase, ribulose-biphosphate carboxylase, malate dehydrogenase, and ATPase from *E. halophila* and *E. halochloris* [12, 13, 14], (Imhoff, unpublished).

Chloride does not accumulate in cells of *E. mobilis* strain 9903, but is progressively excluded at higher salt concentrations. The cytoplasmic concentrations increased only slightly from 54 to 71 mM at external concentrations of 104 to 535 mM (under isoosmotic conditions). This indicates that *E. mobilis* has mechanisms to maintain relatively constant cytoplasmic concentrations of chloride and to exclude chloride from the cytoplasm at high external concentrations [11].

Sulfate is not essential for growth and can be completely omitted from the medium. Its cytoplasmic concentration was always low, irrespective of the salt concentration, but always at low external sulfate concentrations [11].

The cytoplasmic concentration of potassium in *E. mobilis* under standard growth conditions (5% salinity and pH 9.0) as measured by atomic absorption spectroscopy and using an ion sensitive electrode was 300-350 mM. This is equivalent to an accumulation factor of 60-70 in the cytoplasm compared to the extracellular space. Similar accumulation values (up to 100-fold) have been determined in *E. halochloris*, independent of the salinity of the growth medium [14]. These values are not significantly different from those found in many non-halophilic eubacteria.

Organic Compatible Solutes

^{13}C-NMR studies of Galinski [15, 16] have shown that glycine betaine, trehalose and ectoine are accumulated as major compatible solutes within *Ectothiorhodospira halochloris* and other *Ectothiorhodospira* species. This could also be demonstrated for *E. mobilis* strain 9903 (Imhoff, unpublished results), [8]. The results imply the ability of the considered bacteria to synthesize glycine betaine from acetate and carbonate [15]. Glycine betaine also may be taken up by the cells, accumulated in the cytoplasm, and thereby support osmotic adaptation, as found in a number of other bacteria (see [8]). Addition of low concentrations of glycine betaine (5 mM) to the media stimulated growth of all *Ectothiorhodospira* strains at salt concentration above the growth optimum. Growth of *E. halophila* strain 9624 was stimulated only at salt concentrations of more than 27.5% (optimum at 25%). At all other salinities no effect or a slight inhibition were noticed. Growth of *E. halophila* strain 9630 was stimulated at more than 15% salinity (growth optimum). Growth stimulation of *E. mobilis* strain 9903 by glycine betaine was observed at salinities of more than 10% (optimum at 5%).

Growth stimulation by glycine betaine may point to the inability to readily adapt osmotically to these conditions.

If growth stimulation by betaine indicates the inability of the cells to adapt well osmotically in its absence and their ability for betaine uptake (and this is likely), then our results demonstrate that *E. mobilis* 9903 and *E. halophila* 9624 are osmotically well adapted even at salt concentrations slightly higher than their growth optimum. In consequence, this optimum is not osmotically determined, but growth rates at higher salt concentrations cease due to other reasons. Under the same assumptions, osmotic adaptation at optimum salinity of *E. halophila* strain 9630 is only reasonable and growth inhibition at higher salinities is due to limited osmotic adaptation.

Transport of Ions

Cells of various *Ectothiorhodospira* species were used to measure the influence of light/dark changes on the transport of sodium, potassium and protons (Imhoff et al., in preparation). Potassium was always accumulated during light phases but excreted in the dark. Transport of sodium and protons was dependent on pH and sodium concentration and indicated the presence of an active sodium/proton antiporter. At alkaline exterior pH, illumination after a dark phase induced alkalinization of the medium and extrusion of sodium from the cells. After prolonged illumination the cells slowly started to acidify the medium. This may indicate that proton extrusion is a primary event of photosynthesis, but that the high activity of a proton/sodium antiporter overshadows this activity under certain conditions. When the cytoplasmic sodium content has decreased and/or the cytoplasmic pH is acidified, the activity of this antiporter ceases and net proton movement is observed. In the dark sodium ions and protons move along their chemical gradients, sodium moves into the cytoplasm and protons move out. The activity and regulation of ion transport systems in *Ectothiorhodospira* is far from being clear. The available data demonstrate, however, the ability of these cells to maintain low cytoplasmic ion concentrations even at extremely high external salinities.

STRUCTURAL ADAPTATION

To maintain essential physiological functions at high salt concentrations cellular structures have to adapt to changes of the external salinity. These changes concern a single strain at various salt concentrations as well as different strains at their optimum growth conditions. Whether or not the salinity influences the net charge of the cell membrane we measured salt influences on binding of ions to these structures and analyzed proteins and lipids of the membranes. Cell fractions of various *Ectothiorhodospira* strains preferentially bind sodium, but not chloride. Under the experimental conditions, about 2200 nmol/mg of sodium was bound in the membrane fraction and 770 nmol/mg protein in the soluble protein fraction of *E. halochloris*. The binding of chloride was comparably low in both fractions (10-40 nmol/mg protein) and independent from the growth salinity (J. F. Imhoff and B. Meyer, unpublished results). The binding capacity for sodium ions in membrane proteins from *Rhodospirillum rubrum* (no salt requirement), *E. mobilis* (5% salinity required), and two *E. halophila* strains (15% and 25% salinity required) significantly increased with the salt requirement of the bacterial strain.

Polar Lipid Composition

Major components of polar lipids of *Ectothiorhodospira* species are PG, CL, PC, PE and PA. Proportions of other individual polar lipids were less than 1% of the total. In their growth permitive salinity range the three strains were cultivated in the presence of radioactively labelled phosphate or acetate; the lipids were extracted, separated on TLC and quantitated in a liquid scintillation counter. Experiments with radioactively labelled carbon and phosphorus yielded approximately identical results. Most significant salt-dependent changes were observed in the proportions of PG, CL and PE. With increasing salinity we observed in all three strains a strong increase in the proportion of PG and a decrease in that of PE and CL. The absolute content of PE was much lower in the extremely halophilic strains, which accordingly, showed a much more pronounced reduction of the content of CL at higher salinities. The change of the proportions of anionic (PG, CL, PA) to zwitterionic lipids (PE, PC) was remarkable. The content of anionic lipids increased with higher salinities, leading to an increased surplus of negative charges. In *E. mobilis* this charge surplus doubled over the salt range from 2.5 - 15% (B. Thiemann and J. F. Imhoff, in preparation).

ACKNOWLEDGEMENTS

Financial support by the Deutsche Forschungsgemeinschaft is gratefully acknowledged.

REFERENCES

[1] P. Caumette, R. Baulaigue and R. Matheron, *System. Appl. Microbiol.* 10: 284 (1988)

[2] J. F. Imhoff, Halophilic phototrophic bacteria, *in* "Halophilic Bacteria", Vol. I. F. Rodríguez-Valera, ed., CRC Press, Boca Raton, Florida, (1988)

[3] J. F. Imhoff, H. G. Sahl, G. S. H. Soliman and H. G. Trüper, *Geomicrobiology J.* 1: 219 (1979)

[4] J. F. Imhoff, *Adv. Space Research* 6: 299 (1986)

[5] G. Drews, *Arch. Microbiol.* 130: 325 (1981)

[6] H. Nissen and I. D. Dundas, *Arch. Microbiol.* 138: 251 (1984)

[7] F. Rodríguez-Valera, A. Ventosa, G. Juez and J. F. Imhoff, *Microbial Ecol.* 11: 107 (1985)

[8] J. F. Imhoff, *FEMS Microbiol. Rev.* 39: 57 (1986)

[9] J. F. Imhoff, The genus *Ectothiorhodospira*, *in* "Bergey's Manual of Systematic Bacteriology", Vol. 3, J. T. Staley, M. P. Bryant, N. Pfennig, J. G. Holt, eds. Williams & Wilkins, pp. 1654-1658 (1989)

[10] W. D. Grant and H. Larsen, Extremely halophilic archaeobacteria, *in* "Bergey's Manual of Systematic Bacteriology", Vol. 3, J. T. Staley, M. P. Bryant, N. Pfennig, J. G. Holt, eds. Williams & Wilkins, pp. 2216-2219 (1989)

[11] J. F. Imhoff and T. Riedel, *J. Gen. Microbiol.* 135: 237 (1989)

[12] F. R. Tabita and B. A. McFadden, *J. Bacteriol.* 126: 1271 (1976)

[13] B. Meyer and J. F. Imhoff, *J. Gen. Microbiol.* 135: 2829 (1989)

[14] E. Galinski, Diploma thesis, University of Bonn, F.R.G. (1980)

[15] E. Galinski and H. G. Trüper, *FEMS Microbiol. Lett.* 13: 357 (1982)

[16] E. Galinski, H.-P. Pfeiffer and H. G. Trüper, *Eur. J. Biochem.* 149: 135 (1985)

LIPOIC ACID AND DIHYDROLIPOAMIDE DEHYDROGENASE

IN HALOPHILIC ARCHAEOBACTERIA

Michael J. Danson[1], David W. Hough[1], Nataraj Vettakkorumakankav[2]
and Kenneth J. Stevenson[2]

[1]Department of Biochemistry
University of Bath
Bath, U.K.

[2]Department of Biological Sciences
University of Calgary
Calgary
Alberta, Canada

ABSTRACT

We have discovered the presence of dihydrolipoamide dehydrogenase (DHLipDH) in the halophilic archaeobacteria, despite the fact that these organisms lack the multienzyme complexes with which this enzyme is associated in eubacteria and eucaryotes. We will discuss (a) the discovery, purification and characterisation of the halophilic DHLipDH, (b) the detection of the substrate (reduced lipoic acid) in halophilic archaeobacteria, (c) the discovery of DHLipDH and lipoic acid in several other archaeobacterial genera, and (d) the function of DHLipDH in the archaeobacteria and its evolution with respect to the multienzyme complexes of eubacteria and eucaryotes.

INTRODUCTION

On the basis of rRNA sequence comparisons, it has been proposed that archaeobacteria are a phylogenetically distinct group of organisms, as distant from the eubacteria as they are from the eucaryotes [1]. This proposal is supported by a considerable amount of molecular-biological [2] and biochemical [3] data, although, using the same rRNA sequences, Lake [4] has challenged the monophyletic nature of the archaeobacteria.

General and Appliea Aspects of Halophilic Microorganisms
Edited by F. Rodriguez-Valera, Plenum Press, New York, 1991

One of our main interests in the archaeobacteria concerns the comparative enzymology of their central metabolic pathways [5, 6]. A major difference between the archaeobacteria and the other two kingdoms involves the conversion of the 2-oxo acids, pyruvate and 2-oxoglutarate, to their corresponding acyl-CoA thioesters. In eubacteria and eucaryotes, 2-oxo acid dehydrogenase multienzyme complexes catalyse these reactions, each complex consisting of three associated enzymes (a decarboxylase, an acyl-transferase and dihydrolipoamide dehydrogenase), with a covalently bound lipoic acid playing a central catalytic role [7]. Archaeobacteria do not possess these complexes, but have the much simpler 2-oxo acid oxidoreductases, which do not use lipoic acid [8].

It was therefore an unexpected discovery to find the enzyme, dihydrolipo-amide dehydrogenase (DHLipDH), in the halophilic archaeobacteria [9]. DHLipDH catalyses:

$$\text{Dihydrolipoamide} + \text{NAD}^+ \rightleftharpoons \text{Lipoamide} + \text{NADH} + \text{H}^+$$

This was the first report in any species of DHLipDH in the absence of the 2-oxo acid dehydrogenase complexes. This communication will discuss our current ideas on the structure, function and evolution of this enzyme in the halophilic and other archaeobacteria.

DIHYDROLIPOAMIDE DEHYDROGENASE FROM HALOPHILIC ARCHAEOBACTERIA

We have found the enzyme DHLipDH in both classical and alkaliphilic halophiles [9]. Enzymic activity is optimal at approximately 2M NaCl, it exhibits a hyperbolic dependence of velocity on both dihydrolipoamide and NAD^+ concentrations, and it is specific for NAD^+. We have purified to homogeneity the DHLipDH from *Halobacterium halobium* [10] and have shown it to be a dimeric flavoprotein with a polypeptide chain M_r of 58,000 (±3,000). These properties are remarkably similar to those of DHLipDHs from the eubacterial and eucaryotic 2-oxo acid dehydrogenase complexes, a similarity which extends to the catalytic mechanism. That is, through the chemical modification of the halobacterial DHLipDH with the dithiol-specific trivalent arsenicals [10], we have evidence that, as in non- archaeobacterial DHLipDHs, the enzymic activity involves the alternate oxidation and reduction of an intrachain disulphide bond (Figure 1)

A comparison of the amino acid composition of the *H. halobium* enzyme [10] with those of the DHLipDHs from non-archaeobacterial species is consistent with the trend that halophilic proteins contain a significantly higher frequency of acidic and the borderline hydrophobic residues, and a lower frequency of basic amino acids [11]. It is thought that the acidic residues increase stabilization of the protein by effectively competing with salt for the available water, thereby retaining an appropriate hydration shell at high salt concentrations.

With the intention of obtaining the whole sequence of the protein at the DNA level, we have attempted to determine an N-terminal amino acid sequence from which to design oligonucleotide probes. This has not yet been possible as the N-terminus

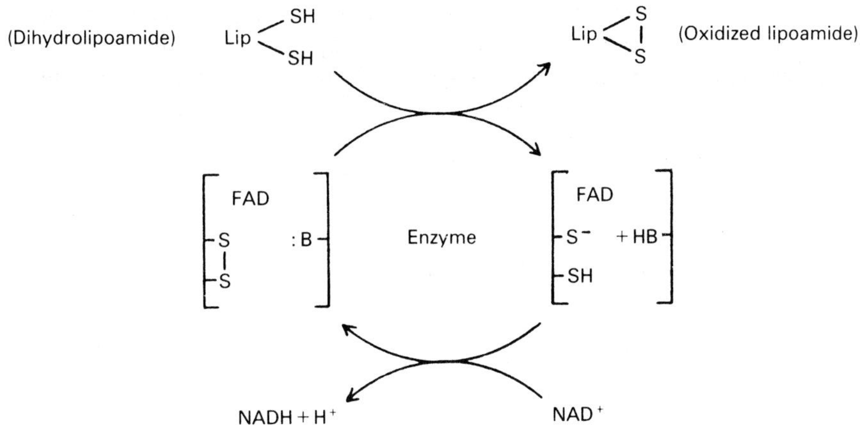

Figure 1. The catalytic mechanism of dihydrolipoamide dehydrogenase.

appears to be blocked. However, we have recently purified the DHLipDH from another halophilic archaeobacterium, *Haloferax volcanii*. The properties of this enzyme are similar to those of the DHLipDH from *H. halobium*, except that the N-terminus is free. We have determined an N-terminal sequence which is aligned with the sequences of the human and *Escherichia coli* DHLipDH N-terminal amino acids (Figure 2). Good homology is observed from residues 13-18, the region which is the start of the FAD-binding domain in the eucaryotic and eubacterial enzymes.

```
                          1         5         10        15

Human liver           A D Q P I D A D V T V I G S G   P
H. volcanii       V V G D I A T G T X L L V I G A G  (P)
E. coli               S T E I K T Q V V V L G A G   P
```

Figure 2.   N-terminal sequences of DHLipDHs from human liver, *H.volcanii* and *E. coli*. Identical residues between sequences are boxed.

THE PRESENCE OF LIPOIC ACID IN HALOPHILIC ARCHAEOBACTERIA

The data strongly suggest that we have discovered a true DHLipDH in the halophilic archaeobacteria, even though the multienzyme complexes to which it is attached in non-archaeobacterial species are absent. To probe its function, it is essential to determine if the presumed substrate, the reduced form of lipoic acid, is present in these halophiles. This has now been achieved in two independent laboratories. Using a bioassay, Noll and Barber [12] have found that *H.volcanii* contains up to 0.8 μg of lipoic acid per g (dry weight) of cells, a value approximately 25% of that in *Escherichia coli*. At the same time, we developed a modified gas chromatographic-mass spectrometric (g.c.-m.s.) procedure to detect lipoic acid [13] (Figure 3). Cell hydrolysis in HCl releases protein-bound lipoic acid and, after extraction into benzene and reduction with $NaBH_4$, the dihydrolipoic acid so generated,

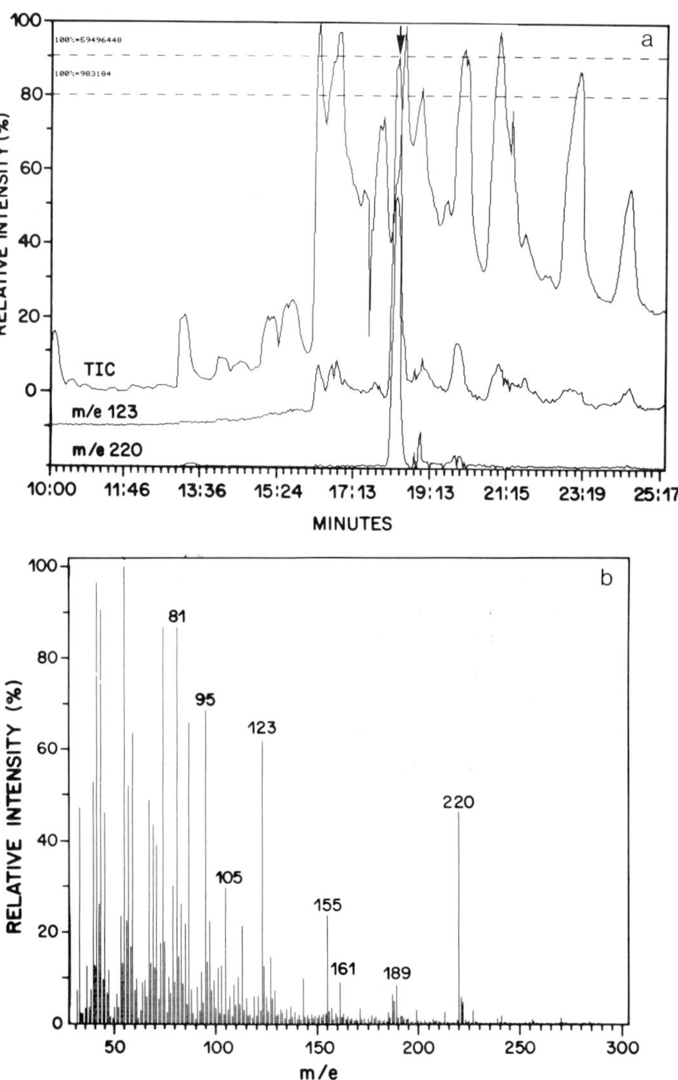

Figure 3 (a) and (b). G.c.-m.s. of oxidised lipoic acid methyl ester from *H.halobium*. (a) gas chromatogram monitored by mass spectrometry. Profiles are of total ion (TIC), m/e=220 (molecular ion) and m/e=123 (a fragmentation ion) as a function of time. (b) Mass spectrum of the peak eluted at 18.3 min of the gas chromatogram (indicated by arrow).

is isolated by covalent chromatography on dithiol-specific p-aminophenyl -arsenoxide-agarose. Elution is achieved with 2,3- dimercaptopropane -1-sulphonic acid and, after extraction into benzene, the dihydrolipoic acid is allowed to O_2-oxidise

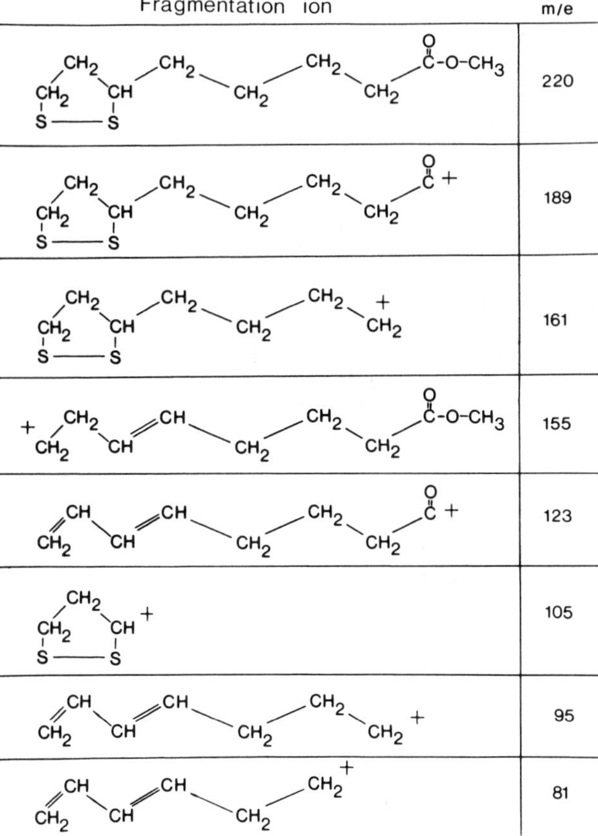

Figure 3. (c)   Assignment of molecular and fragmentation ions.

to the disulphide form. The isolated lipoic acid is converted to the methyl ester by reaction with diazomethane and detected by g.c.- m.s. The mass spectrum shows the characteristic molecular ion (m/e=220) and seven fragmentation ions which, along with those ions retaining the two sulphur ions, allows the definitive detection of lipoic acid. With this methodology, we have identified the presence of lipoic acid in *H. halobium* (Figure 3).

It remains a priority to determine if the lipoic acid is protein bound and, if it is, to identify the nature of the carrier.

LIPOIC ACID AND DIHYDROLIPOAMIDE DEHYDROGENASE IN OTHER ARCHAEOBACTERIA

The archaeobacteria comprise three phenotypes: the extreme halophiles, the thermophiles and the methanogens (Figure 4). They share the common feature that they do not possess the 2-oxo acid dehydrogenase complexes [8] and therefore it was of interest to determine if DHLipDH and/or lipoic acid are present in the other members of this group of organisms.

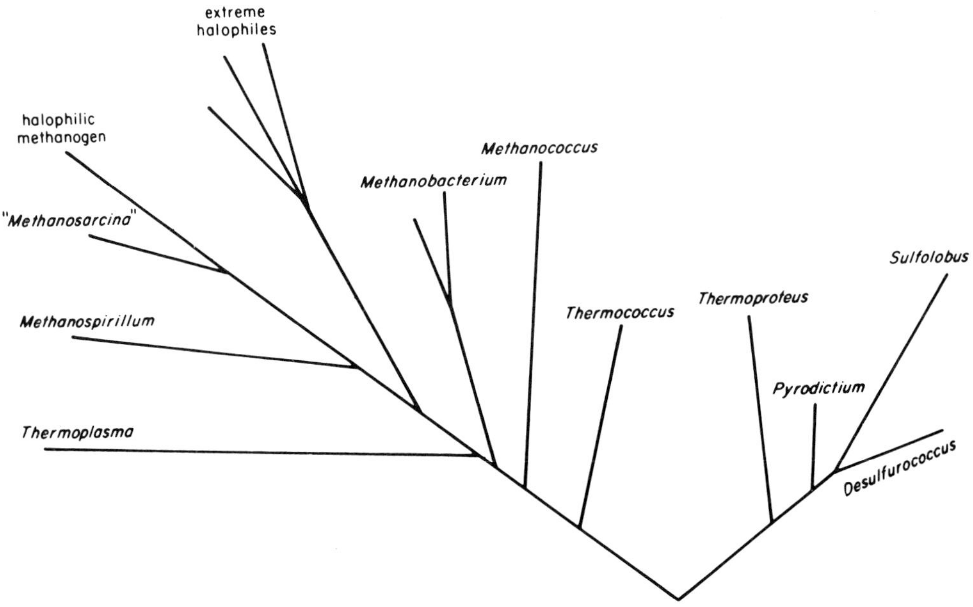

Figure 4.   Archaeobacterial phylogenetic tree based on 16S rRNA sequence comparisons. The tree was constructed by Woese [1] and is reproduced with permission.

The data are not yet complete. We have found DHLipDH in the thermoacidophilic archaeobacterium *Thermoplasma acidophilum* [14] and its M_r value and kinetic and catalytic properties are characteristic of those documented for the halophilic enzyme. However, the enzyme appears to be absent from the phenotypically similar thermoacidophile *Sulfolobus acidocaldarius*. We have also found DHLipDH in the methanogen *Methanosarcina barkeri*; the specific activity is low and further studies are required to characterise the enzyme.

Interestingly, the presence of lipoic acid coincides with the limited enzyme data. The cofactor has been found in *Methanobacterium thermoautotrophicum* but it is absent (or in very low amounts) in *Sulfolobus solfataricus* and the genotypically related thermophiles *Pyrodictium occultum* and *Thermoproteus tenax* [12]. Lipoic acid is present in *Thermococcus celer* but we have no data regarding the enzyme in this organism. Unfortunately, *Tp. acidophilum* has not yet been investigated for its cofactor content but, as we know the DHLipDH is present [14], the search for lipoic acid in this organism is a priority.

The correlation of the lipoic acid and DHLipDH data with the phylogenetic positions of the organisms is noteworthy. Figure 4 shows the generally accepted relationships within the archaeobacteria (their phylogenies with respect to eucaryotes and eubacteria is still a contentious issue [1, 2, 4]); that is, they comprise two main divisions - the methanogens, extreme halophiles, *Thermoplasma* and *Thermococcus* in one and the remaining sulphur-dependent thermophiles in the other. DHLipDH and/or

126

lipoic acid have been detected in all the phenotypes of the former division but neither have yet been discovered in the latter. Further analyses to test this correlation are in progress.

EVOLUTIONARY AND FUNCTIONAL CONSIDERATIONS

Kerscher and Oesterhelt [8] have suggested that the 2-oxoacid: ferredoxin oxidoreductases existed before the divergence of the archaeobacteria, eubacteria and eucaryotes and that the 2-oxoacid dehydrogenase complexes evolved in the eubacterial line after the development of oxidative phosphorylation. Eubacteria retaining their anaerobic niches would still use the oxidoreductases, and eucaryotes would gain the dehydrogenase complexes via the endosymbiotic origin of mitochondria. Our data would suggest that DHLipDH is also an ancient enzyme, originally having a function independent of the 2-oxo acid dehydrogenase complexes, but being recruited into them in the eubacterial lineage.

Can one account for the phylogenetic distribution of DHLipDH and lipoic acid? At present the data are limited, but it should be noted that Lake [4] proposes that the methanogen-halophile branch of the archaeobacteria is related to the eubacteria (both groups possess the enzyme and cofactor) whereas the sulphur-dependent archaeobacterial branch (lacking DHLipDH and lipoic acid) is related to the ancestral eucaryotic lineage which originally lacked mitochondria.

The function of DHLipDH and lipoic acid in the absence of the 2-oxo acid dehydrogenase complexes is also still a matter of speculation [5, 15]. Although the system was originally discovered in the halophilic archaeobacteria, additional information has come from organisms outside this group. *Trypanosoma brucei*, the causative agent of African Sleeping Sickness, lacks a functional mitochondrion in the bloodstream form. Thus it lacks the 2-oxo acid dehydrogenase complexes, but we have discovered that it possesses DHLipDH, attached to the parasite's plasma membrane [16]. We have presented evidence [14] that the *Tp. acidophilum* DHLipDH is also membrane associated and, although we have not yet gained comparable physical evidence for a similar situation in the halophilic archaeobacteria, the highly hydrophobic N-terminal sequence determined for the enzyme from *H.volcanii* suggests that this possibility ought to be investigated.

Little information is available on the function of the DHLipDHs in the absence of the 2-oxoacid dehydrogenase complexes [reviewed in 15]. An involvement in transport or signal transduction is a possibility if the enzyme is located at or near the plasma membrane, and future research must be focussed in this direction. It is clear that the archaeobacteria, especially the extreme halophiles, are ideal systems in which to study this enzyme; they are the organisms in which the non-complexed DHLipDH was discovered and it is felt that from them will come valuable structural, functional and evolutionary information on this dehydrogenase in all three evolutionary lines of descent.

127

ACKNOWLEDGEMENTS

This work was supported by a NATO grant for International Collaboration in Research to MJD and KJS. A grant to KJS from the Natural Sciences and Engineering Research Council of Canada is gratefully acknowledged. NV was supported by the University of Calgary, Canada.

REFERENCES

[1]   C. R. Woese, Bacterial evolution. *Microbiol. Rev.* 51:221 (1987).

[2]   A. T. Matheson and P.P. Dennis (eds), Molecular biology of archaebacteria. *Can. J. Microbiol.* 35, pp 1-244 (1989).

[3]   C. R. Woese and R. S. Wolfe, (eds) Archaebacteria *in* "The Bacteria". Vol.8. Academic Press, London (1985).

[4]   J. A. Lake, Origin of the eucaryotic nucleus: eucaryotes and eocytes are genotypically related. *Can. J. Microbiol.* 35:109 (1989).

[5]   M. J. Danson, Archaebacteria: the comparative enzymology of their central metabolic pathways. *Adv. Microbiol. Physiol.* 29:165 (1988).

[6]   M. J. Danson, Central metabolism of the archaebacteria: an overview. *Can J. Microbiol.* 35:58 (1989).

[7]   R. N. Perham, L. C. Packman and S. E. Radford, 2-Oxo acid dehydrogenase multi-enzyme complexes: in the beginning and halfway there. *Biochem. Soc. Symp.* 54:67 (1987).

[8]   L. Kerscher and D. Oesterhelt, Pyruvate: ferredoxin oxidoreductase - new findings on an ancient enzyme. *Trends Biochem. Sci.* 7:371 (1982).

[9]   M. J. Danson, R. Eisenthal, S. R. Hall and D. L. Williams, Dihydrolipoamide dehydrogenase from halophilic archaebacteria. *Biochem. J.* 218:811 (1984).

[10]   M. J. Danson, A. McQuattie and K. J. Stevenson, Dihydrolipoamide dehydrogenase from halophilic archaebacteria: purification and properties of the enzyme from *Halobacterium halobium. Biochemistry* 25:3880 (1986).

[11]   H. Eisenberg and E. J. Wachtel, Structural studies on halophilic proteins, ribosomes and organelles of bacteria adapted to extreme salt concentrations. *Ann. Rev. Biophys. Biophys. Chem.* 16:69 (1987).

[12]   K. M. Noll and T. S. Barbar, Vitamin contents of archaebacteria. *J. Bacteriol.* 170:4315 (1988).

[13]   K. J. Pratt, C. Carles, T. J. Carne, M. J. Danson and K. J. Stevenson, Detection of bacterial lipoic acid: a modified gas-chromatographic mass-spectrometric procedure. *Biochem. J.* 258:749 (1989).

[14]   L. D. Smith, S. J. Bungard, M. J. Danson and D. W. Hough, Dihydrolipoamide dehydrogenase from the thermoacidophilic archaebacterium *Thermoplasma acidophilum. Biochem. Soc. Trans.* 15:1097 (1987).

[15]   M. J. Danson, Dihydrolipoamide dehydrogenase: a new function for an old enzyme? *Biochem. Soc. Trans.* 16:87 (1988).

[16]   M. J. Danson, K. Conroy, A. McQuattie and K. J. Stevenson, Dihydrolipoamide dehydrogenase from *Trypanosoma brucei.* Characterisation and cellular location. *Biochem. J.* 243:661 (1987).

NITRATE REDUCTION IN THE EXTREMELY HALOPHILIC BACTERIA

Lawrence I. Hochstein

Ames Research Center Moffett Field, CA 94035 U.S.A.

ABSTRACT

Membranes prepared from *Haloferax denitrificans, Haloferax mediterranei* and a *Haloferax* strain designated as Baja-12 contained a nitrate reductase that was stable as well as most active in the absence of added NaCl (60 mM NaCl). Nitrate reduction was competitively inhibited by azide and chlorate; the latter also was a substrate. Cyanide inhibited the reduced form of the enzyme in an inhibition antagonized by nitrate. Membranes prepared from cells grown in the presence of tungstate did not reduce nitrate suggesting that the enzyme was a molybdenum-containing protein. The similarity of this enzyme to other prokaryotic nitrate reductases suggests a eubacterial origin of the enzyme.

INTRODUCTION

Little is known concerning the enzymology of denitrification in the extremely halophilic bacteria. Although a putatively denitrifying *Halobacterium, Halobacterium marismortui*, was first described in 1940 [1], more than a decade passed before Marquez and Brodie [2] were to describe a partially purified nitrate reductase (NAR) from an organism isolated from the Great Salt Lake and also designated as *H. marismortui* [3]. Enzyme activity was salt-dependent and characterized by an extremely high temperature optimum [2]. Subsequently, Werber and Mevarech [4] reported on a NAR from another extreme halophile isolated from the Dead Sea [5] and also designated as *H. marismortui* [6]. The induction of NAR activity in this *H. marismortui* was unusual in that it occurred some 80 hours after the organism was first exposed to nitrate and reached a maximum value after nitrite reductase activity was in decline [4].

We have isolated several extremely halophilic bacteria, including *Haloferax denitrificans* [7] originally described as *Halobacterium denitrificans* [8], which only grew anaerobically in the presence of either nitrate or nitrite. Growth was accompanied by the production of nitrite, nitrous oxide and dinitrogen unequivocably establishing that these organisms were denitrifiers. We subsequently demonstrated that *H. mediterranei* and *H. marismortui* also grew anaerobically and produced nitrite, nitrous oxide, and dinitrogen [9]. The availability of denitrifying extreme halophiles distinct from *H. marismortui* suggested an opportunity to examine the halobacterial enzymes unequivocally associated with denitrification. Here we describe several properties of the membrane-bound NAR which are consistent with it being a dissimilatory enzyme and present evidence that the enzyme does not exhibit the salt-dependent properties usually associated with the enzymes from the extremely halophilic bacteria.

MATERIALS AND METHODS

Low aeration cultures of *H. denitrificans* (ATCC 35960), *H. mediterranei* (ATCC 33500), and strain Baja-12, were grown in nitrate-containing complex medium [8]. The medium (700 ml in 2 litre flasks) was incubated at 37°C on a gyrotary shaker operated at 100 RPM. Cells were suspended in 20 mM HEPES-3M NaCl-10 mM MgCl$_2$-100 μM dithiothreitol pH 7.4 buffer (H buffer) and disrupted by passage through a French Pressure Cell operated at 5,000 psi. Any intact cells and debris were removed by centrifugation at 5,900 x g for 30 min at 4°C. The supernatant was decanted and centrifuged at 250,000 x g for 90 min at 4°C. The resulting supernatant was designated as the cytoplasmic fraction; the sediment was made up in H buffer and constituted the membrane fraction. NAR activity was assayed at 37°C in a reaction mixture that contained the following additions (in μmol) in a total volume of 2.5 ml: Tris pH 7.0 buffer (100); potassium nitrate (20); methyl viologen (0.5); dithionite, from a 230 mM stock solution neutralized with NaOH (23); and enzyme. The reaction was initiated by the addition of nitrate following preincubation of the other components for 2 min at 37°C and terminated after 10 min by agitating the reaction mixture with a Vortex mixer. Nitrite was determined by a diazotization procedure [10]. Protein was determined in the presence of 1% sodium dodecylsulfate by the bicinchoninic acid procedure [11] using bovine serum albumin (Fraction V) as the protein standard.

RESULTS

Although NAR activity is assumed to be repressed by oxygen, induction can take place at low oxygen concentrations [12, 13, 14]. The *Haloferax* nitrate reductases were induced in cultures in which there was sufficient oxygen to support carotenoid biosynthesis (Figure 1). Under these conditions, the doubling time was 4hr which was similar to the value observed when cells were grown with more efficient aeration [8]. However, the cell yield was significantly less. The doubling time and cell yield were unaffected by the presence of nitrate indicating that the cells were not using nitrate as an oxidant. Furthermore, induction of NAR activity, as judged by the accumulation of nitrite, took place as the culture entered the maximum stationary phase. Tungstate, which had no effect on growth, inhibited the accumulation of nitrite, and membranes prepared from cells grown in the presence of tungstate had no detectable NAR activity. This is presumptive evidence for the presence of a molybdo-enzyme in which an inactive tungstate form of the enzyme is produced [15]. The NAR produced in low aeration cells had a lower specific activity than the enzyme from anaerobically-grown

130

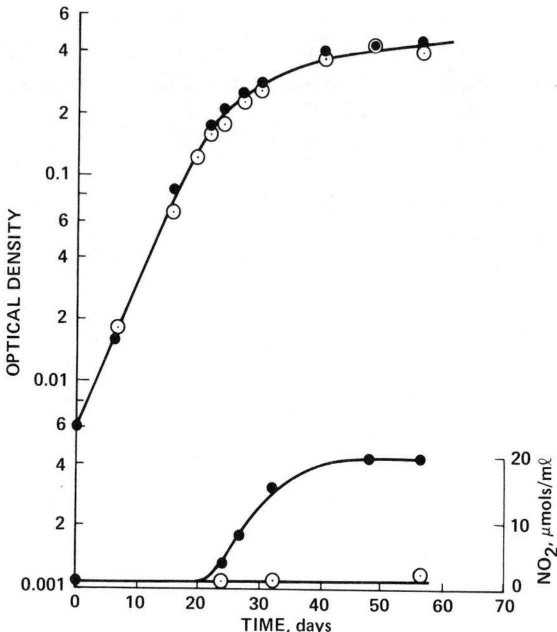

Figure 1. Induction of NAR activity during low-aeration growth. a) *H. denitrificans* was grown in complex medium containing nitrate with (open circles) or without 3mM sodium tungstate (closed circles). Growth in the absence of nitrate was similar and is not shown. b) Nitrite accumulation during growth in medium with (open circles) or without 3 mM sodium tungstate (closed circles).

Table 1. Requirements for nitrate reductase activity

Conditions	NO$_2$ (+ ΔNO$_2$/10 min)
Complete	40
– methyl viologen	3
– nitrate	<1
– dithionite	<1
– membranes	<1

cells. However the properties of the two types of enzyme upon purification were comparable suggesting they are similar [F. Lang and L. I. Hochstein, unpublished data].

NAR activity was located in the membrane and cytoplasmic fractions, with about 70% of the activity associated in the former fraction. The requirements for NAR activity in membranes from *H. denitrificans* are shown in Table 1. While a significant amount of nitrite was produced in the absence of methyl viologen, relatively little, if any, nitrite was produced in the absence of nitrate, dithionite, or the membrane fraction. Similar results were obtained with membranes from the other organisms. The most rapid rate of nitrite accumulation occurred between pH 7.0 and 7.4; however, nitrite accumulation was 90% and 82% of maximum at pH 6.4 and pH 8, respectively. The membrane-bound nitrate reductases from *H. denitrificans* (Figure 2a), *H. mediterranei* and Baja-12 were active over a wide range of salt concentrations. At 37°C, the maximum rate of nitrite accumulation took place in the absence of added salt (NaCl = 60 mM) and decreased at higher salt concentrations. Replacing NaCl with KCl did not materially affect the results. The enzyme was most active when assayed at 85°C in the presence of 4 M NaCl. Since methyl viologen could not be maintained in a reduced state at temperatures greater than 65°C in the presence of 60 mM NaCl, we were unable to determine the optimum temperature for the enzyme at this salt concentration. NAR activity was remarkably stable in the absence of added NaCl. When membranes from *H. denitrificans* were diluted in buffer lacking NaCl, no loss of activity was observed after 1 week (Figure 2b). The nitrate reductases from *H. mediterranei* and Baja-12 were also stable when incubated in the absence of added NaCl. For comparative purposes, the inset to Figure 2b shows the inactivation of a membrane-bound NADH dehydrogenase which is a more typical response. This NADH dehydrogenase [16] was solubilized when membranes were exposed to sub-optimal concentrations of NaCl. When the membranes from *H. denitrificans* were incubated in 50 mM HEPES pH 7.4 buffer that was 60 mM with respect to NaCl, as much as 98% of the membrane-bound NAR was subsequently found in the soluble fraction accompanied by about 50% of the protein originally associated with the membranes. When soluble, the enzyme remained stable when stored in the presence of 60 mM NaCl. Essentially identical results were obtained with the NAR from Baja-12.

NAR activity was competitively inhibited by azide and chlorate whereas cyanide was a non-competitive inhibitor. The enzyme from *H. denitrificans* catalyzed the chlorate-dependent oxidation of methyl viologen radical, a reaction that was inhibited by azide. Preincubating membranes at any concentration of cyanide increased the extent of inhibition. In addition, cyanide was more effective in the presence of dithionite suggesting that the inhibitor reacted with the reduced form of the enzyme. Nitrate was found to protect the enzyme against cyanide inhibition (Table 2) which was unexpected since cyanide was a non-competitive inhibitor with respect to nitrate. This suggests that the enzyme may undergo a conformational change upon combination with nitrate so that the cyanide binding site becomes unavailable.

DISCUSSION

The NARs from the 3 extreme halophiles examined in this study were similar. All were most active and stable at very low concentrations of NaCl while the differences in their kinetic parameters did not appear significant (Table 3). The NAR from the extreme halophiles was membrane-bound, reduced chlorate, and was competitively inhibited by azide, properties characteristic of NARs associated with

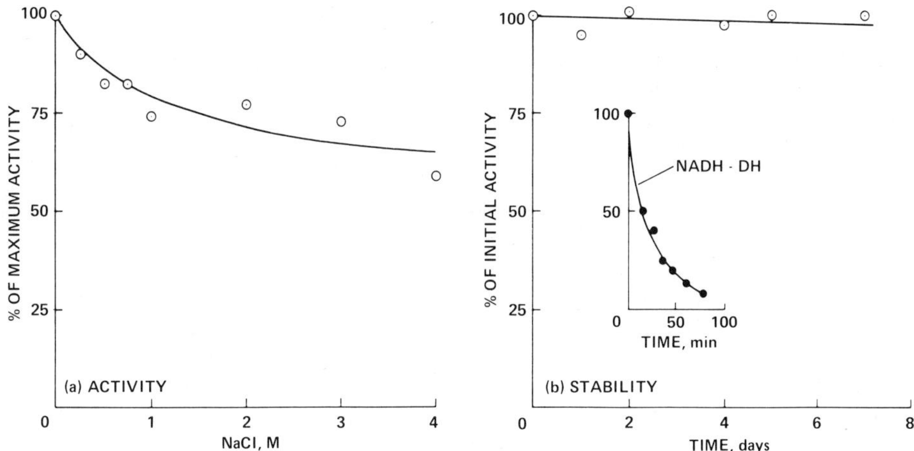

Figure 2. The effect of NaCl on the membrane-bound NAR from *H. denitrificans*.
  a) Activity. The membrane fraction was assayed in reaction mixtures
    containing the indicated concentration of NaCl. Replacing NaCl with KCl
    had no significant effect. 100% activity is equal to the production of 22
    nmols of nitrite/10 minute.
  b) Stability. The membrane fraction (5 μg protein) was incubated in 50 mM
    HEPES pH 7.2 buffer. 100% is equal to the production of 19 nmols of
    nitrite/10 min. Insert. The stability of a membrane-bound NADH
    dehydrogenase in buffer containing 100 mM NaCl (Hochstein, unpublished
    data)

Table 2. The effect of nitrate on cyanide inhibition

Preincubate with	$+\Delta$nMols NO$_2$		% Inh.
	-KCN	+KCN	
methyl viologen	21	7	69
nitrate	23	17	23

Membranes containing 7 μg of protein were incubated
at room temperature in reaction mixtures either lacking
nitrate or methyl viologen. After 10 min, nitrate or
methyl viologen were added to start the reaction which
was terminated after 10 min. When present, the final
concentration of KCN was 500 μM.

$$\textbf{Table 3. } \text{Comparison of various NARs}$$

Parameter	Organism		
	S_1[a,b]	R_4[a,b]	Baja-12[b]
Specific activity[c]	0.38	0.55	0.22
$K_m(NO_3)$ mM	0.27	0.16	0.16
Inhibition (K_i)			
Azide μM	2	1	1
chlorate, mM)	1	0.60	0.50
cyanide, mM)	0.28	0.16	0.18
Nitrate protection	+	nd	+
Active in low salt	+	+	+
Stable in low salt	+	+	+
Reduce chlorate	+	nd	nd

[a] S^1 (*H. denitrificans*), R_4 (*H. mediterranei*)
[b] N. Mushell and L. I. Hochstein (unpublished data)
[c] μmols/min/mg protein.

denitrification [17]. As in the case of the NAR from *Escherichia coli* [18], the halobacterial enzyme was inhibited by cyanide; was more sensitive in the dithionite-reduced form; and nitrate protected the enzyme against cyanide. The apparent K_m values for nitrate compared very well with the apparent K_m of 0.2 mM for the NAR from *H. marismortui* [4], and were within the range of apparent K_m values reported for other nitrate reductases [19], although considerably less than the one for the NAR from the moderately halophilic bacterium, *Paracoccus halodenitrificans* [20]. Paradoxically, the most unusual property of the haloferaxal NAR was the absence of salt-dependent properties. Most enzymes from the extremely halophilic bacteria are activated by relatively high salt concentrations [21]. In the case of the enzyme from *H. denitrificans*, maximum nitrate reduction took place in the absence of added salt. While NaCl appeared to inhibit NAR activity, this inhibition may be more apparent than real. Membranes also brought about the disappearance of nitrite and did so most actively at high salt concentrations. Thus, the amount of nitrite detected when the assay was carried out at higher salt concentrations could reflect the increased disappearance of nitrite rather than an inhibition of nitrite production. The absence of salt dependence for NAR activity was contrary to the observation of Marquez and Brodie [2] who reported that the NAR had an absolute requirement for NaCl, a requirement that was temperature dependent. On the other hand, Werber and Mevarech [4], who also observed a temperature dependent salt optimum, reported considerable NAR activity in the presence of 85 mM NaCl. Although these discrepancies could reflect the use of different organisms, this does not seem likely since the nitrate reductases obtained from *H. mediterranei* and Baja-12 were also most active in the absence of added salt. It may be that the NARs from *H. denitrificans* and the two strains of *H. marismortui* are functionally distinct so that the former has a dissimilatory role, the latter a different function. The anomolous differential sequence of induction of the nitrate and nitrite reductases in the "Dead Sea strain" of *H. marismortui* [4] appears inconsistent with a dissimilatory role for that enzyme. It may be significant that the nitrate reductases described in this study were obtained from *Haloferax* species whereas *H. marismortui* has recently been shown to belong to the genus *Haloarcula* [22].

The nitrate reductases from *H. denitrificans*, *H. mediterranei* and Baja-12 were unusually stable in the absence of added NaCl. While most proteins from extremely halophilic bacteria are stabilized by high concentrations of NaCl [21], stability in the absence of NaCl is not unknown. The purification procedure described by Marquez and Brodie [2] involved dialyzing the enzyme against a buffer solution 0.17 M with respect to NaCl. From this one might infer that this NAR was also stable at low salt concentrations. Purple membranes are also stable in the absence of added salt, a property used for their isolation [23]. The superoxide dismutase from *H. cutirubrum* [24] and the dihydrolipoamide dehydrogenase from *H. halobium* are also stable in the absence of salt [25].

The N-terminal region of the halobacterial superoxide dismutase exhibits considerable homology with eubacterial superoxide dismutases which suggests that the superoxide dismutase may have arisen prior to the divergence of the Archaeobacteria and Eubacteria. If nitrate respiration appeared after the appearance of oxygen respiration [26], it seems unlikely that NAR activity preceded the eubacterial-archaebacterial divergence. An intriguing possibility is that the enzyme found in the extreme halophiles originated from a non-halophilic bacterium via lateral gene transfer.

ACKNOWLEDGEMENTS

This research was supported by a grant from the NASA Exobiology Program on the early evolution of life.

REFERENCES

[1]   B. Elazari-Volcani, *in,* "Bergey's Manual of Determinative Bacteriology, 7th Edition", R. S. Breed, E. G. D. Murrary and N. R. Smith, eds, pp. 207-212, Williams and Wilkens, Baltimore (1957)

[2]   E. D. Marquez and A. F. Brody, The effect of cations on the heat stability of a halophilic nitrate reductase, *Biochim. Biophys. Acta* 321: 84 (1973)

[3]   R. Nachum, Studies of the extracellular amylase isolated from an extremely halophilic bacterium, *Ph.D.Dissertation*, University Southern California, University Microfilms, Ann Arbor Michigan 70: 5225 (1969)

[4]   M. M. Werber and M. Mevarech, Induction of a dissimilatory reduction pathway of nitrate in halobacterium of the Dead Sea, *Arch. Biochem. Biophys.* 186: 60 (1978)

[5]   M. Ginzburg, L. Sachs and B. Z. Ginzburg, Ion metabolism in a Halobacterium. I. Influence of age of culture on intracellular concentrations, *J. Gen. Physiol.* 55: 187 (1970)

[6]   M. Ginzburg, Ion metabolism in a Halobacterium. I. Influence of age of culture on intracellular concentrations, *in* "Energetics and structure of halophilic microorganisms", S.R. Caplan and M. Ginzburg, eds, pp. 561–583, Elsevier-North Holland Biomedical Press, New York (1978)

[7]   B. J. Tindall, G. A. Tomlinson and L. I. Hochstein, Transfer of *Halobacterium denitrificans* (Tomlinson, Jahnke and Hochstein) to the genus *Haloferax* as *Haloferax denitrificans* comb. nov., *Int. J. Syst. Bacteriol.* 39: 359 (1989)

[8]   G. A. Tomlinson, L. L. Jahnke, and L. I. Hochstein, *Halobacterium denitrificans* sp. nov. an extremely halophilic denitrifying bacterium, *Int. J. Syst. Bacteriol.* 36: 66 (1986)

[9]   R. Mancinelli and L. I. Hochstein, The occurrence of denitrification in extremely halophilic bacteria. *FEMS Microbiol. Lett.* 35: 55 (1986)

[10]   M. K. Showe, and J. A. Demoss, Localization and regulation of synthesis of nitrate reductase in *Escherichia coli*, *J. Bacteriol.* 95: 1305 (1968)

[11]   P. K. Smith, R. I. Krohn, G. T. Hermanson, A. K. Mallia, F. H. Gartner, M. D. Provenzano, E. K. Fujimoto, N. M. Goeker, B. J. Olson and D. C. Klenk, Measurement of protein using bicinchoninic acid, *Anal. Biochem.* 150: 76 (1985)

[12]   K. Calder, K. A. Burke and J. Lascelles, Induction of nitrate reductase and membrane cytochromes in wild type and chlorate-resistant *Paracoccus denitrificans*, *Arch. Microbiol.* 126: 149 (1980)

[13]   L. I. Hochstein, M. Betlach and G. Kritikos, The effect of oxygen on denitrification during steady state growth of *Paracoccous halodenitrificans*, *Arch. Microbiol.* 137: 74 (1984)

[14]   P. Justin and D. P. Kelly, Metabolic changes in *Thiobacillus denitrificans* accompanying the transition from aerobic to anaerobic growth during continuous chemostat culture, *J. Gen. Microbiol.* 107: 131 (1988)

[15]   H. G. Enoch and R. K. Lester, Effects of molybdate, tungstate,and selenium compounds on formate and dehydrogenase and other enzyme systems in *Escherichia coli*, *J. Bacteriol.* 110: 1032 (1972)

[16]   L. I. Hochstein and B. P. Dalton, Studies of a halophilic NADH dehydrogenase. I. Purification and properties of the enzyme. *Biochim. Biophys. Acta* 302: 216 (1983)

[17]   A. H. Stouthammer, Biochemistry and genetics of nitrate reductase in bacteria, *Adv. Microbiol. Physiol.* 14: 315 (1976)

[18]   M. W. W. Adams and L. E. Mortenson, The effect of cyanide and ferricyanide on the activity of the dissimilatory nitrate reductase of *Escherichia coli*, *J. Biol. Chem.* 257: 1791 (1982)

[19]   A. Craske and S. J. Ferguson, The respiratory nitrate reductase from *Paracoccus denitrificans*. Molecular characterisation and kinetic properties, *Eur. J. Biochem.* 158: 429 (1986)

[20]   J-P. Rosso, P. Forget and F. Pichinoty, Les nitrate-reductases bacteriennes. Solubilization, purification et proprietes de l'enzyme a de *Micrococcus halodenitrificans, Biochim. Biophys. Acta* 321: 443 (1973)

[21]   J. Lanyi, Salt-dependent properties of proteins from extremely halophilic bacteria, *Bacteriol. Rev.* 38: 272 (1974)

[22]   A. Oren, M. Ginzburg, B. Z. Ginzburg, L. I. Hochstein, and B. E. Volcani, *Haloarcula marismortui (Volcani)* sp. nov. nom. rev., an extremely halophilic bacterium, *Int. J. Syst. Bacteriol.* 40: 209 (1990)

[23]   R. Henderson, The purple membrane from *Halobacterium halobium, Ann. Rev. Biophys. Bioeng.* 6: 87 (1977)

[24]   B. P. May and P. P. Dennis Superoxide dismutase from the extremely halophilic archaebacterium *Halobacterium cutirubrum, J. Bacteriol.* 169: 1417 (1987)

[25]   M. J. Danson, J. A. McQuattie and K. J. Stevenson, Dihydrolipoamide dehydrogenase from halophilic archaebacteria: purification and properties of the enzyme from *Halobacterium halobium, Biochemistry* 25: 3880 (1986)

[26]   E. Broda, The history of inorganic nitrogen in the biosphere, *J. Mol. Evol.* 7: 87 (1975)

RETINAL-OPSIN-DEPENDENT DETECTION OF SHORT-WAVELENGTH ULTRAVIOLET RADIATION (UV-B), AND ENDOGENOUS BIAS ON DIRECTION OF FLAGELLAR ROTATION IN TETHERED *HALOBACTERIUM HALOBIUM* CELLS

Gottfried Wagner, Torsten Rothärmel and Bernhard Traulich

Membran- und Bewegungsphysiologie
Botanisches Institut 1
Justus-Liebig-Universität
Senckenbergstrasse 17-25
D-6300 Giessen, Federal Republic of Germany

SUMMARY

The UV-B sensory pigment in *Halobacterium halobium* has been characterized in our investigations in the photoenergetic pigments-deficient mutant strain Flx-03 l.c., with mutant strain L-07 as the all-*trans* retinal-dependent control. Evidences in favour of sensory rhodopsin I (SR-I) are: (a) *H. halobium*, mutant strain Flx-03 l.c., shows threshold value of UV-B avoidance response much the same as reported for photoreceptor wild type cells. (b) *H. halobium*, mutant strain L-07, deficient in retinal synthesis, shows no UV-B avoidance response under the standard conditions, unless reconstituted by exogenous all-*trans* retinal. (c) UV-B avoidance response shows strong photochromism under bichromatic irradiation, consistent with the reported properties of SR-I but unknown for SR-II. (d) Membrane vesicles of *H. halobium*, mutant strain Flx-03 l.c., prepared of cells in the late exponential growth phase, show SR-I as the only photochromic pigment detectable. With tethered cells of the *H. halobium* mutant strain Flx-03 l.c., the bias in direction of flagellar rotation has been discriminated as a function of (a) continuous orange sensory light, (b) continuous blue sensory light, (c) arginine, (d) sodium acetate or (e) potassium cyanate. Similar to attractant light conditions, high energy situations of the cell result in preferred CW rotation while low energy situations, similar to repellent light conditions, result in preferred CCW rotation of the flagellum. Thus, rotational sense of the halobacterial flagellum is a variable function of the two parameters: (i) external stimulation, (ii) cellular competence.

INTRODUCTION

In *Halobacterium halobium*, two sensory pigments are known to receive the

General and Applied Aspects of Halophilic Microorganisms
Edited by F. Rodriguez-Valera, Plenum Press, New York, 1991

information of solar light and to channel its message via steps of amplification to the flagellar motor, namely sensory rhodopsin I (SR-I) and sensory rhodopsin II (SR-II) [1]. The range of wavelengths, monitored by this archaebacterium, reaches from short ultraviolet radiation of 280 to 320 nm (UV-B), versus long ultraviolet radiation of 320 to 380 nm (UV-A) to visible light of 380 to at least 600 nm [2]. In comparison, the ground state absorption maximum of photochromic SR-I is at 587 nm with the physiologically most active intermediate state at 373 nm; the ground state absorption maximum of SR-II is at 480 nm [1]. Thus, the question remains in *H. halobium*, by what sensory pigment the short ultraviolet radiation of 280 to 320 nm is detected. As reported here, SR-I appears to be the physiological UV-B sensory pigment, possibly via intramolecular energy transfer from aromatic amino acid residues of the SR-I apoprotein to the 373 nm-intermediate state of the holopigment [3].

A question closely related to the halobacterial UV-B sensory pigment is that of the cellular mechanism of UV-B avoidance response, i.e. the overall "colour discrimination" which is exercised by this archaebacterial cell. Observation of tethered halobacteria has permitted discrimination of rotational sense of the flagellum in the *clockwise* versus *counter clockwise* mode under statistical conditions (CW/CCW; [3]). As reported here, repelled *H. halobium* preferentially persists in the CCW mode while attracted *H. halobium* preferentially persists in the CW mode [4].

MATERIALS AND METHODS

H. halobium mutant strain Flx-03 l.c. of *low carotenoid* content was used, deficient in bacteriorhodopsin (BR) and halorhodopsin (HR) but positive for SR-I and SR-II [1]. For control experiments, the *H. halobium* mutant strain L-07 was used, which synthesizes rhodopsin-like pigments only upon addition of exogenous all-*trans* retinal [5]. The cells were grown as described [3], and used for the experiments three days after inoculation (late exponential growth phase).

The experiments were done with tethered cells [3]: after short heating (5 min at 60°C), 1.5 ml culture of late exponential growth phase cells were squeezed gently through 23 gauge needles (0.6 mm outer diameter) of two syringes joined by polyethylene tube (Braun-Melsungen, FRG). Immediately after this flagellar shearing procedure, the cells were centrifuged with slow acceleration at finally 6,000*g for 2 min (Micro-Rapid, Hettich, Tuttlingen, FRG) and allowed within 1 h to swim into peptone-free motility medium [3], enriched with 50 mmol/l arginine [6]. Rotating single cells, tethered to the surface of etched cover-glass (4 mol/l NaOH for 24 h), were observed with a microscope (ICM 405, Zeiss, Oberkochen, FRG with quartz optics) under semi-anaerobic conditions at 37°C over several hours without arginine. Rotational sense of the tethered cells was inverted in order to reflect rotational sense of the flagellum. For the results of Figs. 6 to 8, a custom-made laminar flow-through coverglass-cell with 200 µl internal volume was used; the response of tethered cells was monitored for 1 h before and after medium exchange, done by a flow of 500 µl/min x 9 min each.

Stimulating light was obtained from a 500 W Xenon Cermax lamp (ILC, Sunnyvale, Ca., USA), or a 75 W Xenon lamp (Osram, München, FRG), filtered through the interference filters UV-M-IL 282 nm or UV-PAL 362 nm or FIL 589 nm (Schott, Mainz, FRG) and passing the objective lenses. For data collection, background light from a 60 W tungsten lamp was used, passing the condensor and a bandpassfilter with peak transmittance at 700 nm (half bandwidth 670 to 800 nm; B&W, Wiesbaden, FRG). The irradiance of background light (0.7 Wm^{-2}, and of stimulating light applied via a timer-coupled electronic photoshutter (Melles Griot, Rochester, N.Y., USA), were measured in place of the tethered cells by means of a silicon photoelement (BPX 79, Siemens, München, FRG) calibrated against a thermistor bolometer (Radiometer 65 A, YSI Kettering, Yellow Springs, Ohio, USA), or a UV-detector (AUV 222, UDT, Hawthorne, Ca., USA). The cellular responses were monitored by an infrared-sensitive camera SN 76 with high resolution monitor (both Grundig, Fürth, FRG) and video-recorded on U-matic (VO-5.800P, Sony, Tokyo, Japan) with VTG-33 timer (For.A, Tokyo, Japan).

Absorption spectroscopy in membrane vesicles of cells of *H. halobium*, mutant strain Flx-03 1.c., in the late exponential growth phase, was done as used before [7].

RESULTS AND DISCUSSION

Statistical variability of spontaneous reversal period is logarithmically Gaussian distributed (Figure 1; [8, 2, 3]). Therefore, mean values for reversal periods of tethered cells, separated in CW or in CCW direction, are given here as geometric means, and each data point represents the mean of at least 30 reversal periods with S.E.M. (= ± 10% or less).

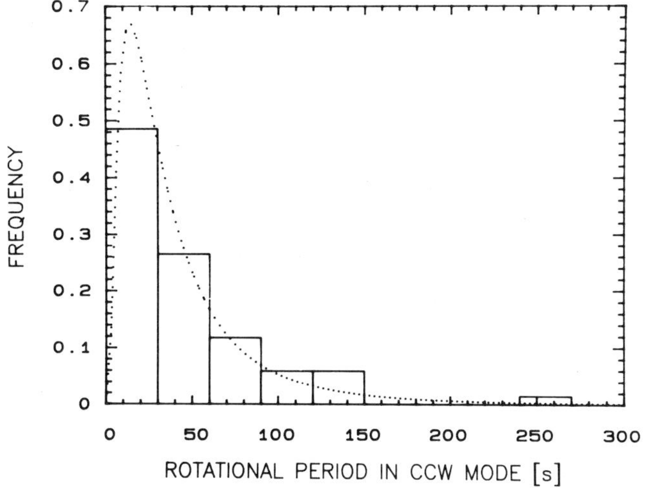

Figure 1. Distribution of the length of flagellar spontaneous rotational periods in CCW direction of tethered cells of *H. halobium*, mutant strain Flx-03 1.c., appears logarithmically Gaussian (histogram), as evidenced by the underlaid function (dotted line). Similarly, number and velocity of rotational CCW or CW periods show a logarithmically Gaussian distribution (not shown). (T. Rothärmel, unpublished data).

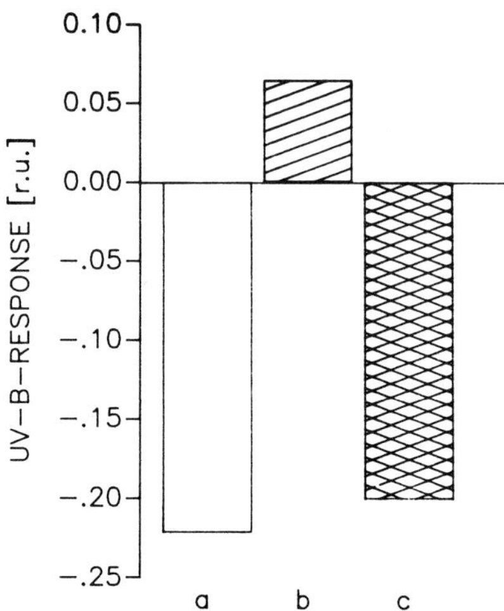

Figure 2. UV-B pulse irradiation by 0.09 W m^{-2} * 250 ms of tethered cells leads: in *H. halobium*, mutant strain Flx-03 l.c. to abbreviated reversal period **(a)**; in *H. halobium*, mutant strain L- 07 before reconstitution with all-*trans* retinal to slight extension of the spontaneous reversal period **(b)**; in *H. halobium*, mutant strain L-07 after this reconstitution, to UV-B avoidance response similar to Flx-03 l.c. **(c)**; see also **(a)**. The data are given in relative units, normalized to the spontaneous reversal period. (B. Traulich and T. Rothärmel, unpublished data).

Application of a UV-B pulse close to the threshold value, i.e. 0.09 W m^{-2} * 250 ms (Figure 2) leads to a significant avoidance response in *H. halobium*, mutant Flx-03 l.c., with abbreviated reversal period by 22% relative to the spontaneous reversal period, while retinal-deficient *H. halobium*, mutant L-07, kept responding spontaneously within an error of ± 7%; the avoidance response of L-07 could be reconstituted by culture in exogenous all-*trans* retinal (Figure 2). This indicates the UV-B response being dependent on a retinal-protein, either the reported SR-I or SR-II. Photochromism of the UV-B response under varied background light conditions [9, 3] compatible with the photochromic SR-I intermediate state 373 nm, but not reported for SR-II [1], favours SR-I as the UV-B sensory pigment. Furthermore, membrane vesicles of late exponential cells of *H. halobium*, mutant strain Flx-03 l.c., prepared in parallel to the experiments of Figure 2, showed SR-I as the only detectable photochromic pigment (Figure 3). Intramolecular energy transfer from aromatic amino acid residues of the SR-I apoprotein to the holopigment appears possible.

Upon increase of the UV-B photon fluence from 0.1 μmol quanta * m^{-2} (energy fluence of 0.04 W m^{-2} * 250 ms) to 1.75 μmol quanta * m^{-2} (energy fluence of 0.7 W m^{-2} * 250 ms), *H. halobium*, mutant strain Flx-03 l.c. showed a fluence-dependent transient pattern of behavioural response (Figure 4): the maximum

142

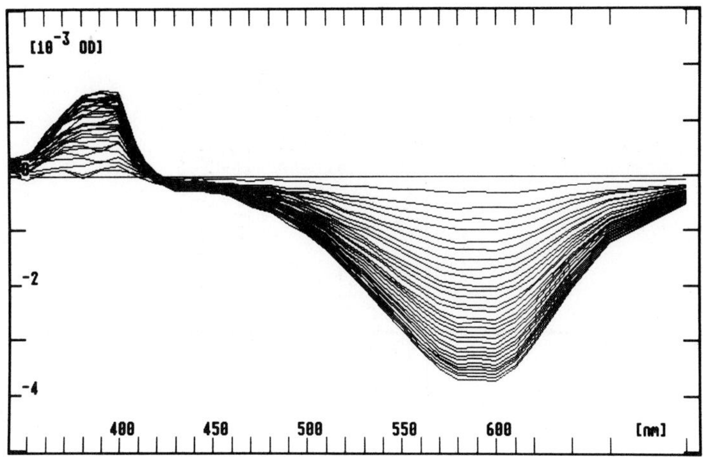

Figure 3.   Absorbance change of membrane vesicles of *H. halobium*, mutant strain Flx-03 l.c., normalized to 1 mg total membrane protein. The spectra were recorded in 10 ms time intervals between 0 ms and 300 ms after turning on the white light used simultaneously for actinic action and for monitoring absorption changes. Membrane vesicles were suspended in 4 mol/l NaCl, and sample concentrations were 1.5 OD at 400 nm. (R. Uhl and G. Wagner, unpublished data).

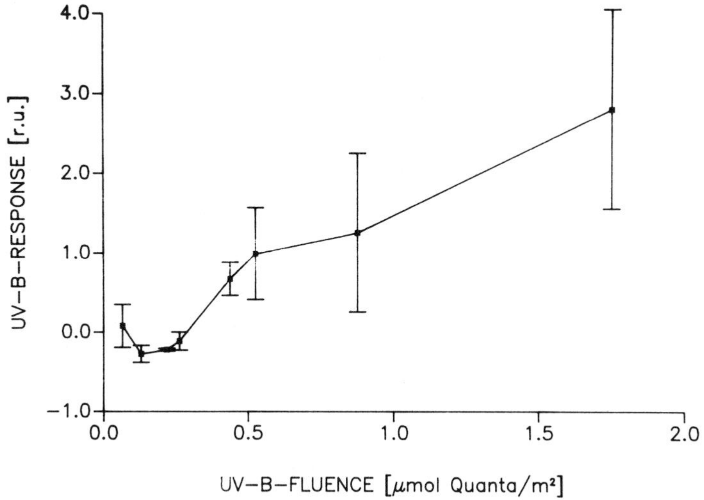

Figure 4.   UV-B response of tethered cells of *H. halobium*, mutant strain Flx-03 l.c., leads to a fluence-dependent transient pattern of response with the most of avoidance response at the UV-B fluence of 0.2 μmol quanta $*$ m^{-2}, followed by attractant response at higher UV-B fluences. The data are given in relative units, normalized to the spontaneous reversal period. (B. Traulich, unpublished data).

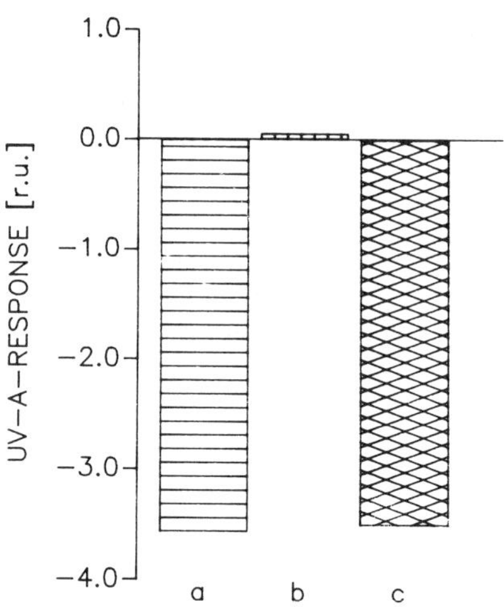

Figure 5. UV-A avoidance response of tethered cells at a standard fluence of 1.7 μmol quanta * m^{-2} leads: in *H. halobium*, mutant strain Flx-03 l.c. unexposed to UV-B pre-irradiation, to greatly abbreviated reversal period **(a)**; in *H. halobium*, mutant strain Flx-03 l.c. after pre-exposure to a UV-B fluence of 10 μmol quanta * m^{-2}, to fully "bleached" photosensory response **(b)**; in *H. halobium*, mutant strain L-07 after reconstitution with all-*trans* retinal, to UV-A avoidance response similar to Flx-03 l.c. before "bleaching" **(c)**; see also Figures 2 **(c)** and 5 **(a)**. The data are given in relative units, normalized to the spontaneous reversal period. (B. Traulich and T. Rothärmel, unpublished data).

avoidance response is seen close to 0.2 μmol quanta * m^{-2}, followed by a ramp of attractant response up to a measured almost three-fold extended reversal period. When UV-B avoidance response is "bleached" by pre-exposure of *H. halobium*, mutant strain

Flx-03 l.c., to a fluence of 10 μmol quanta * m^{-2}, the standard UV-A avoidance response of a fluence of 1.7 μmol quanta * m^{-2} was fully "bleached" as well (Figure 5; see also the control of reconstituted *H. halobium* cells, mutant strain L-07).

A set of experiments was designed to specify the biased rotational sense of tethered *H. halobium* cells, mutant Flx-03 l.c., under different environmental conditions. The metabolic substrates arginine or sodium acetate [6, 10] caused a preference in flagellar CW rotation with strongly suppressed CCW rotation, while potassium cyanide as an inhibitor of cytochrome oxidase provided the inverse result, i.e. strongly suppressed CW rotation (Figure 6). Similar to arginine or sodium acetate, continuous orange sensory light caused a preference of flagellar CW rotation (Figure

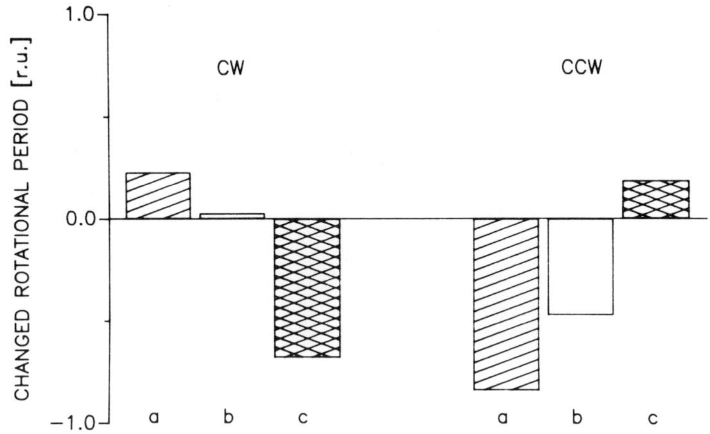

Figure 6. In tethered cells of *H. halobium*, mutant strain Flx-03 l.c., metabolic substrates, and respiratory inhibitor, lead to typical bias in flagellar rotational sense: 50 mmol/l arginine **(a)** or 20 mmol/l sodium acetate **(b)** suppresses CCW rotation while 2.5 mmol/l potassium cyanate **(c)** suppresses CW rotation. The data are given in relative units, normalized to the sum of rotations per spontaneous reversal period in CW or in CCW direction. (T. Rothärmel, unpublished data).

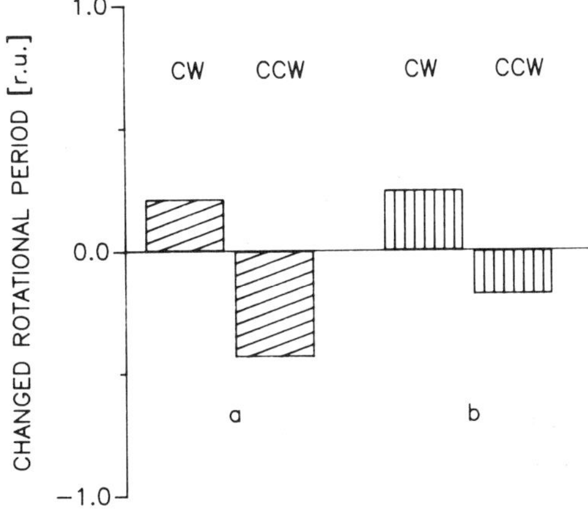

Figure 7. Bias in rotational velocity in tethered cells of *H. halobium*, mutant strain Flx-03 l.c., in the presence of 50 mmol/l arginine **(a)** or in the presence of continuous 0.7 Wm^{-2} orange light from FIL 589 nm **(b)**. Both parameters stimulate the CW response while the CCW response is suppressed. The data are given in relative units, normalized to the rotational velocity per spontaneous reversal period in CW or in CCW direction. (T. Rothärmel and B. Traulich, unpublished data).

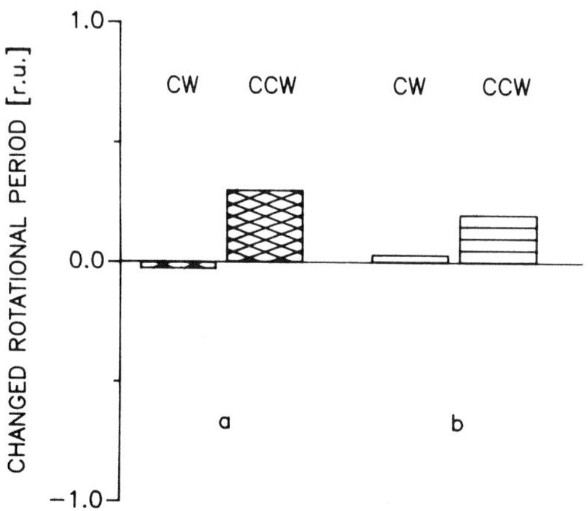

Figure 8. Bias in rotational velocity in tethered cells of *H. halobium*, mutant strain Flx-03 l.c., in the presence of 2.5 mmol/l potassium cyanate (**a**) or in the presence of 5.0 Wm^{-2} continuous blue light from BP 347 nm (**b**). Both parameters stimulate the CCW response while the CW response remains uninfluenced. The data are given in relative units, normalized to the rotational velocity per spontaneous reversal period in CW or in CCW direction. (T. Rothärmel and B. Traulich, unpublished data).

7); continuous blue sensory light caused a preference of flagellar CCW rotation, similar to potassium cyanide (Figure 8). Thus besides external stimulus conditions, rotational sense of the halobacterial flagellum seems greatly influenced by the cellular state of energy.

REFERENCES

[1]   J. L. Spudich and R. A. Bogomolni, Sensory rhodopsins of Halobacteria, *Ann. Rev. Biophys. Biophys. Chem.* 17:193 (1988)

[2]   A. Schimz and E. Hildebrand, Photosensing and processing of sensory signals in *Halobacterium halobium*, *Bot. Acta*, 101:111 (1988)

[3]   B. Traulich, Aktionsspektroskopische Untersuchungen zur Photoenergetik und zur Photosensorik von *Halobacterium halobium*, Dissertation, Univ. Giessen, FR Germany (1989)

[4]   T. Rothärmel, Rotationsverhalten des Geiβelmotors von *Halobacterium halobium* unter vercshiedenen Reizen und Energiezuständen der Zelle, Diploma, Univ. Giessen, FR Germany (1988)

[5]   G. Wagner, D. Oesterhelt, G. Krippahl and J. K. Lanyi, Bioenergetic role of halorhodopsin in *Halobacterium halobium* cells, *FEBS Lett.* 131:341 (1981)

[6]   R. Hartmann, H.-D. Sickinger and D. Oesterhelt, Anaerobic growth of halobacteria, *Proc. Natl. Acad. Sci. USA* 77:3821 (1980)

[7]   J. Otomo, W. Marwan, D. Oesterhelt, H. Desel and R. Uhl, Biosynthesis of the two halobacterial light sensors P_{480} and sensory rhodopsin and variation in gain of their signal transduction chains, *J. Bacteriol*, 171: 2155 (1989)

[8]   W. Marwan and D. Oesterhelt, Signal formation in the halobacterial photophobic response mediated by a fourth retinal protein (P_{480}), *J. Mol. Biol.* 195:333 (1987)

[9]   B. Traulich and G. Wagner, Photochromic properties of the UV-B sensory pigment in *Halobacterium halobium*, BPT-Report 6/86, ISSN 0176/0777.

[10]   A. Schimz and E. Hildebrand, Chemo-sensory responses of *Halobacterium halobium*, *J. Bacteriol.* 140:749 (1979)

STEPS IN THE PHOTOSENSORY SIGNAL CHAIN OF *HALOBACTERIUM HALOBIUM*

Eilo Hildebrand and Angelika Schimz

Institut für Biologische Informationsverarbeitung
Kernforschungsanlage Jülich
D-5170 Jülich, FRG

KEYWORDS

Photobehaviour, sensory transduction, sensory signal processing, signal integration, cellular oscillator, G-protein, protein methylation, sensory adaptation, archaebacteria.

ABSTRACT

Halobacteria swim by means of polarly inserted flagella and reverse their swimming direction about every 10 s. They detect light intensity changes through different retinal-containing pigments, and the evoked cellular signals modulate the interval length between two reversals. Several functional elements in the photosensory signal chain have been deduced by behavioural studies or could be identified biochemically. They are: a G-protein, a signal integrator, an endogeneous oscillator, which controls the reversal pattern, a flagellar switch, and methyl-accepting proteins, which are involved in the extinction of sensory signals and thereby control adaptation. The latter step is closely related to the integrator.

INTRODUCTION

Halobacterium halobium carries sensory photosystems which enable the organism to detect light intensity changes in different spectral regions and to orient itself in an indirect way with respect to light. Halobacteria gather in an environment of high fluence in the visible to allow light-energy harvesting by bacteriorhodopsin (BR) and halorhodopsin (HR), and avoid damaging ultraviolet irradiation.

The cells swim by means of two polarly inserted bundles of flagella equally well in both directions of their long axis [1]. About every 10 s they spontaneously reverse their swimming direction [2, 3]. Small deviations during the reversal events lead to a three-dimensional random walk, and it is the activation of the sensory photosystems which biases this movement pattern in an appropriate manner.

Increasing light intensity in the green to red range lengthens the interval between two reversals whereas in the UV to blue region it shortens the interval. Decreasing light intensity has the opposite effects on the interval length [2, 4, 5, 6].

Halobacteria respond to step-like changes of light intensity as well as to temporal light gradients [7]. Most probably the organism detects light intensity changes when swimming up or down an intensity gradient in its natural environment and will swim for a longer duration in the favourable direction and for a shorter time in the unfavourable direction, and will thus find the optimal environment.

Besides sensory photoreceptors a number of successive molecular events is involved in the light-dependent behaviour of this unicellular organism. Several attempts have been started to identify the receptors and to detect and characterize following steps in the photosensory pathway.

PHOTORECEPTOR PIGMENTS

It is well established that the sensory pigments of *Halobacterium* are retinal-protein complexes like the energy-converting pigments, bacteriorhodopsin (BR) and halorhodopsin (HR) [8, 9]. P-480, named also sensory rhodopsin II(sR II), has been identified as the photoreceptor of the blue-sensitive photosystem [6, 10, 11, 12]. There is, however, some uncertainty about the pigments which underlie the long-wavelength (green to red) and the ultraviolet photosystem [13, 14]. It has been argued that "slowly cycling rhodopsin", sR 587, triggers the response in the long-wavelength range, and that its long-living photointermediate, S 373, should be responsible for the reaction at 370 nm [15]. This pigment therefore has been named sensory rhodopsin I (sR I). However, action spectra in the long-wavelength range show three peaks at 565, 590 and 610 nm, the one at 565 nm being dominant (Hildebrand and Schimz, unpublished).

Two observations clearly point to a photochromic system as proposed by Spudich and Bogomolni [15]: (1) Sensitivity to UV is increased by visible background light. (2) UV background light converts the attractant response to long-wavelength stimuli into a repellent response and *vice versa*. This effect can be best explained by the assumption that under these conditions the UV-intermediate is formed by the stimulus and determines the response, because the UV-system is about 50 times as sensitive as the long-wavelength one. In a photochromic system as proposed, one would expect 587 nm to be the most efficient wavelength to convert the ground state into the S 373 intermediate, and thereby to generate the highest sensitivity in the UV. However, the most efficient wavelength to generate the highest responsiveness in the UV was found to be 510 nm (Hildebrand and Schimz, unpublished).

Although most recent results support the existence of a photochromic system, it has to be shown that it is actually represented by sR I.

150

RESPONSE REGULATION BY AN ENDOGENEOUS OSCILLATOR

Although the switching of flagellar rotation from clockwise (CW) to counter-clockwise (CCW) and *vice versa* is not strictly periodic, but shows a broad distribution of interval lengths [3], we postulate an endogeneous clock which controls the flagellar switch. The arguments for such an oscillator are the following: (1) The responsiveness to photostimuli is not constant throughout an interval between two reversal events, but changes in a characteristic sawtooth-shaped manner [3, 16]. (2) Certain stimulation programs induce phenomena which are interpreted as a route from limit cycle oscillation to chaotic oscillation [7, 17]. Periodic stimuli or temperal exponential light gradients lead, depending on their frequency or steepness, respectively, to the following behavioural patterns: (i) to new predictable periodicities of reversals, e.g. phase-locking, (ii) to period doubling (bifurcation from a fixed interval length into two alternating ones), and (iii) to deterministic chaotic oscillation. Our experimental results which indicate chaos show similarities to those obtained by a simple model calculation of periodic perturbation of a limit cycle oscillation, when a certain amount of noise is added [18, 19]. The observed random fluctuations in the interval lengths may thus be a consequence of stochastic events at a later step in the transduction chain.

The interval length can be influenced by extracellular changes of the calcium concentration and by addition of dibutyryl-cyclic GMP. These results led us to assume that cGMP and calcium may be antagonistically acting parts of the oscillator. IBMX and fluoroaluminat show effects which are in accordance with the involvement of a phosphodiesterase and a G-protein [20]. Hydrolysis of cGMP measured in vitro was found to be stimulus-dependent and was activated by activators of G-proteins and reduced by IBMX. An α-subunit of a G-protein of an apparent MW of 59 kDa could be detected immunologically [21]. The localization of the G-protein in the sensory pathway is still an open question. In analogy to other known G-proteins we postulate a functional relation to the receptor proteins.

THE FLAGELLAR SWITCH

Our oscillator hypothesis assumes that a certain substance, e.g. cGMP or calcium, periodically rises and decreases in its concentration, and, at a critical level, triggers a flagellar switch, which in turn changes the rotational sense of the flagellar motor [3]. The random fluctuations of the interval length may be a consequence of certain properties of the flagellar switch or motor.

Two models have been presented which describe the frequency distribution of interval lengths and correlate it to the switching of flagellar rotation. A four-state model of the motor, which includes an autocatalytical step, describes quantitatively the interval distribution [11]. The second model claims three states of the switch, a non-reversing (N), a reversing (R), and an inactive (I) one, which are related to each other by first-order kinetic processes [22].

Both models explain the switching mainly as a stochastic phenomenon. Our oscillator hypothesis grounds on periodic activities as the fundamental processes which are superposed by shocastic events. Stochastic transitions between different states of the switch or motor may well attribute for the random fluctuations of interval lengths.

The nature of the switch is presently unknown. Possibly it is part of the motor or at least closely related to it. Recently a switching factor could be isolated and it turned out to be fumarate [23].

SIGNAL INTEGRATION PRIOR TO THE OSCILLATOR

Signals from different photosensory inputs and from chemosensory receptors are integrated [5]. Particular stimulation programs led us to propose an integration step which lies before the oscillator in the sensory pathway [24].

For about 500 ms after a reversal event the cell is absolutely refractory to a repellent stimulus, i.e. the interval between two reversals is not influenced [3]. We have shown that refractoriness is closely related to properties of the oscillator [20]. Even under these conditions a repellent signal is formed, since we found it to be integrated with an attractant signal. Signals are integrated only up to a certain delay between stimuli, which we call "signal lifetime". The signal lifetime at the integrator is about 1.3 s for repellents [24] and 4 s for attractants [25] and does not depend on the photosystems nor on the stimulus strength. The repellent signal declines rapidly during the first 100 ms and slowly thereafter [25]. We assume that the output signals of the integrator act on the oscillator which is an additional site of integration for successively evoked signals.

METHYL-ACCEPTING PROTEINS AND ADAPTATION

Like in eubacteria, proteins could be identified in halobacteria which, depending on stimulation by attractants or repellents, are methylated or demethylated. As in eubacteria the methyl donor is S-adenosyl-methionine [[26, 27]. Methyl-accepting proteins in eubacteria were found to be related to sensory adaptation. A similar function could be expected in halobacteria.

Methylated protein bands of a MW between 65 and 135 kDa were found in *H. halobium* by gel electrophoresis [27, 28] and fluorography [29]. In vitro experiments show that methylation of these proteins is enhanced at reduced calcium concentration while demethylation is reduced under these conditions, which indicates that the methyl-transferase as well as the -esterase depends on calcium [27].

Demethylation was also measured by the amount of methanol formation and, surprisingly, the rate of demethylation increases upon both kinds of photostimuli, attractants as well as repellents [29]. Methylation is strongly inhibited in the presence of homocysteine. Under these conditions the signal lifetime at the integrator increases 2- to 3-fold in the case of attractants as well as of repellents. Similarly extended signal lifetimes were found in a mutant strain, M 160, which shows enhanced demethylation [20, 25].

From this close relationship between methylation and signal lifetime we have concluded that methylation is the molecular device to terminate the sensory signal at the integrator, and allows the cell to adapt. The signal lifetimes correspond well to the range during which the cells adapt to new steady light conditions after having received an attractant or repellent stimulus.

Since calcium effectively controls the enzymatic reactions of protein methylation and demethylation, and since reduction of external calcium shortens the signal lifetimes [20, 25], we have proposed that changes in the cellular free calcium concentration may be the signal, which, in a feed-back loop, triggers the extinction of the sensory (excitatory) signal at the integrator, which, of course, should be closely related to the methyl-accepting proteins [13].

FINAL REMARKS

Several steps in the photosensory pathway of halobacteria could be identified or deduced by a thorough analysis of the behavioural pattern under appropriate stimulation programs. The biochemical identification and characterization is presently under study. Goals for the near future are to show periodic changes in the concentration of certain cellular components as e.g. cGMP and/or calcium, and to unravel the functional relationship between the receptors and the G-protein as well as the feed-back loop for adaptation.

ACKNOWLEDGEMENT

Our work was supported by the Deutsche Forschungs-gemeinschaft (SFB 160 and grant Hi 114/1-2).

REFERENCES

[1]   M. Alam and D. Oesterhelt, Morphology, function and isolation of halobacterial flagella, *J. Mol. Biol.* 176: 459 (1984)

[2]   E. Hildebrand and A. Schimz, Photosensory behavior of a bacteriorhodopsin-deficient mutant, ET-15, of *Halobacterium halobium. Photochem. Photobiol.* 37: 581 (1983)

[3]   A. Schimz and E. Hildebrand, Response regulation and sensory control in *Halobacterium halobium* based on an oscillator. *Nature*, 317: 641 (1985)

[4]   E. Hildebrand and N. Dencher, Two photosystems controlling behavioural responses of *Halobacterium halobium. Nature*, 257: 46 (1975)

[5]   J. L. Spudich and W. Stoeckenius, Photosensory and chemosensory behavior of *Halobacterium halobium, Photobiochem. Photobiophys.* 1: 43 (1979)

[6]   T. Takahashi, H. Tomioka, N. Kamo and Y. Kobatake, A photosystem other than PS 370 also mediates the negative phototaxis of *Halobacterium halobium. FEMS Microbiol. Lett.* 28: 161 (1985)

[7]   A. Schimz and E. Hildebrand, Periodicity and chaos in the response of *Halobacterium* to temporal light gradients. *Eur. Biophys. J.* 17: 237 (1989)

[8]   N. A. Dencher and E. Hildebrand, Sensory transduction in *Halobacterium halobium*: Retinal pigment controls UV-induced behavioral response. *Z. Naturforsch.* 34c: 841 (1979)

[9]   A. Schimz, W. Sperling, P. Ermann, H. J. Bestmann and E. Hildebrand, Substitution of retinal by analogues in retinal pigments of *Halobacterium halobium*. Contribution of bacteriorhodopsin and halorhodopsin to photosensory activity. *Photochem. Photobiol.* 38: 417 (1983)

[10]   E. K. Wolff, R. A. Bogomolni, P. Scherrer, B. Hess and W. Stoeckenius, Color discrimination in halobacteria: Spectroscopic characterization of a second receptor covering the blue-green region of the spectrum. *Proc. Natl. Acad. Sci. USA.* 83: 7272 (1986)

[11]   W. Marwan, and D. Oesterhelt, Signal formation in the halobacterial photophobic response mediated by a fourth retinal protein (P-480). *J. Mol. Biol.* 195: 333 (1987)

[12]   P. Scherrer, K. McGinnis and R. A. Bogomolni, Biochemical and spectroscopic characterization of the blue-green photoreceptor in *Halobacterium halobium*. *Proc. Natl. Acad. Sci. USA* 84: 402 (1987)

[13]   A. Schimz and E. Hildebrand, Photosensing and processing of sensory signals in *Halobacterium halobium*. *Bot. Acta.* 101: 111 (1988)

[14]   J. L. Spudich and R. A. Bogomolni, Sensory rhodopsins of halobacteria. *Ann Rev. Biophys. Biophys. Chem.* 17: 193 (1988)

[15]   J. L. Spudich and R. A. Bogomolni, Mechanism of colour discrimination by a bacterial sensory rhodopsin. *Nature*, 312: 509 (1984)

[16]   E. Hildebrand and A. Schimz, Role of the response oscillator in inverse responses of *Halobacterium halobium* to weak light stimuli. *J. Bacteriol.* 169: 254 (1987)

[17]   A. Schimz and E. Hildebrand, Entrainment and temperature dependence of the response oscillator in *Halobacterium halobium*. *J. Bacteriol.* 166: 689 (1986)

[18]   L. F. Olsen and H. Degn, Chaos in biological systems. *Quart. Rev. Biophys.* 18: 165 (1985)

[19]   W. M. Schaffer, S. Ellner and M. Kot, Effects of noise on some dynamical models in ecology. *J. Math. Biol.* 24: 479 (1986)

[20]   A. Schimz and E. Hildebrand, Effects of cGMP, calcium and reversible methylation on sensory signal processing in halobacteria. *Biochim. Biophys. Acta.* 923: 222 (1987)

[21]   A. Schimz, K.-D. Hinsch and E. Hildebrand, Enzymatic and immunological detection of a G-protein in *Halobacterium halobium*. *FEBS Lett.* 249: 59 (1989)

[22]   D. A. McCain, L. A. Amici and J. L. Spudich, Kinetically resolved states of the *Halobacterium halobium* flagellar motor switch and modulation of the switch by sensory rhodopsin, *Int. J. Bacteriol.* 169: 4750 (1987)

[23]   W. Marwan, F. W. Schäfer and D. Oesterhelt, Signal transduction in *Halobacterium* depends on fumarate, *EMBO J.* 9: 355 (1990)

[24]   E. Hildebrand and A. Schimz, Integration of photosensory signals in *Halobacterium halobium*. *J. Bacteriol.* 167: 305 (1986)

[25]   E. Hildebrand and A. Schimz, The lifetime of photosensory signals in *Halobacterium halobium* and its dependence on protein methylation. *Biochim. Biophys. Acta*, (in press) (1990)

[26]   A. Schimz, Methylation of membrane proteins is involved in chemosensory and photosensory behavor of *Halobacterium halobium*. *FEBS Lett.* 125: 205 (1981)

[27]   A. Schimz, Localization of the methylation system involved in sensory behavior of *Halobacterium halobium* and its dependence on calcium. *FEBS Lett.* 139: 283 (1982)

[28]   S. I. Bibikov, V. A. Baryshev and A. N. Glagolev, The role of methylation in the taxis of *Halobacterium halobium* to light and chemoeffectors. *FEBS Lett.* 146: 255 (1982)

[29]   M. Alam, M. Lebert, D. Oesterhelt and G. L. Hazelbauer, Methyl-accepting taxis proteins in *Halobacterium halobium*. *EMBO J.* 8: 631 (1989)

MODE OF ACTION OF HALOCINS H4 AND H6: ARE THEY EFFECTIVE

AGAINST THE ADAPTATION TO HIGH SALT ENVIRONMENTS?

I. Meseguer, M. Torreblanca and F. Rodriguez-Valera

Departamento de Genética Molecular y Microbiologia
Universidad de Alicante
Campus de S. Juan, Apartado. 374
Alicante, Spain

ABSTRACT

Halocins H4 and H6, two bacteriocins produced by halobacterial strains, are proteins able to kill other halobacterial strains than those which produce them. Although their physico-chemical features are quite distinct, their modes of action seem to be similar. Both halocins induce morphological changes and lysis in sensitive cells, affect light-induced pH changes and inhibit α-aminoisobutiric acid transport. All these factors lead us to suppose that the target of both halocins must be located at membrane level, affecting one or more of the mechanisms which take part in the complex machinery of regulation and maintenance of the electrochemical gradients steady state through the membrane. The fact that halobacteria live in extremely aggressive media considerably increases the effectiveness of substances such as halocin H4 and H6 to kill salt-dependent cells.

INTRODUCTION

The members of the Halobacteriaceae family produce a group of bacteriocin-like substances which have been named halocins [1,2]. The production of proteins which kill closely related organisms, i.e. bacteriocins, is commonplace among eubacteria [3,4]. With yeasts a similar case occurs with the so-called killer-toxins [5]. Since the halobacteria are members of the third urkingdom, the Archaebacteria [6,7], this kind of behaviour appears to be of an uncommon universality. Two halocins, termed H4 and H6, have been studied [8 - 11]. Halocin H4 is a protein of 28.000 MW which requires salts in the medium for stability [8] like many halobacterial enzymes [12], and halocin H6 is a heat-stable non-salt-dependent protein of 32.000 MW [10]. Both halocins bind to the target cell and cause death and lysis by exponential kinetics

General and Applied Aspects of Halophilic Microorganisms
Edited by F. Rodriguez-Valera, Plenum Press, New York, 1991

[9,10]. The modes of action of these halocins have been studied, observing that lysis is preceded by progressive deformation of the cells, which become spherical and without any apparent cytoplasm, i.e. ghost cells [9,11].Several bacteriocins which cause morphological changes in sensitive cells have been described. The appearance of deformations in the affected cells strongly indicates that the bacteriocin affects the cell envelope, either the wall or the cellular membrane. Bacteriocins whose target is the cell wall either induce the formation of protoplasts, as with megacin A, identified as a phospholipase [13], or spheroplasts, like colicin M which inhibits murein biosynthesis [14]. Other bacteriocins cause morphological alterations without affecting the wall, as for instance staphylococcin C55 [15], streptococcin A-TT22 [16], boticin E-S5$_1$ [17]. In the case of halocins H4 and H6, the facts that (a) the wall remains apparently intact after treatment, and (b) wall-less spheroplasts are sensitive to the halocins action [11], seem to indicate that their primary target is the cytoplasmic membrane.

In a previous work we demonstrated that halocin H4 affected both the light-induced pH changes of a cell suspension, and α-aminoisobutiric acid transport [9]. In this work we show how halocin H6 causes effects similar to those produced by halocin H4, indicating that the mechanisms of action of the two substances must be similar in spite of their significantly different physico-chemical characteristics.

MATERIAL AND METHODS

Bacterial strains: The producer strain of Halocin H4 was *Haloferax mediterranei* ATCC 33500 and that of halocin H6 was *Haloferax gibbonsii* strain Ma 2.39 [18]. As sensitive strains, *Halobacterium halobium* NRC817 and *Halobacterium halobium* P-mutant [19] were used.

Media, growth conditions and purification of halocins: Media and growing culture conditions were as described previously [8, 10]. The cultures at the beginning of the stationary phase were tangencially filtered twice, using the Pellicon system (Millipore), the first time to eliminate cells and large particles, using 0.45 μm pore size filters, the second to eliminate high molecular weight impurities using ultrafilters which retain molecules weighing over 100.000 Daltons. Tangencial ultrafiltration by the same system was used to concentrate halocins H4 and H6, using ultrafilters which retain molecules over 10.000 Daltons. The halocins were purified from this concentration by hydroxyl-apatite column chromatography, using a $PO_4^{=}$ gradient of 10mM to 120mM in 2.5 M NaCl for halocin H4 and of 10mM to 400mM for halocin H6. Concentrating the maximum activity fractions, H4 was obtained with a satisfactory degree of purity, but H6 still contained a large amount of impurities. Therefore the concentrate of the maximum activity fractions from the hydroxylapatite column were refiltered through a Sephadex G-50 column using tris-ClH 0.05M pH 6.7 buffer as eluent. Concentrating the maximum activity fractions then obtained produced halocin H6 sufficiently pure for use in this work. The degree of purity of the concentrates was determined by SDS-polyacrylamide gel electrophoresis, activity was measured as described previously [9] and expressed as arbitrary units per ml (AU/ml).

pH changes induced by light: For these experiments *Halobacterium halobium* P-mutant [19] was used. This mutant is sensitive to both halocin H4 and H6. pH changes were measured and registered as described [9] in a test tube containing 3ml (magnetically stirred) cell suspension (1mg prot/ml in 25% SW) placed in a thermostatized chamber at 37°C. Illumination was provided by a self-focusing 250 watt quartz-halogen lamp (ENH), through a 2cm thick water filter, a "hot mirror" and OG-530 filters. Final concentration of halocin H4, when added, was 400 AU/mg of protein and 410 AU/mg protein of halocin H6. DCCD was added in ethanol solution at a final concentration of 20 μM. Protein was determined by the Lowry method [20].

Transport of α-aminoisobutyric acid: The transport was measured by the filtration method using $|^{14}C|$ α-aminoisobutyricacid. Two different aspects were studied with this compound, the uptake and release of AIB by treated and untreated cells.

For the first, 1μCi/ml (final concentration) $|^{14}C|$ AIB were added to a suspension of *Halobacterium halobium* NRC 817 (1mg protein/ml), and 0.1ml samples were taken at 5 minute intervals. The samples were filtered using HA 0.45 μ Millipore filters and washed three times with 1ml 25% SW. The filters were dissolved with 1ml Cellosc¹ve (Sigma), and after adding 4ml scintillation liquid (Ready-Solv HP, Beckman) were counted in a Beckman LS-2800 scintillation counter.

The release of $|^{14}C|$ AIB was studied using a cell suspension (1mg prot/ml) previously incubated for two hours with this compound (1μCi/ml final concentration),

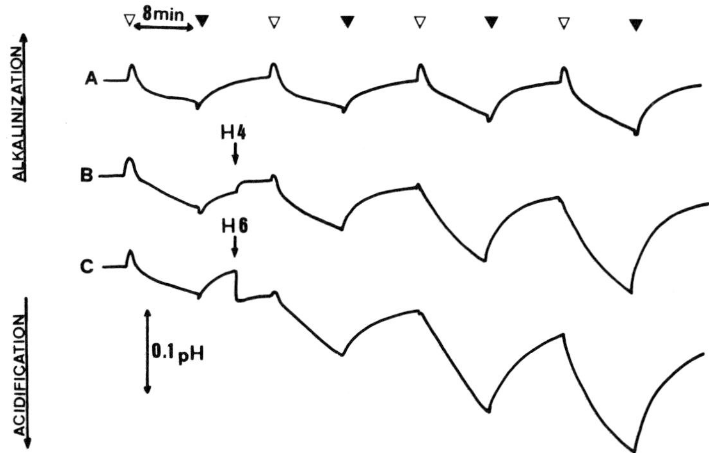

Figure 1.   Effect of halocins H4 and H6 on light-induced pH changes of a P-mutant cell suspension (1mg prot./ml). A) untreated control cells; B) cells treated with halocin H4 (400 AU/mg protein); C) cells treated with halocin H6 (410 AU/mg protein). The addition of halocin is indicated by arrows. ∇ indicates light switched on, and $\blacktriangledown$ light switched off. (See Materials and Methods).

then centrifuged and resuspended in an equal volume of 25% SW. 0.1ml samples taken at 5 minute intervals were filtered and counted as above. Halocin H6 was added at final concentrations of 410 or 840 AU/mg protein.

RESULTS AND DISCUSSION

In a previous work we showed that halocin H4 affected the light-induced pH changes of a P-mutant cell suspension [9]. With a suspension of untreated cells, a series of perfectly constant pH changes can be produced in the medium in response to alternating periods of light and darkness. These changes, consistent with those described in the literature [21], start with a rapid but brief alkalinization when illumination commences, followed by a slower and more prolonged acidification. When the light is switched off the process is reversed, first acidification and then alkalinization tending towards restoration of the initial pH (Figure 1A). When we added halocin H4 or H6 to a suspension of this type, the pH changes were modified as shown in

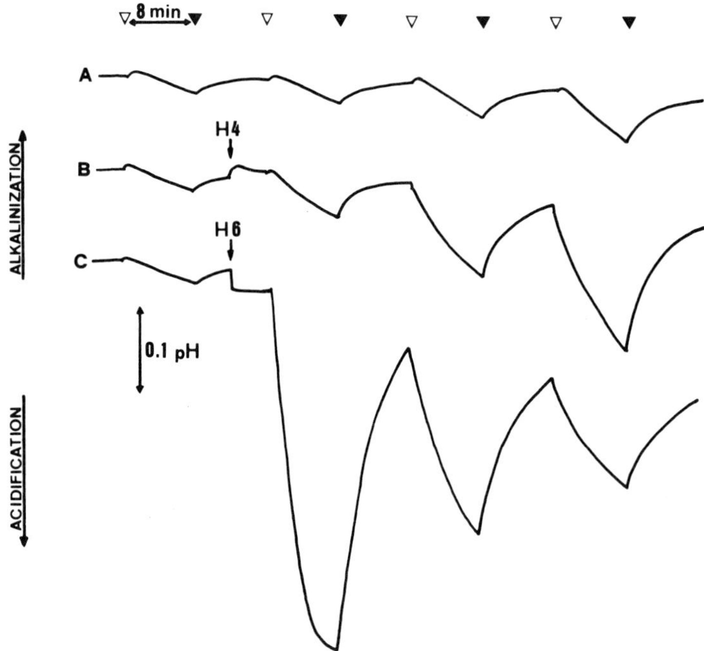

Figure 2. Effect of halocins H4 and H6 on light-induced pH changes of a DCCD treated P-mutant cell suspension (1mg prot./ml., 20 μm DCCD). A) DCCD treated cell control; B) cells treated with DCCD and halocin H4 (400 AU/mg prot.); C) cells treated with DCCD and halocin H6 (410 AU/mg prot.). The addition of halocin is indicated by arrows. ∇ indicates light switched on, and $\blacktriangledown$ light switched off. (See Materials and Methods).

160

Figures 1B and 1C respectively. On the one hand both the alkalinization on exposure to light and the acidification in darkness tended to disappear, and on the other the acidification with illumination and alkalinization in darkness intensified, although not sufficiently to restore the initial pH. Moreover, these effects were intensified in proportion with longer periods of halocin treatment. When the cells are treated simultaneously with halocin and DCCD, an ATPase inhibitor which has recently been shown to inhibit the Na^+/H^+ antiporter also in these microorganisms [22], a synergistic effect occurs, with more accentuated pH decrease during illumination and increase in darkness (Figure 2, B and C).

These results clearly indicate that the halocins either directly or indirectly affect the proton interchange through the membrane. Either the halocins favour the exit of protons during illumination, or they impede their return into the cell. It could be concluded that the halocins excercise some kind of effect on the light-dependent proton-pump, i.e. bacteriorhodopsin, but this is improbable since the halocins effect is not light-dependent, and they are equally effective against sensitive cells in darkness. Another explanation for the effects observed could be that they inhibit one or more of the mechanisms involved in the reentry of protons into the cells. To date two mechanisms responsible for this function in halobacterial membrane have been described: ATPase and Na^+/H^+ antiporter. Finally, the possibility must still be considered that halocins H4 and H6 affect some other membrane mechanism involved in the interchange of other ions such as K^+ or Cl^-, and thus indirectly affect the H^+ interchange. Whilst it is known that halorhodopsin is an inward-directed light-

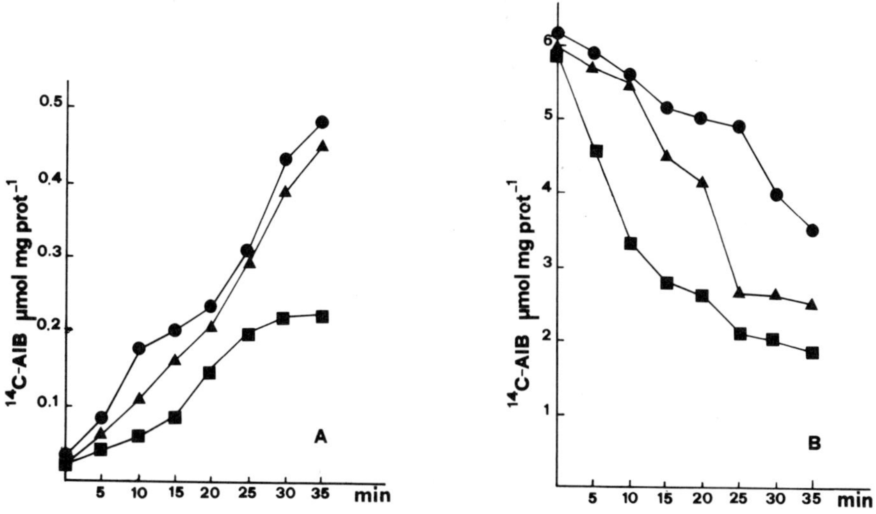

Figure 3. Effect of halocin H6 on [^{14}C |AIB uptake (A) and release (B), measured by the filtration method as described in Materials and Methods. (●) untreated control cells; (▲) treated with 410 AU/mg prot.; (■) treated with 820 AU/mg prot.

dependent chloride pump, the mechanism responsible for this function in darkness is not yet known but several authors claim that such mechanism must exist in these organisms. Although there is no direct evidence regarding the mechanism affecting K^+, the existence of a uniport system to ensure the entrance of this ion against the gradient is postulated.

Resuming, any of these mechanisms could well be the halocins primary target, but the inhibition of AIB transport produced by halocin H6 (Figure 3A) and previously observed with halocin H4 [9], seems to point to the Na^+/H^+ antiporter, since the aminoacid transport in these microorganisms has been described as a Na^+ gradient-dependent symport system [23], which is itself generated by the Na^+/H^+ antiporter. Moreover some authors have theorized that this system and the AIB transport system may have a common subunit [24]. Nevertheless, there is not yet sufficient evidence to discount the other possibilities mentioned.

Studies of other bacteriocins whose primary target is at membrane level lead to the supposition that the bacteriocins action forms pores in the membrane through which a free interchange of ions with the medium takes place, quickly destroying the ion gradients [13]. In the case of Halocins H4 and H6 this possibility is ruled out because if such pores existed the light-induced pH gradients could not be formed and the release of AIB from treated cells would be more rapid than that observed in Figure 3B.

With the aim of clarifying these facts, we have commenced a study into the effect of these halocins on intracellular pH and volume, membrane potential and proton-motive force. Preliminary results indicate that these parameters do suffer alteration on halocin treatment of the cells, but not drastically.

What was quite evident from the combined total of these results is that both halocins in some way affect the mechanisms involved in maintaining ion gradients. As halobacteria inhabit media with Na^+ concentrations in the order of 3-4M, to compensate for the strong external osmotic pressure they have developed a mechanism which enables them to achieve an internal concentration of K^+ up to one thousand times that of the medium. This mechanism cannot function independently from the other ions most abundant in these media, principally Na^+ and Cl^-. This means that halobacteria must have a complex, efficient machinery for maintaining the steady state of the ion gradients through the membrane. Logically therefore, if halocins H4 and H6 affect one or more of those mechanisms the cells would lose their defence against the extreme environment they inhabit. It is clear also that this machinery would be an eminently suitable primary target for bacteriocin-like substances in view of their role in the competition for the same ecological niche.

ACKNOWLEDGEMENTS

This work was supported by grants PR84/1099 and BT11/86 of the CAICYT (Spanish Ministry of Education and Science). M.Torreblanca was in receipt of a fellowship from the Generalitat Valenciana. The secretarial assistance of K. Hernandez is gratefully acknowledged.

REFERENCES

[1]   F. Rodriguez-Valera, G. Juez and D. J. Kushner, Halocins: salt-dependent bacteriocins produced by extremely halophilic rods. *Can. J. Microbiol.* 28:151 (1982).

[2]   I. Meseguer, F. Rodriguez-Valera and A. Ventosa, Antagonistic interactions among halobacteria due to halocin production. *FEMS Microbiol. Lett.* 36:177 (1986).

[3]   P. Reeves, The Bacteriocins. *Bacteriol. Rev.* 29:24 (1965).

[4]   J. R. Tagg, A. S. Dajani and L. W. Wannamaker, Bacteriocins of gram-positive bacteria. *Bacteriol. Rev.* 40:722 (1976).

[5]   R. G. E. Palfree and H. Bussey, Yeast killer toxin: purification and characterization of the protein toxin from *Saccharomyces cerevisiae. Eur. J. Biochem.* 93:487 (1979).

[6]   C. R. Woese and G. E. Fox, Phylogenetic structure of the prokaryotic domain: the primary kingdoms. *Proc. Natl. Acad. Sci. USA* 74:5088 (1977),

[7]   C. R. Woese, Archaebacteria and cellular origins: an overview. *Zbl. Bakt. Hyg. I. Abt. Orig.* C3:1 (1982).

[8]   I. Meseguer and F. Rodriguez-Valera, Production and Purification of halocin H4 *FEMS Microbiol Lett.* 28:177 (1985).

[9]   I. Meseguer and F. Rodriguez-Valera, Effect of halocin H4 on cells of *Halobacterium halobium. J. Gen. Microbiol.* 132:3061 (1986).

[10]   M. Torreblanca, I. Meseguer and F. Rodriguez-Valera, Halocin H6, a bacteriocin from *Haloferax gibbonsii. J. Gen. Microbiol.* 135:2655 (1989).

[11]   M. Torreblanca, I. Meseguer and F. Rodriguez-Valera, Effects of Halocin H6 on the morphology of sensitive cells. Biochem and Cell Biol. (In press).

[12]   J. K. Lanyi, Salt dependent properties of proteins from extremely halophilic bacteria. *Bacteriol. Rev.* 38:272 (1974).

[13]   J. Konisky, Colicins and other bacteriocins with established modes of action. *Ann. Rev. Microbiol.* 36:125 (1982).

[14]   K. Schaller, J. V. Höltje and V. Braun, Colicin M is an inhibitor of murein biosynthesis. *J. Bacteriol.* 152:994 (1982).

[15]   C. C. Clawson and A. S. Dajani, Effect of bacteriocidal substance from *Staphylococcus aureus* on group A Streptococci. *Infect. Immun.* 1:491 (1970).

[16]   J. R. Tagg, E. A. Phil and A. R. McGiven, Morphological changes in a susceptible strain of *Streptococcus pyogenes* treated with Streptococcin A. *J.Gen. Microbiol.* 79:167 (1973).

[17]   J. S. Ellison, C. F. T. Mattern and W. A. Daniel, Structural changes in *Clostridium botulinum* type E after treatment with boticin $S5_1$. *J. Bacteriol.* 108:526 (1971).

[18]   M. Torreblanca, F. Rodriguez-Valera, G. Juez, A. Ventosa, M. Kamekura and M. Kates, Classification of non alkaliphilic halobacteria based on numerical taxonomy and polar lipid composition, and description of *Haloarcula,* gen. nov. and *Haloferax*, gen. nov. *Syst. Appl. Microbiol.* 8:89 (1986).

[19]   G. Juez and F. Rodriguez-Valera, A mutant of *H. halobium* with constitutive production of bacteriorhodopsin. *FEMS Microbiol. Lett.* 23:167 (1984).

[20]   O. H. Lowry, N. J. Rosebrough, A. L. Farr and R. J. Randall, Protein measurements with the folin phenol reagent. *J. Biol. Chem.* 193:265 (1951).

[21]   E. P. Bakker, H. Rotemberg and R. Caplan, An estimation of the light-induced electrochemical potential difference of protons across the membrane of *Halobacterium halobium. Biochim. Biophys. Acta.* 440:557 (1976).

[22]   N. Murakami and T. Konishi, Mechanism of function of Dicyclohexylcarbodiimide-sensitive Na^+/H^+ Antiporter in *Halobacterium halobium*: pH effect. Archiv. *Biochem. Biophys.* 2:515 (1989).

[23]   R. E. MacDonald, R. V. Greene and J. K. Lanyi, Light-activated amino acid transport systems in *Halobacterium halobium* envelope vesicles: role of chemical and electrical gradients. *Biochemistry*, 16:3227 (1977).

[24]   T. A. Krulwich, Na^+/H^+ Antiporters. *Biochim. Biophys. Acta.* 726:245 (1983).

BIOENERGETICS OF *HALOBACTERIUM HALOBIUM* AND OF

H. MARISMORTUI

Ben-Zion Ginzburg

Plant Biophysical Laboratory
Botany Department
The Hebrew University
Jerusalem
Israel

ABSTRACT

Changes in ATP have been followed *in vivo* by P-NMR. Uncouplers (protonophores and ionophores), were shown to inhibit ATP synthesis, growth and respiration, but had little or no effect on pH gradients either in *H. halobium* or in *H. marismortui*. The K gradient, which was dissipated in *H. halobium* when uncouplers were used at 10 times the concentrations that inhibited metabolic reactions, remained intact in *H. marismortui* with uncouplers at 1000 times these concentrations. The electrical conductance of the membranes of *H. marismortui* was found to be 2000 s/cm^2, one million times larger than that of *H. halobium*. These observations are not consistent with the chemiosmotic paradigm. It is suggested that enhanced interactions of water molecules are involved in maintaining large ionic gradients. Evidence for the existence of an enhancement of this sort comes from NMR, calorimetric and dielectric measurements.

H. halobium and *H. marismortui* belong to the Archaeobacteriaceae and thus are low in the evolutionary hierarchy. In other words, they are primitive organisms. The *H. marismortui* membranes possess unusual features and a question arose as to whether the biochemical and physiological mechanisms of this organism are more primitive than those of more advanced organisms.

When anaerobic preparations of *H. marismortui* supplied with glycerol and oxygen were examined *in vivo* by P-NMR, ATP was found to be synthesized at a reasonable rate [1,2]. It should be pointed out that the methods used to measure ATP

General and Applied Aspects of Halophilic Microorganisms
Edited by F. Rodriguez-Valera, Plenum Press, New York, 1991

Table 1. Ionic composition of halophilic bacteria

| Bacterial species | State of Bacteria | millimoles/kg water | | | |
| | | K$^+$ | | Na$^+$ | |
		out	in	out	in
H. halobium	Logarithmic	5	4500	3500	850
	Starved	5	1400	3500	3300
H. marismortui	Logarithmic	2	3700	3500	2000
	Starved	2	3000	3500	1100

Table 2. Effect of uncouplers on bioenergetic parameters of halophilic bacteria.

	Uncoupler		*H. halobium*	*H. marismortui*
Effect on ATP in vivo (NMR)	FCCP TTFB	3μM 3μM	Decrease by 90% from level with oxygen	Decrease by 90% from level with oxygen
Effect on growth	FCCP TTFB	1μM 2μM	complete inhibition	complete inhibition
Effect on endogenous respiration (oxygen uptake)	FCCP TTFB	3μM 3μM	50% inhibition 50% inhibition	50% inhibition 50% inhibition
Effect on glycerol respiration stimulation (oxygen uptake)	FCCP TTFB Nigericin Valinomycin Valinomycin + FCCP	10μM 10μM 6μM 3μM 3μM 2μM	100% inhibition 100% inhibition 100% inhibition No effect No effect	100% inhibition 100% inhibition 100% inhibition No effect No effect
Effect on pH difference	FCCP TTFB	30-50μM 30-50μM	Small decrease Small decrease	No effect No effect
Effect on cellular K	FCCP TTFB	3μM 30μM 30μM	No effect K disappeared after 10 h as above	No effect No effect, even at 1mM, after 48 hours

by NMR with halophilic bacteria can give only approximative values. Rates of ATP synthesis and the concentrations of ATP and Pi detected resembled those of *H. halobium*. This latter has already been studied by conventional methods [3, 4, 5] and can be used as a reasonable control for our NMR observations, the first of their kind for halophilic microorganisms. As can be seen in Table 1, both species are rich in K^+. The inner pH, which is about one unit higher than that of the outside, was deduced from the chemical shift of inorganic phosphate. There is some uncertainty about the pH value which is very sensitive to ionic species at high ionic concentrations. When the cells of *H. marismortui* are exposed to low concentrations of uncouplers (1 - 3 μM), both protonophores (FCCP, TTFB) and ionophores (nigericin or valinomycin), ATP disappeared (Table 2). The protonophores inhibited growth of the organisms at even lower concentrations. Interestingly enough, the uncouplers inhibited respiration, both endogenous and that induced by addition of glycerol, over the same range of concentrations. There was a very small effect on pH in *H. halobium*, but only at uncoupler concentrations 10 times higher, and no effect on the pH of *H. marismortui*. The above-mentioned inhibitors had remarkably little effect on cell K^+; in *H. halobium* dissipation of cell K^+ was observed only 5 - 8 h after the onset of treatment, and only at higher concentrations than were needed for dissipation of ATP. In *H. marismortui* no effect on cell K^+ was found, not even after 48 h with 1 mM protonophores. Ionophores were equally ineffective. This confirms former similar data obtained by other methods [6,7].

Figure 1 shows a comparison between the dielectric dispersions of *H. halobium* and *H. marismortui*. The low-frequency dispersion is the so-called ß-dispersion associated with the cell membrane. The extent of the dispersion of *H. marismortui* is relatively small, the reason being the high electrical conductance of

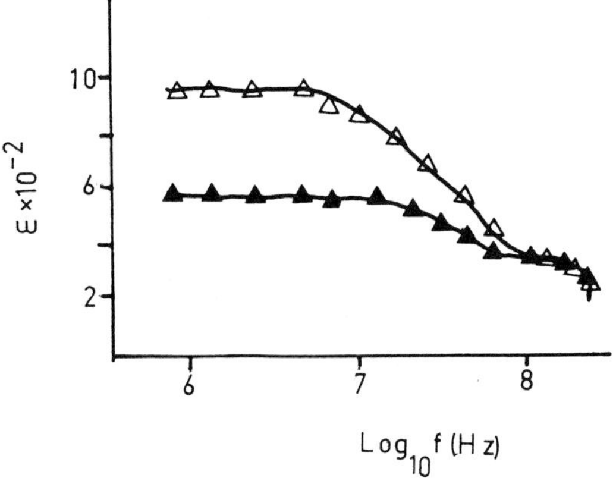

Figure 1. Variation of permitivity (ϵ) as function of frequency for suspensions of *H. halobium* (Δ) and *H. marismortui* ($\blacktriangle$) at volume fractions of 0.65. Measurements were made at 25°C [10]

the cell membrane, which was calculated to be 2000 s/cm^2, or at least six orders of magnitude higher than that of *H. halobium*, or of any other known cell-membrane. Such very high values are consistent with former measurements which revealed the very high permeability of the membrane of *H. marismortui* to small organic molecules such as fructose and sucrose [8] and to the small ions Na^+, Cl^-, K^+ [6, 7]. Conductances of typical energy-coupling membranes reported in the literature [9] do not exceed 2.10^{-4} s/cm^2; even the addition of gramicidin to thylakoid vesicles does not raise the membrane conductance above 0.15 s/cm^2 [9]. We have here, therefore, direct evidence for bioenergetically competent cells with cell membranes with a very high conductance indeed.

Neither the effects of uncouplers nor the electrical conductivity of the membrane of the *H. marismortui* are consistent with the chemiosmotic paradigm.

Table 3. Molar heat of dehydration of pellets of cells

	Molar heat of dehydration (KJ/mole water)	
		Excess
Human red blood cells	40.6 ± 1.13	-0.45 (a)
H. halobium	46.2 ± 0.96	1.9 (b)
H. marismortui	50.4 ± 0.85	5.9 (b)
Water	41.05 ± 0.92	
KCl solution, 3M	44.3 ± 0.35	

(a) above water; (b) above KCl solution.

Table 4. Molar heat of dehydration (Joules per mole water) of 3 different populations of cell-water molecules in *H. marismortui* (see text).

	Population		
	a	b	c
% Heat	46.6	27.4	25.8
Heat of dehydration	44.51	63.03	
Excess (a)		18.0	

(a) above KCl solution

If the membrane of *H. marismortui* does not function as an efficient barrier to the ions K^+ and H^+, what is the mechanism which maintains high gradients of these ions across the membrane?

The dynamic properties of intracellular water molecules of *H. marismortui* revealed by the NMR measurements, can be summarised as follows. Firstly, the translational diffusion coefficient, D, of the cell water is of the order of 10^{-8} $cm^2.s^{-1}$. This is at least three orders of magnitude smaller than the value for pure water, or of water in other biological cells. Secondly, the protons of the cell water molecules exchange rapidly, at a rate much above $100\,s^{-1}$. This exchange involves macromolecular protons including those of acidic aminoacids which are present in large amounts in *H. marismortui* proteins. Thirdly, rotational motions of the cell water involves processes that have a reorientation rate slower than $10^8\ s^{-1}$, at least three orders of magnitude smaller than the rotational diffusion rate of pure water, which is about $10^{11}s^{-1}$, at

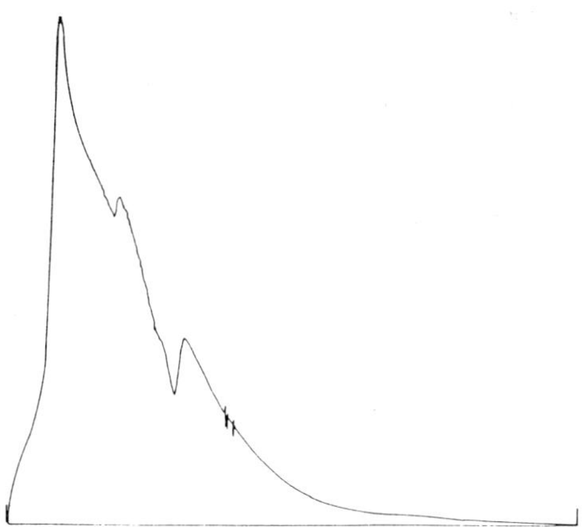

Figure 2.   Thermogram of the dehydration of *H. marismortui* pellet (isothermal mode).

20°C. Lastly, the cell water in *H. marismortui* may exist in possibly two, or even three different environments. About 45% of the cell water freezes at -20°C together with the extracellular water. The NMR signal that remained after freezing indicated the presence of two more water environments, characterized by a difference in NMR relaxation [11].

The average molar heat of dehydration of *H. marismortui* is 50.4 KJ.mole^{-1} H_2O. This is 9 KJ.mole^{-1} H_2O more than that of the heat of evaporation of pure water measured under equivalent conditions, and 6 KJ.mole^{-1} H_2O more than the heat of dehydration of concentrated solutions of NaCl of KCl (Table 3). In comparison, the molar heat of dehydration of a pellet of red blood cells (RBC) is almost that of pure water. RBC can be considered to be a relatively concentrated protein solution, mainly of haemoglobin, and lacking the high ion-concentration of halophilic bacteria. Thus, the excess heat of dehydration, if it is compared to pure water in the case of RBC, and to that of salt solution in the case of the bacteria, expresses as a first approximation the water-water interaction. It should be noticed that the molar heat of dehydration given in Table 3 is an average value for the pellet-water. As at least 30% of this water is outside the bacteria, the excess molar heat of dehydration is even larger than indicated in Table 3. Closer inspection of Figure 2 gives more information on the thermal events of the dehydration of the bacteria. We can discern three events, indicative of the existence of three populations of water within the bacteria, with different intensities of water-water interactions. The first, which comprises about 50% of the heat of the total process, has a water-water interaction similar to that of concentrated salt solution (3.5M NaCl) and must refer to extracellular water (25 - 30% of total water) plus about one-half of the intracellular water. To the remaining cell-water (populations b+c) can be assigned an excess of heat of 18 KJ mole^{-1}. This is equivalent to an addition of one hydrogen bond per water molecule and is consistent with the model suggested by NMR-proton measurements with 2, or possibly 3 populations of water molecules within the cells. The calorimetric results of the *H. halobium* suggest that the population of enhanced water-water interactions is much smaller, only one-third of that of *H. marismortui* [12]. The results are consistent with experimental findings that less than one-third of cell K^+ is retained by starved *H. halobium* whereas starved *H. marismortui* retain most of their K^+ for very long periods of time (Tables 1, 2). The distribution of Na^+, fructose, sucrose and insulin between the cell water and the outer solution is consistent with this model of the cell [8].

Figure 1 shows a comparison between the dielectric dispersions of *H. halobium* and *H. marismortui*. The low-frequency dispersion is the so-called ß-dispersion associated with the cell membrane. We observe here also the beginning of another dispersion at higher frequencies. The instrument was limited in its frequency range and the dispersion could not be observed in its entirety [10]. Recent measurements (Morgan and Bone, personal communication) of cell suspensions by TDR (Time Domain Reflectometry) enable measurements to be extended to higher frequencies. These results show very clearly two dispersions in *H. halobium*, the ß-dispersion and another at a much higher frequency, the characteristic frequency at about 500 MHZ. This second dispersion is completely new to biology, and has not yet been seen in any other living cell. It could be due to the existence of a water-protein-K^+-H^+ complex, where there exist <u>enhanced</u> water-water interactions, in other words a second water environment.

Recently we found a dramatic narrowing of the inorganic P lines in NMR spectra when the temperature of bacterial suspensions of either species was lowered. This observation indicates an exchange mechanism between two environments, which can be due either to the presence of paramagnetic ions or is consistent with the model discussed above. It could also be that a combination of the two effects is responsible for the phenomenon discussed [1, 2].

Thus the second water environment could be responsible for maintaining the large K^+, Na^+ and H^+ gradients across the highly conductive membrane of *H.*

170

marismortui. These gradients could bring about a localised conduction of the ions through the ATP synthetase system and thus ATP synthesis.

ACKNOWLEDGEMENTS

Most of the research on the halophilic bacteria has been done in joyful cooperation with Dr. Margaret Ginzburg. I would like to give my thanks Dr. Iain D. Campbell for the opportunity to do enjoyable NMR work in the Biochemistry Department of Oxford University, and to him and to Dr. P. Quirk for the use of some of our unpublished work. I would like to thank Drs. Morgan and Bone from U.C.N.W. in Bangor, for fruitful cooperative work with TDR.

REFERENCES

[1]   P. Quirk, B. Z. Ginzburg, I. D. Campbell, Abstract in *Fifth European Bioenergetics Conference*, p.159 (1988)

[2]   P. Quirk, B. Z. Ginzburg, I. D. Campbell, in preparation.

[3]   A. Danon, W. Stoeckenius, *Proc. Natl. Acad. Sci. USA*, 71: 1234 (1974)

[4]   H. Michel, D. Oesterhelt, *Biochemistry*, 19:4607 (1980)

[5]   G. Wagner, H. Michel, D. Oesterhelt, *Eur. J. Biochem.* 89: 169 (1978)

[6]   M. Ginzburg, L. Sachs, B. Z. Ginzburg *J .Memb. Biol.* 5:78 (1971a)

[7]   M. Ginzburg, B. Z. Ginzburg, D. C. Tosteson, *J. Memb. Biol.* 6:259 (1971b)

[8]   H. Morgan, M. Ginzburg, B. Z. Ginzburg, *Biochem. et Biophys. Acta.* 924:54 (1987)

[9]   M. Ginzburg, B. Z. Ginzburg, *Biochem. et Biophys. Acta.* 584: 398 (1979)

[10]   W. M. Arnold, B. Wendt, U. Zimmerman, R. Kornstein, *Biochem. et Biophys. Acta.* 813: 117 (1985)

[11]   H. T. Edzes, Chap. 6: NMR studies of water in a Halobacteria, in Ph. D. Thesis, University of Groningen (1977)

[12]   B. Z. Ginzburg, *Thermochem. Acta.* 46: 249 (1981)

THE RESPONSE OF HALOPHILIC BACTERIA TO HEAVY METALS

Joaquín J. Nieto

Department of Microbiology and Parasitology
Faculty of Pharmacy
University of Sevilla
Sevilla 41012
Spain

ABSTRACT

The susceptibility of a great number of halophilic strains (145 extremely halophilic archaeobacteria and 308 moderate halophiles), including both culture collection strains and fresh isolates from widely differing geographical areas, to arsenate, cadmium, chromium, cobalt, copper, lead, mercury, nickel, silver and zinc ions, were surveyed by using an agar dilution method. The halobacterial culture collection strains showed different susceptibilities clustering into five groups. In contrast, the halococci collection strains exhibited similar responses to the majority of the heavy metals tested. The behaviour of the moderately halophilic collection strains was only similar with respect to six metal ions. Concerning the fresh isolates, the metal-susceptibility levels of the moderate halophiles were, in general, very heterogeneous among the four taxonomic groups as well as within the strains included in each group. All the halophilic strains were sensitive to mercury and silver and a great fraction of them were also sensitive to zinc. On the other hand, most of them were tolerant of lead and chromium. Different patterns of metal susceptibility were found for the rest of metals. A range of concentrations for defining metal-tolerance in halophilic bacteria for additional studies are proposed.

INTRODUCTION

Among the halophilic bacteria, the extremely halophilic archaeobacteria (which grow best in media with 2.5 - 5.2 M salt) and the moderate halophiles (which grow optimally in media containing 0.5 - 2.5 M salt) [1] constitute the most important groups of microorganisms adapted to live in hypersaline environments. Despite extensive studies with regard to their taxonomy [2, 3], physiology [1, 4] and ecology [5], very little attention has been focused on their response to potential inhibitors such as heavy metals, let alone their ability to develop heavy metal resistances. However, such information might be desirable under several points of view. First, some of the possible metal-resistant halophilic bacteria could be used as bioassay indicator organisms in saline aquatic polluted environments [6]. Secondly, the isolation of these kinds of strains might provide useful information about the ecological role of these halophilic microorganisms in heavy metal biotransformations in natural environments. Thirdly, the possible heavy metal resistances could be used both in halophilic archaeobacteria and moderate halophiles as genetic markers, since the lack of suitable markers for these microorganisms when studying genetic transference mechanisms is well known. In addition, since most metal resistance in bacteria is encoded by genes located in plasmids, the detection of such plasmids would be very useful to elucidate the above mentioned genetic processes.

The purpose of our studies in the last three years was to determine the natural susceptibility levels of a large number of halobacteria, halococci and moderate halophiles to ten heavy metals which are common industrial pollutants, in order to be able to discriminate metal-tolerant from metal-sensitive strains, and therefore isolate metal-resistant strains from polluted environments. An additional aim was to check if the well-known differences among these halophilic groups were reflected in their responses to the heavy metals tested. In this paper, we summarize the most noticeable data obtained in these studies [7 - 10].

A total of 453 halophilic strains (68 halobacteria, 77 halococci and 308 moderate halophiles), including both culture collection strains and fresh isolates from widely differing hypersaline habitats located in Spain, were selected for these studies. All strains were grown in a medium composed of basal salt solution with a final concentration of ca. 25% (for archaeobacteria) or 10% (for moderate halophiles) (wt/vol) to which 0.5% (wt/vol) Difco yeast extract was added. The composition of the basal salt solution has been described elsewhere [11, 12]. The pH was always adjusted to 7.2. Incubation was done at 37°C in an orbital shaker at 200 strokes/min. The toxicity of the 10 heavy metals was determined by an agar dilution method [13]. The metals tested were provided from standard commercial sources as sodium arsenate, lead acetate, silver nitrate, cadmium chloride, cobalt chloride, cupric sulphate, mercuric chloride, nickel sulphate, zinc sulphate and potassium chromate. A very wide range of metal concentrations were used for susceptibility testing: 0.005, 0.01, 0.05, 0.1, 0.5, 1, 2.5, 5. 10, 20 and 40 mM. Agar plates with 20 ml of the growth medium and a determined metal concentration were inoculated with 10^4 and 10^5 microorganisms from exponentially growing cultures. Plates were read after incubation at 37°C for 2 days (for moderate halophiles), 10 days (for halobacteria) or 20 days (for halococci). In all cases the Minimal Inhibitory Concentration (MIC) was determined. The MICs for all strains were tested in at least three different experiments.

Table 1 shows the MICs of the 10 heavy metals tested against some of the collection strains used in our studies. Amongst the halobacteria, these strains can be

Table 1. MICs of 10 metal ions against some representatives of the halophilic collection strains used in these studies.

Microorganism	MIC(mM) of									
	Ag	As	Cd	Co	Cr	Cu	Hg	Ni	Pb	Zn
Halobacteria										
H. mediterranei ATCC 33500	0.5	20	2.5	1	5	1	0.01	2.5	20	0.5
H. volcanii DS2	0.5	20	0.5	1	5	1	0.01	2.5	20	0.5
H. hispanicum ATCC 33960	0.5	10	0.5	1	5	1	0.01	2.5	10	0.5
H. vallismortis ATCC 29715	0.5	10	0.5	1	5	1	0.01	2.5	10	0.5
H. gibbonsii ATCC 33959	0.5	10	0.5	1	5	1	0.01	2.5	10	0.5
H. saccharovorum ATCC 2952	0.05	10	0.5	0.5	2.5	2.5	0.01	1	20	0.5
H. halobium CCM 2090	0.05	10	0.5	0.5	2.5	2.5	0.01	1	20	0.5
H. trapanicum NCMB 767	0.05	10	0.5	0.5	2.5	2.5	0.01	1	5	0.5
H. salinarium CCM 208	0.05	20	0.1	0.5	2.5	1	0.01	1	5	0.5
H. californiae" ATCC 33799	0.05	10	0.1	0.5	1	2.5	0.05	1	10	0.5
H. sinaiiensis" ATCC 33800	0.05	10	0.05	0.5	2.5	1	0.05	1	5	0.05
Halococci										
H. morrhuae CCM 859	0.1	10	1	1	2.5	2.5	0.05	2.5	1	0.5
pH. morrhuae NCMB 757	0.1	10	2.5	0.5	2.5	2.5	0.05	2.5	2.5	0.5
H. morrhuae strain Delft	0.5	5	1	0.5	2.5	2.5	0.05	2.5	10	0.5
H. morrhuae CCM 537	0.1	10	1	1	2.5	2.5	0.05	2.5	2.5	0.5
H. saccharolyticus P423	0.1	10	2.5	1	2.5	2.5	0.05	2.5	1	0.5
Moderate halophiles										
D. halophila CCM 3662	0.05	10	1	2.5	2.5	1	0.05	5	5	0.5
F. halmephilum CCM	0.05	10	2.5	1	2.5	0.5	0.05	2.5	2.5	05
H. elongata ATCC 33173	0.0	20	5	2.5	1	2.5	0.05	2.5	5	0.5
H. subglaciescola UQM 2962	0.05	5	5	2.5	2.5	2.5	0.05	2.5	2.5	0.5
M. albus CCM 3517	0.05	1	0.5	0.5	0.5	1	0.05	1	2.5	0.1
M. halophilus NRCC 14033	0.05	1	0.5	1	1	0.5	0.05	2.5	2.5	0.5
M. halobius CCM 2591	0.05	5	0.5	1	5	2.5	0.05	5	2.5	1
"P. halosaccharolytica" CCM 2851	0.05	20	1	2.5	2.5	2.5	0.05	5	5	0.5
V. costicola NCMB 701	0.5	1	0.5	1	2.5	1	0.05	1	2.5	0.1

grouped into five major groups on the basis of the similar MICs of all the metals tested, with the strains *H. mediterranei* and *H. volcanii* showing, in general, the highest MIC values and, in contrast, "*H. californiae*" and "*H. sinaiiensis*" with the lowest MICs. The response to Cd and Pb ions was heterogeneous [8]. A greater homogeneity was observed with the halococci collection strains except for *H. morrhuae* strain Delft, which exhibited the highest tolerance levels to Ag and Pb ions of all these strains but also the highest sensitivity to As ions [9]. Finally, concerning the moderately halophilic collection strains, these showed similar responses to Ag, Hg, Ni, Pb and Zn ions, except for *M. albus*, which showed, in general, the lowest tolerance. On the other hand, the responses to As, Cd, Co, Cr and Cu were very heterogeneous. Only *H. elongata* and "*P. halosaccharolytica*" showed MICs higher than 10 mM against As ion and only three strains (*H. elongata, H. subglaciescola* and *F. halmephilum*) showed MICs higher than 1mM against Cd. It is noticeable that the highest tolerance to Ag and the lowest tolerance to Zn ions was exhibited by *V. costicola* [9].

With respect to the response of all halobacteria to the metals tested, a great heterogeneity was found, since not even 50% of strains showed the same MIC values for the metals tested, except for mercury and copper (70 and 78% of the strains had the same MICs, respectively). Highest MIC values were found with respect to As and Pb ions (34% of strains were tolerant up to 10 mM of this ion). Otherwise, a high sensitivity to mercury (all strains were inhibited by 0.05 mM) and silver was detected. The susceptibility of halobacteria to mercury has been previously used as a "classical" phenotypic feature in some taxonomic studies [14, 15]. However, it must be pointed out that the concentration of $HgCl_2$ used in those reports was higher than 0.1 mM, that is, a concentration which inhibited all strains tested.

The responses of the halococci examined were very homogeneous to Cu, Cr, Ni, Hg and Zn ions, since more than 75% of the strains showed the same MIC values while only 68% of them exhibited the same values for Ag ions. In contrast, great heterogeneity was found in the MICs obtained for As, Cd, Pb and Co ions. Halococci showed greater tolerance levels to As, Cd and Ni ions than did rod-shaped halobacteria; except for Co and Pb, halococci were slightly more metal-tolerant than extremely halophilic rods. These small differences in metal-tolerance might be caused by the well-known differences in structures of cellular envelopes between halococci and halobacteria. Amongst the moderately halophilic eubacteria, different susceptibility levels were found among the five different groupings (*V. costicola* strains, *D. halophila* strains, *Acinetobacter* sp. strains, *Flavobacterium* sp. strains and the Gram-positive cocci) as well as within the strains belonging to the same group, with the following exceptions: *V. costicola*, for Ag Cd, Cu, Ni and Pb ions; *Flavobacterium* sp. and moderate halophilic cocci for Ni and Zn; *Acinetobacter* sp. for As ions; *D. halophila* for Pb, Ag, As and Co ions. In contrast, all these eubacteria showed a very homogeneous sensitivity to mercury. On the contrary, six different MIC values were found in the cases of cadmium and chromium. This fact might suggest that moderately halophilic eubacteria exert very heterogeneous behaviour in relation with their individual natural susceptibility levels to the ten heavy metals tested. The strains of *D.halophila* were the most homogeneous of all these strains since more than 80% of them exhibited the same MICs for Ag, As, Co, Hg and Pb metal ions. In any case, it seems clear that the

176

response to heavy metals should definitely not be used as a taxonomic feature in this group of moderate halophiles.

It is well known that unlike antibiotic resistance, which is evaluated using therapeutic doses, there are no "standard" acceptable metal concentrations used by all researchers to specify metal resistance. Indeed, metal salts and microbiological medium components can interact in ways which make data interpretation difficult [16, 17]. Thus, the toxicity of some metals could result from metal complex formation rather than from individual metal ions. Therefore, such interactions must always be taken into account. Anyway, for comparative purposes, we have chosen those concentrations that have been employed in testing media containing yeast extract in previous studies carried out with eubacteria [6, 18]. Thus, those strains which were not inhibited by 10 mM As; 1 mM Ag, Cd, Co, Cr, Cu, Ni, Pb and Zn; and 0.1 mM hg, were regarded as tolerant. It is very important to point out that these percentages have been obtained in media with 0.5% yeast extract and any further comparison study with halophilic bacteria must be done with this same yeast extract concentration in order to obtain significant results. We have found in moderate halophiles that a reduction in the concentration of this medium component caused a noteworthy increase in all metal

Table 2. Percentages of tolerance of the different groupings of halophilic bacteria to ten heavy metals.

Taxonomic grouping	Metal ion									
	Ag	As	Cd	Co	Cr	Cu	Hg	Ni	Pb	Zn
Extremely Halophilic archaeobacteria										
Halobacteria	0	15	9	25	72	80	0	45	100	0
Halococci	0	33	30	8	82	90	0	83	78	0
Moderately Halophilic Eubacteria										
V. costicola	0	44	0	14	88	0	0	91	100	0
Acinetobacter sp.	0	92	88	83	79	88	0	100	100	67
D. halophila	0	92	89	100	100	35	0	100	100	0
Gram-positive cocci	0	8	0	61	90	63	0	100	100	2
Flavobacterium sp.	0	46	14	28	68	32	0	100	100	0

toxicities while a higher yeast extract concentration resulted in lessened toxicities of the majority of the metals tested [12]. Our results are in agreement with previous reports that showed the influence of some components of culture media on the availability of the toxic heavy metal ions for the microbes [16, 17].

As can be seen in Table 2, all the halophilic strains tested were sensitive to silver and mercury and the majority of them were also sensitive to zinc. Therefore, the sensitivity of halophilic bacteria to these metal ions seems to be a general feature of them. On the other hand, a great tolerance of lead, chromium and nickel (except for halobacteria) was also found. Amongst the archaeobacteria, only a significant difference in nickel tolerance was detected, although it seems that halococci exhibit higher general tolerance levels than halobacteria. Concerning the moderate halophiles, the strains belonging to *V. costicola* showed the highest sensitivity to copper, cobalt and cadmium (as Gram-positive cocci in this case) of all moderate halophiles tested. In contrast, the strains of *Acinetobacter* sp. and *D. halophila* were the most metal-tolerant since more than 65% of them demonstrated tolerance of 8 and 6 metals, respectively. Onishi et al. [19] carried out a study on the cadmium-tolerance of 41 strains of halophilic eubacteria with different salt requirements, reporting that about half of the 31 moderate halophiles tested were cadmium-tolerant. However, the concentration of cadmium chloride used by them to define metal-tolerance was of 0.23 mM; using our "standard" concentration, only seven strains would have been scored as cadmium-tolerant.

From these results, we conclude that a new range of heavy metal concentrations are needed in order to distinguish with greater confidence between sensitive and metal-tolerant halophilic strains, and to be able to isolate heavy-metal-resistant strains from environmental sources. For this purpose, we propose those metal concentrations which prevented growth, in our experimental conditions, of at least 90% of all strains studied for each of the three major groups of halophilic bacteria (halobacteria, halococci and moderate halophiles). These metal concentrations are as follows (in mM): As, 20; Ag, 1; Cd, 2.5; Co, 2.5; Cr, 10; Cu, 2.5; Hg, 0.1; Ni, 5; Pb, 20; Zn, 2.5, for halobacteria. As, 20; Ag, 1; Cd, 2.5; Co, 2.5; Cr, 5; Cu, 2.5; Hg. 0.1; Ni, 5; Pb, 5; and Zn, 2.5, for halococci; and As, 20; Ag, 1; Cd, 5; Co, 2.5; Cr, 20; Cu, 2.5; Hg. 0.1; Ni, 5; Pb, 10; and Zn, 2.5, for moderately halophilic eubacteria.

ACKNOWLEDGEMENTS

This work was supported by grants from the Dirección General de Investigación Científica y Técnica and from the Junta de Andalucía.

REFERENCES

[1]   D. J. Kushner, The Halobacteriaceae, *in* "The bacteria", Vol. 8, C. R. Woese and R. S. Wolfe, eds., Academic Press, London (1985)

[2]   G. Juez, Taxonomy of extremely halophilic archaebacteria, *in* "Halophilic bacteria", Vol. 1, F. Rodriguez-Valera, ed., CRC Press, Boca Raton, Florida (1988)

[3]   A. Ventosa, Taxonomy of moderately halophilic heterotrophic eubacteria, *in* "Halophilic bacteria", Vol. 1, F. Rodriguez-Valera, ed., CRC Press, Boca Raton, Florida (1988)

[4]   D. J. Kushner and M. Kamekura, Physiology of halophilic eubacteria, *in* "Halophilic bacteria", Vol. 1, F. Rodriguez-Valera, ed., CRC Press, Boca Raton, Florida (1988)

[5]   F. Rodriguez-Valera, Characteristics and microbial ecology of hypersaline environments, *in* "Halophilic bacteria", Vol. 1, F. Rodriguez-Valera, ed., CRC Press, Boca Raton, Florida (1988)

[6]   J. T. Trevors, K. M. Oddie and B. H Belliveau, *FEMS Microbiol. Rev.*, 32: 39 (1985)

[7]   M. T. García, J. J. Nieto, A. Ventosa and F. Ruiz-Berraquero, *J. Appl. Bacteriol.* 63: 63 (1987)

[8]   J. J. Nieto, A. Ventosa and F. Ruiz-Berraquero, *Appl. Environ. Microbiol.* 53: 1199 (1987)

[9]   J. J. Nieto, C. G. Montero, A. Ventosa and F. Ruiz-Berraquero, *System. Appl. Microbiol.* 12: 116 (1989)

[10]   J. J. Nieto, R. Fernández-Castillo, A. Ventosa, E. Quesada and F. Ruiz-Berraquero, *Appl. Environ. Microbiol.* 55: 2385 (1989)

[11]   M. C. Gutiérrez, M. T. García, A. Ventosa, J. J. Nieto and F. Ruiz-Berraquero, *J. Appl. Bacteriol.* 61: 67 (1986)

[12]   A. Ventosa, E. Quesada, F. Rodriguez-Valera, F. Ruiz-Berraquero and A. Ramos-Cormenzana, *J. Gen. Microbiol.* 128: 1959 (1982)

[13]   J. A. Washington, II, and V. L. Sutter, Dilution susceptibility test: agar and macrobroth dilution procedures, *in* "Manual of clinical microbiology", 3rd ed., American Society for Microbiology, Washington, C.D. (1980)

[14]   R. R. Colwell, C. D. Litchfield, R. H. Vreeland, L. A. Kiefer and N. E. Gibbons, *Int. J. Syst. Bacteriol.* 29: 379 (1979)

[15]   F. Rodriguez-Valera, G. Juez and D. J. Kushner, *System. Appl. Microbiol.* 4: 369 (1983)

[16]   S. Ramamoorthy and D. J. Kushner, *Microb. Ecol.* 2: 162 (1975)

[17]   H. Babich and G. Stotzky, *Crit. Rev. Microbiol.* 8: 99 (1980)

[18]   T. V. Riley and B. J. Mee, *Antimicrob. Agents Chemother.* 22: 889 (1982)

[19]   H. Onishi, T. Kobayashi, N. Morita and M. Baba, *Agric. Biol. Chem.* 48: 2441 (1984)

BIOCHEMICAL CHARACTERIZATION OF DIHYDROFOLATE

REDUCTASE OF *HALOBACTERIUM VOLCANII*

Tal Zusman and Moshe Mevarech

Department of Microbiology
George S. Wise Faculty of Life Sciences
Tel Aviv University
Tel Aviv, 69978
Israel

ABSTRACT

The enzymatic activity of the enzyme dihydrofolate reductase of *Halobacterium volcanii* depends on salt concentration. the following communication describes the effect of salt concentration and pH on the catalytic parameters of this enzyme. In addition, the mode of inhibition of three different inhibitors of DHFR was studied.

INTRODUCTION

The extremely halophilic archaebacteria are prokaryotic microorganisms that are able to grow under conditions of salt concentrations that can reach saturation [1]. These bacteria adapt to such high salt concentrations by maintaining an internal ionic strength equal to or greater than that of the medium [2]. Most halophilic enzymes require, for their activity and stability, salt concentrations in which non-halophilic enzymes are usually inactive and sometimes even insoluble [3]. This phenomenon depends, most probably, on structural differences between halophilic and non-halophilic proteins. Halophilic proteins are found to be highly acidic [4] and are capable of binding unusually high amounts of salt and water molecules [5].

The enzyme dihydrofolate reductase (DHFR) catalyzes the NADPH dependent reduction of dihydrofolate to tetrahydrofolate which is an essential cofactor in the biosynthesis of thymidine, purines and several amino acids [6]. Since DHFR is an

General and Applied Aspects of Halophilic Microorganisms
Edited by F. Rodriguez-Valera, Plenum Press, New York, 1991

essential enzyme in one carbon metabolism it became a target for antibacterial and antitumor agents. Most of these agents are analogues of folic acid. Furthermore, DHFRs from different bacterial and eukaryotic sources can be distinguished from each other on the basis of their relative inhibition by those drugs [7]. In the following communication we describe the effect of folate analogues on the archaebacterial DHFR of *Halobacterium volcanii* (h-DHFR).

In a previous communication we have described the isolation of the enzyme and its coding gene. This enzyme whose activity and stability depend on salt concentration, exhibits excess of acidic residues [8]. The following communication presents detailed analysis of the dependence of the kinetic properties of this enzyme on pH and salt concentration.

MATERIALS AND METHODS

Chemicals:

Dihydrofolic acid (DHF), NADPH and folate analogues were purchased from Sigma. All the salts employed were of analytical grade.

The enzyme: Halophilic DHFR (h-DHFR) was purified from the overproducing strain *H. volcanii* WR215 [9] by ammonium sulfate mediated chromatography as described elsewhere [8].

Enzymatic assay:

Enzymatic activity was measured by following the decrease in the absorbance at 340 nm during the first two minutes immediately after diluting 2.5 μg of enzyme into 1 ml assay mixture at 25ºC. Under standard conditions the assay mixture contained (final concentrations): 2 M KCl, 25 mM 2-(N-morpholine) ethanesulfonic acid (MES) pH 6, 0.05 mM DHF and 0.08 mM NADPH. When the Michaelis constants were determined the reaction mixture was buffered with 0.1 M K-phosphate pH 7. The concentration of NADPH was determined spectrophotometrically using an extinction coefficient of 6.2 mM^{-1} cm^{-1} at 340 nm. The concentration of DHF was determined enzymatically by measuring the amount of NADPH oxidized in a reaction that contained a large excess of NADPH versus DHF. The amount of DHF was calculated using a molar absorption change, at 340 nm, of 12.3 [10].

Inhibition of h-DHFR by folate analogues:

Measurements were performed after 3 minutes preincubation of the enzyme with the inhibitor. The K_i values were calculated from the double reciprocal plots of velocity vs. DHF concentrations. The concentrations of folic acid and methotrexate were determined spectrophotometrically using extinction coefficients of 27 mM^{-1} cm^{-1} at 282 nm and 22 mM^{-1} cm^{-1} at 302 nm respectively.

RESULTS

The effect of salt concentration on the kinetic parameters

The K_m values of DHF and NADPH at pH 7 were found to be 5×10^{-5} M and 8.5×10^{-6} M respectively. The effect of salt concentration on the K_m and V_{max} for DHF is demonstrated in Figure 1. Whereas the K_m is not affected by salt concentration, the V_{max} decreases as the concentration of KCl decreases.

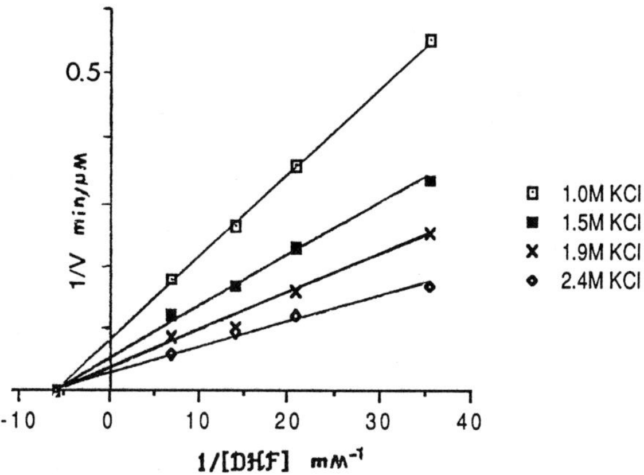

Figure 1. The effect of salt concentration on the kinetic parameters.

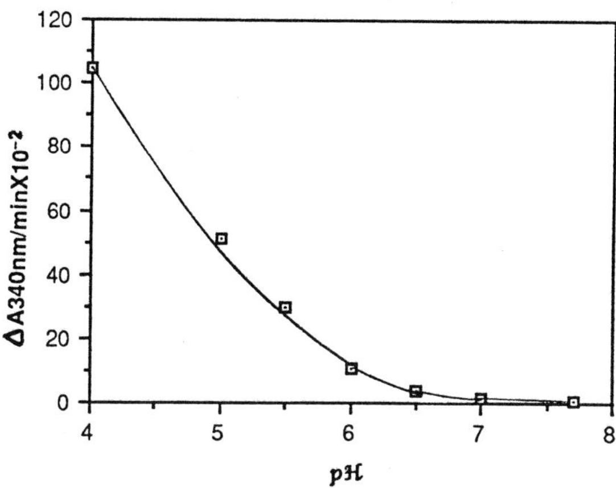

Figure 2. The effect of pH on enzymatic activity.

The effect of pH on h-DHFR activity

The effect of pH on the enzymatic activity was determined and the results are given in Figure 2. The enzymatic activity exhibits a considerable increase when the pH decreases. At pH values of 5 and below the enzyme becomes unstable (half life of ≈15 minutes at pH 4 and ≈30 minutes at pH 5), and therefore the activity measurements were performed shortly after diluting the enzyme into the assay mixture. When calculating the enzymatic activity at this low pH range a correction was introduced to account for the instability of the substrates.

Change in salt dependency as a function of pH

The activity and stability of h-DHFR are affected by salt concentration [8]. The dependence of the enzymatic activity on high salt concentration is in itself a function of pH as can be seen in Figure 3. The dependence of the enzymatic activity on salt concentration decreases as the pH decreases. At pH 4 the enzymatic activity is, virtually, independent of salt concentration.

Inhibition of h-DHFR by DHF and its analogues

Substrate inhibition by DHF was observed at concentrations greater than 0.069 mM as can be seen in Figure 4. NADPH, on the other hand, does not inhibit the enzyme. Folic acid cannot replace the DHF in the oxidation of NADPH over the pH range of 4-8. However, folic acid clearly inhibits the reduction of DHF as indicated in Table 1. Methotrexate is a known inhibitor of eukaryotic and eubacterial DHFRs, on the other hand trimethoprim is more efficient in inhibiting eubacterial DHFRs [7]. The effects of these inhibitors on *H. volcanii*-DHFR is summarized in Table 1.

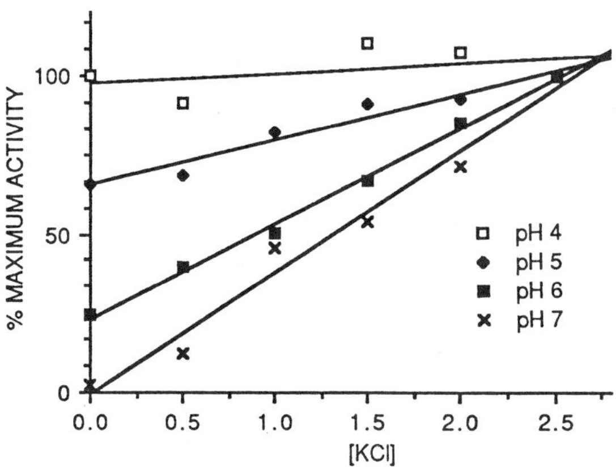

Figure 3. The effect of pH on salt dependency.

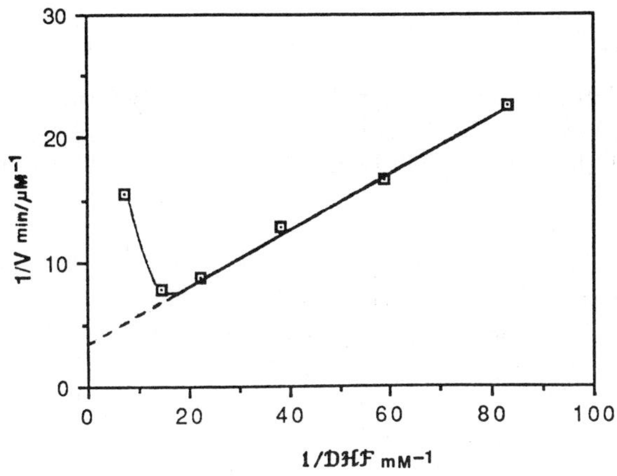

Figure 4. Substrate inhibition by DHF.

Table 1. Mechanisms of inhibition by folate analogues

Substrate analogue	Type of reversible inhibition	K_i value
		M
Folic acid	competitive	2.3×10^{-4}
Methotrexate	non-competitive	5.0×10^{-8}
Trimethoprim	competitive	6.7×10^{-6}

DISCUSSION

In spite of the resemblance of the structure of h-DHFR to the structures of DHFRs from non-halophilic sources, its function is fully adapted to the intracellular high salt concentration of the halobacterium. In order to be able to explain salt adaptation in terms of the effect of salt concentration on the individual rate constants of the catalytic reaction, a detailed kinetic scheme for h-DHFR should be first established. Such a scheme was suggested for the DHFR of *E. coli* by Fierke et al [11]. This scheme was derived from measurements of the various rate constants leading to the different binary and ternary complexes as well as to the hydride transfer reaction and dissociation of the products. The scheme suggests that at low pH the rate limiting step is the dissociation of tetrahydrofolate from the termary complex of FH_4-E-NADPH. At high pH, however, the rate limiting step is the hydride transfer reaction. This hydride transfer reaction has a pK_a of 6.5 reflecting ionization of a single group in the active site. This group is Asp27 which upon protonation transfers the proton to N-5 of the pteridine ring of dihydrofolate that in turn catalizes the hydride transfer.

If this kinetic scheme holds true also for h-DHFR it is possible that the effect of salt on the catalytic activity is by changing the binding or dissociation rate constants of the substrates or the products, or by having an indirect effect on the pK_a of the Asp29 (equivalent to Asp27 of *E. coli*-DHFR), through which the hydride transfer reaction is affected. Such an indirect effect of salt concentration was observed by Russel and Fersht [12] in the enzyme subtilisin. In that system unshielded surface charge affects the pK_a of a single amino acid (His 64) whose unprotonated state is essential for activity.

The effect of pH on the enzymatic activity of h-DHFR supports the view that the rate limiting step in the catalytic activity is the hydride transfer reaction. As mentioned above, this reaction is catalyzed by the protonation of Asp27. Unlike *E. coli*-DHFR (e-DHFR) in which Asp27 is followed by two uncharged residues (Leu28 and Ala29), in h-DHFR the equivalent Asp is followed by two lysines [8]. The existence of two positive charges should lower the pK_a of the Asp, and this decrease is reflected in the pH dependence of the enzymatic activity. Perry et al. [13] described a mutant of *E. coli* DHFR which has a Val28 to Arg transition. This change causes a decrease of more than twenty-fold in the activity at pH 7 in comparison with the wild type. Unfortunately, the pH profile of the activity of this mutated enzyme was not given.

On the assumption that salt concentration affects the pK_a of Asp29 and thereby the hydride transfer, it is possible to understand the reduced dependency of the enzymatic activity on salt concentration at lower pH values. This kinetic model is supported by the fact that the binding of DHF is not affected by salt concentration (Figure 1).

The halophilic DHFR resembles the DHFR of eubacteria in many respects; homology in the amino acid sequence [8], as well as its inability to reduce folic acid (Table 1). Similarly to *E. coli*-DHFR, DHF causes substrate inhibition. In addition, the modes of inhibition of folic acid and methotrexate are in agreement with those reported for *E. coli*-DHFR [14]. However, trimethoprim inhibits *E. coli*-DHFR in a noncompetitive manner [7], whereas h-DHFR was found to be inhibited competitively, but with a rather low K_i value (Table 1).

ACKNOWLEDGEMENT

This work was supported in part by the Endowment Fund for Basic Research in Life Sciences: Charles H. Revson Foundation.

REFERENCES

[1]   D. J. Kushner, The Halobacteriaceae, *in* "The Bacteria", Vol. 8, pp. 171-214, C. R. Woese and R. S. Wolfe, eds, Academic Press, New York (1985)

[2]   J. H. B. Christian and J. A. Waltho, Solute concentration within cells of halophilic and non-halophilic bacteria, *Biochim. Biophys. Acta* 65: 506 (1962)

[3]   J. K. Lanyi, Salt dependent properties of proteins from extremely halophilic bacteria, *Bacteriol. Rev.* 38: 272 (1974)

[4]   R. Reistad, On the composition and nature of the bulk protein of extremely halophilic bacteria, *Arch. Microbiol.* 71: 353 (1970)

[5]   G. Zaccai, G. J. Bunick and H. Eisenberg, Denaturation of a halophilic enzyme monitored by small-angle neutron scattering, *J. Mol. Biol.* 192: 155 (1986)

[6]   R. L. Blakley, Chemistry and biochemistry of folates, *in* "Folates and Pteridines", Vol. 1, pp. 191-253, R. L. Blakley and S. J. Benkovic, eds. Wiley & Sons, New York (1984)

[7]   J. J. Burchall and G. H. Hitchings, Inhibitor binding analysis of dihydrofolate reductases from various species, *Mol. Pharmacol.* 1: 126 (1965)

[8]   T. Zusman, I. Rosenshine, G. Boehm, R. Jaenicke, B. Leskiw and M. Mevarech, Dihydrofolate reductase of the extremely halophilic archaebacterium *Halobacterium volcanii*, *J. Biol. Chem.* 264: 18878 (1989)

[9]   I. Rosenshine, T. Zusman, R. Werczberger and M. Mevarech, Amplification of specific DNA sequences correlates with resistance of the archaebacterium *Halobacterium volcanii* to dihydrofolate reductase inhibitors trimethoprim and methotrexate, *Mol. Gen. Genet.* 208: 518 (1987)

[10]   B. L. Hillcoat, P. F. Nixon and R. L. Blakey, Effect of substrate decomposition on the spectrophotometric assay of dihydrofolate reductase, *Anal. Biochem.* 21: 178 (1967)

[11]   C. A. Fierke, K. A. Johnson and S. J. Benkovic, Construction and evaluation of the kinetic scheme associated with dihydrofolate reductase from *Escherichia coli*, *Biochemistry* 26: 4085 (1987)

[12]   A. J. Russel and A. R. Fersht, Rational modification of enzyme catalysis by engineering surface charge, *Nature* 328: 496 (1987)

[13]   K. M. Perry, J. J. Onuffer, N. A. Touchette, C. S. Hendon, M. S. Gittelman, C. R. Mathews, J. T. Chen, R. J. Mayer, K. Taira, S. J. Benkovic, E. E. Howell and J. Kraut, Effect of single amino acid replacements on the folding and stability of dihydrofolate from *Escherichia coli*, *Biochemistry* 26: 2674 (1987)

[14]   S. R. Stone and J. F. Morrison, Kinetic mechanism of the reaction catalyzed by dihydrofolate reductase from *Escherichia coli*, *Biochemistry* 21: 3757 (1982)

PART III

COMPOSITION AND STRUCTURE

POLAR LIPID STRUCTURE, COMPOSITION AND BIOSYNTHESIS

IN EXTREMELY HALOPHILIC BACTERIA

M. Kates and N. Moldoveanu

Department of Biochemistry
University of Ottawa
Ottawa, Ontario, K1N 9B4
Canada

ABSTRACT

The polar lipids of extremely halophilic bacteria are derived from the saturated dialkylglycerol ether, sn-2,3-diphytanylglycerol, abbreviated "archaeol", and consist of phospholipids (archaeol analogues of phosphatidylglycerol, PG, phosphatidylglycerophosphate, PGP, and phosphatidylglycerosulfate, PGS) and glycolipids (sulfated triglycosyl-archaeol, S-TGA-1, triglycosylarchaeol, TGA-2, sulfated diglycosyl-archaeol, S-DGA-1, and others). The polar lipid composition, particularly that of the glycolipids, appears to be correlated with the taxonomic classification of the extreme halophiles on the level of the genera so far distinguished: *Halobacterium, Haloarcula, Haloferax, Halococcus, Natronobacterium* and *Natronococcus*. Biosynthesis of these archaeol analogues of phospholipids and glycolipids proceeds by complex pathways in a multienzyme, membrane-bound system, absolutely dependent on high salt concentration. The first step is the alkylation of a glycerol derivative (presumably dihydroxyacetone) with a C_{20}-isoprenylpyrophosphate. The alkylated product, containing ether-linked C_{20}-isoprenyl groups, serves as precursor for both the isoprenylarchaeol phospholipids and the glycolipids. Stepwise reduction of the isoprenyl chains to phytanyl chains then gives the final saturated archaeol analogues of phospholipids and glycolipids.

INTRODUCTION

The extremely halophilic bacteria, together with the two other archaeobacterial groups, methanogens and thermoacidophiles, are clearly distinguished from the

General and Applied Aspects of Halophilic Microorganisms
Edited by F. Rodriguez-Valera, Plenum Press, New York, 1991

eubacteria with respect to their membrane lipid structures, being derived from a C_{20}-isopranylglycerol diether, *sn*-2,3-diphytanylglycerol diether [1, 2], rather than *sn*-1, 2-diacylglycerol diester as in all other organisms. Variants of this structure containing one or two C_{25}-isopranyl groups are also found in certain extreme halophiles [3, 4]. To overcome the cumbersome nomenclature of these lipids, Nishihara et al. [5] have introduced the trivial name "archaeol" for the diphytanylglycerol diether, modified by appropriate alkyl group designations (e.g., C_{20}, C_{25}), and this nomenclature will be used here.

The polar lipids of extreme halophiles consist of archaeol analogues of phospholipids and glycolipids: the phospholipids are mainly phosphatidylglycerophosphate (PGP) with small amounts of phosphatidylglycerol (PG), phosphatidylglycerosulfate (PGS) and phosphatidic acid (PA); the glycolipids are mainly a sulfated triglycosylarchaeol (S-TGA-1), a sulfated tetraglycosylarchaeol (S-TeGA), a sulfated diglycosyl archaeol (S-DGA-1), and a triglycosylarchaeol (TGA-2), as well as several minor glycolipids [1, 2] (see Figure 1).

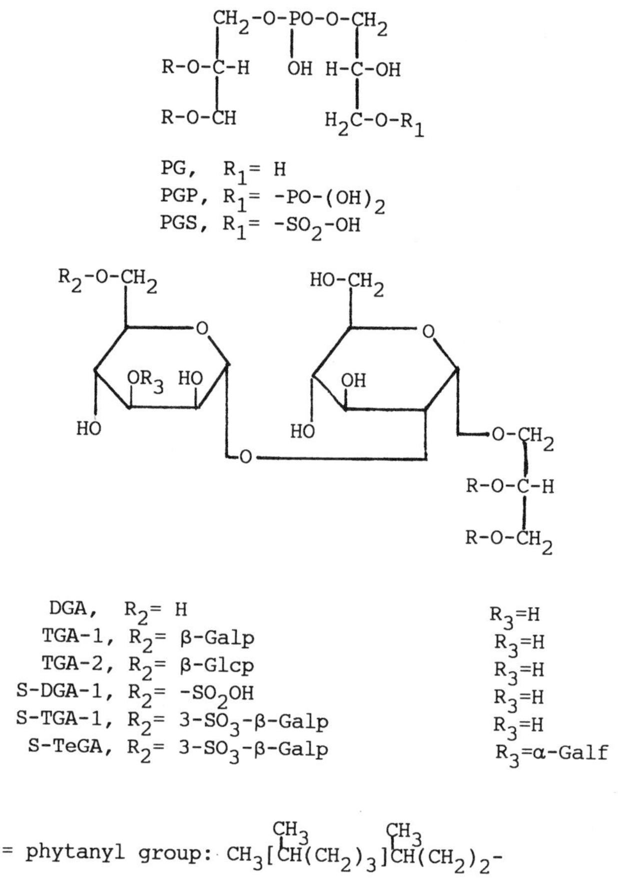

PG, $R_1 =$ H
PGP, $R_1 = -PO-(OH)_2$
PGS, $R_1 = -SO_2-OH$

		$R_3 = $
DGA,	$R_2 =$ H	$R_3 =$ H
TGA-1,	$R_2 = \beta\text{-Gal}p$	$R_3 =$ H
TGA-2,	$R_2 = \beta\text{-Glc}p$	$R_3 =$ H
S-DGA-1,	$R_2 = -SO_2OH$	$R_3 =$ H
S-TGA-1,	$R_2 = 3\text{-SO}_3\text{-}\beta\text{-Gal}p$	$R_3 =$ H
S-TeGA,	$R_2 = 3\text{-SO}_3\text{-}\beta\text{-Gal}p$	$R_3 = \alpha\text{-Gal}f$

R = phytanyl group: $CH_3[CH(CH_2)_3]CH(CH_2)_2-$

Figure 1.   Structures of archaeol phospholipids and glycolipids in extreme halophiles.

The exclusive presence of such unusual lipids in the extreme halophiles raises the question as to what biosynthetic pathways are used for their synthesis and whether there is any correlation between the structure and composition of these lipids and the taxonomic classification of the halophiles. This article will review first the polar lipid composition of various species of extreme halophiles with respect to their taxonomic classification, and then the mechanisms of biosynthesis of the major phospholipids and glycolipids in these organisms.

POLAR LIPID COMPOSITION OF HALOBACTERIALES

Halobacteriaceae

So far, three genera have been clearly distinguished in this family of extreme halophiles: *Halobacterium, Haloarcula*, and *Haloferax* [6]. All species in these genera contain archaeol PGP as the major phospholipid and archaeol PG and PGS as minor phospholipids, with the exception of *Haloferax* species which characteristically lack PGS. Traces of archaeol PA are present in species of all three genera (Table 1).

The glycolipid composition of these three genera is more discriminating and shows a remarkable correlation with taxonomic classification on the level of the genus [6, 7]. Thus, all species of *Halobacteria* examined so far contain S-TGA-1 as major glycolipid and TGA-1, and S-TeGA as minor glycolipids; *Haloarcula* species contain major proportions of TGA-2 and minor proportions of an unidentified diglycosyl-archaeol, DGA-2 (containing glucose and mannose); and *Haloferax* species contain S-DGA-1 as major glycolipid and DGA-1 as minor glycolipid (see Table 1).

However, it should be noted that there are some exceptions to the above correlations which may prove to be useful in establishing new genera of *Halobac-teriaceae*, for example: *Halobacterium sodomense* and two unidentified species of rods, 3.5 rp 4 and 3.1 palp 4, contain an unknown glycolipid (GL-2) instead of the major S-TGA-1 found in *Halobacteria* [6]; *Amoebacter morrhuae* and unidentified species Ma 2.20 which are classified as *Haloarcula*, contain the desulfated TGA-1 instead of TGA-2 and *Halobacterium trapanicum* although containing glycolipid TGA-2 characteristic of *Haloarcula* species, also has TGA-1 and an unidentified glycolipid (GL-1) [6] (Table 1).

Halococcaceae

For many years, only one species of *Halococcus* was recognized, namely *H. morrhuae* [8], but recently, a large number of extremely halophilic, non-alkaliphilic cocci were isolated from several hypersaline habitats in Spain and classified into four phenons (A-D) [9]. One of the strains of phenon D has been classified as a new species *Halococcus saccharolyticus* [10]. Examination of the lipids of selected strains of phenons A - D showed the presence in all of them of PGP and PG (derived from both the C_{20}-C_{20} and C_{20}-C_{25} archaeol), an unidentified phospholipid, traces of PA, a sulfated diglycosyl (Man-Glc) archaeol (S-DGA), three unidentified glycolipids and

Table 1. Distribution of polar lipids in known genera of extreme halophiles.

Genus	PGP	PG	PGS	PA	S-TGD-1	TGD-1	TGD-2	S-TeGD	S-DGD-1	DGD-1
Halobacterium[a]	+++	+	+	tr	++	+	−	+	−	−
Haloarcula[b]	+++	+	++	tr	−	−	++	−	−	−
Haloferax[c]	++	+++	−	tr	−	−	−	−	++	+
Halococcus[d,e]	+++	+	−	tr	+[d]	tr[d]	−	−	+[e]	−
Natronobacterium[f]	+++	+	−	tr	−	−	−	−	−	−
Natronococcus[g]	+++	+	−	tr	−	−	−	−	−	−

Type species: [a] H. cutirubrum, H. halobium, H. salinarium, H. saccharovorum [6,7];

[b] H. marismortui, H. valismortis, H. hispanica, H. californiae,

H. sinaiensis [6,7];

[c] H. mediterranei, H. volcanii, H. gibbonsii [6,7];

[d] H. morrhuae [8], Sarcina literalis and a Sarcina sp. [11];

[e] H. saccharolyticus [10]; contains S-DGD-1, several unidentified glycolipids and

a phosphoglycolipid;

[f] N. pharoanis, N. magadii, N. gregoryi [13,14];

[g] N. occultis [13-15]; contains a cyclic form of PGP.

one unidentified phosphoglyco-lipid; no PGS was detected in any of the strains examined [10]. In contrast, *H. morrhuae* [8] and two extremely halophilic *Sarcina* species [1] [11], while containing PGP and PG and no PGS, appeared to contain S-TGA-1 and TGA-1 and little or no S-DGA (Table 1). Further studies of *Halococci* species are needed to establish the presence of new genera of *Halococci*.

Haloalkaliphiles

This group of halophiles has been classified so far into two genera, *Natronobacterium* and *Natronococcus* [12]. Both genera have a relatively simple polar lipid composition, the main components being the C_{20}-C_{20} and C_{20}-C_{25} archaeol analogues of PGP and PG, and minor amounts of cyclic-PGP, and two unidentified phospholipids; no PGS or glycolipids have been detected [12-15] (Table 1).

BIOSYNTHESIS

Biosynthesis of the archaeol analogues of phospholipids and glycolipids proceeds by complex pathways in a multienzyme, membrane-bound system that is absolutely dependent on 4M salt concentration [2, 16, 17]. Synthesis of the isoprenoid/-isopranoid chains (C_{20}, C_{25}) in the archaeol lipid core occurs via the mevalonate pathway for isoprenoids, also absolutely dependent on 4M salt concentration [17], starting from acetate (and involving lysine [18]), and proceeding to mevalonate, isopentenylpyrophosphate and then to geranylgeranylpyrophosphate (and perhaps partial reduction to phytylpyrophosphate) [2, 17]:

> lysine
> ↓
> acetate →→ mevalonate →→ isopentenyl-PP →→ dimethylallyl-PP →→
> geranyl-PP →→ farnesyl-PP →→ geranylgeranyl-PP [→→phytyl-PP].

Final reduction of geranylgeranylpyrophosphate (or phytylpyrophosphate) to the phytanyl group takes place only after linkage to the glycerol backbone to form an unidentified C_{20}-isoprenyl glycerol ether derivative ("pre-diether") which is the precursor of both phospholipids and glycolipids, as has been shown in biosynthetic studies with intact cells of *H. cutirubrum* [19] (see Figures 2 and 3).

The "pre-diether" may be formed by an unknown mechanism whereby a suitably substituted glycerol derivative, tentatively dihydroxyacetone (DHA), is alkylated with a C_{20}-isoprenylpyrophosphate (geranylgeranylpyrophosphate or phytylpyrophosphate) [17, 19]. The choice of DHA or a substituted DHA as acceptor of the isoprenylgroup is based on the finding that the glycerol moiety of diphytanyl-glycerol undergoes dehydrogenation at C-2 but not at C-1 (or C-3) [20], thus eliminating triose phosphates (DHA-P and glyceraldehyde-P) and glycerol-P as possible acceptors, since these would exchange hydrogen at C-1 by aldoketo or keto-enol isomerizations (see Figure 2). Glycerol alone (or a substituted glycerol) might also

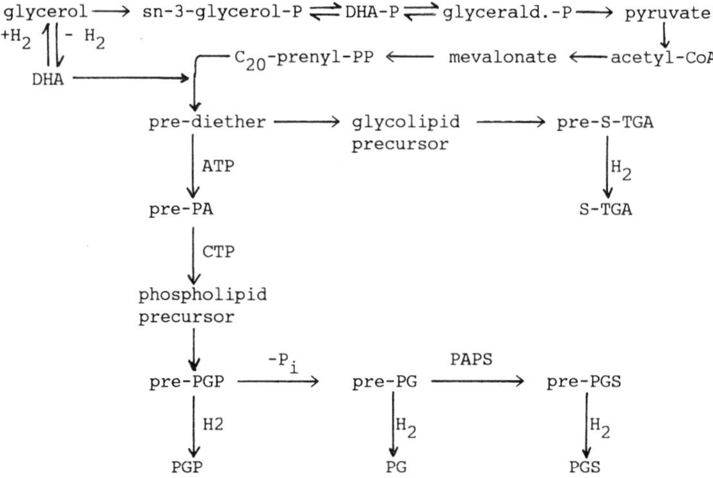

Figure 2. Proposed pathways for biosynthesis of phospholipids and glycolipids in extreme halophiles.

$$H_2C-O-CH_2CH=C-[(CH_2)_2CH=C-]_3-CH_3$$
$$H-C-O-CH_2CH=C-[(CH_2)_2CH=C-]_3-CH_3$$
$$H_2C-O-X$$

pre-diether	X = unidentified
pre-PA	X = $-PO-(OH)_2$
pre-PG	X = $-PO(OH)-O-CH_2CH(OH)CH_2OH$
pre-PGP	X = $-PO(OH)-O-CH_2CH(OH)CH_2-O-PO(OH)_2$
pre-PGS	X = $-PO(OH)-O-CH_2CH(OH)CH_2-O-SO_3-OH$
Phospholipid precursor	X = $-P(=O)(O^-)-O-P(=O)(O^-)-O$-cytidine
Glycolipid precursor	X = unidentified
pre-S-TGA	X = $-Glc-Man-Gal-3-SO_3-$

Figure 3. Structures of isoprenyl precursors of the diphytanyl ether analogues of phospholipids and glycolipids of extreme halophiles.

serve as an acceptor, as has been suggested previously [21], but it would have to undergo dehydrogenation at C-2 at some stage to account for the observed loss of hydrogen at C-2 in phytanylglycerol lipids of *H. cutirubrum* [20].

Evidence supporting the involvement of an isoprenylpyrophosphate as donor of the C_{20}-isoprenyl group is provided by the demonstration that bacitracin, which complexes with polyprenylpyrophosphates [22], is a powerful inhibitor of the biosynthesis of "pre-diether" and of phospholipids and glycolipids in whole cells of *H. cutirubrum* [23].

196

Pulse-labelling studies with whole cells of *H. cutirubrum* using ^{32}P-phosphate or ^{14}C-glycerol as precursors showed that the phospholipids were labelled in the order pre-PA > phospholipid precursor > pre-PGP > pre-PG > pre-PGS [19]. These results strongly suggest the following pathway for biosynthesis of phospholipids in extreme halophiles (see Figures 2 and 3): the pre-diether is phosphorylated with ATP, and the pre-PA formed is then converted to the "phospholipid precursor", most likely cytidinediphosphate archaeol, which is then reacted with glycerophosphate to form pre-PGP; dephosphorylation of pre-PGP by a specific phosphatase would give rise to pre-PG, which could then be converted to pre-PGS by reaction with PAPS. All of the "pre-" phospholipids are subsequently hydrogenated to give the final, saturated archaeol analogues of PGP, PG and PGS.

The pre-diether also serves as the precursor of the glycolipids [19] (Figures 2 and 3), being glycosylated stepwise with glucose, mannose and galactose folowed by sulfation with PAPS to give the pre-S-TGA which is finally reduced to the saturated S-TGA Evidence in support of a stepwise glycosylation and sulfation has been provided by pulse-labelling studies with whole cells of *H. cutirubrum* grown in presence of ^{35}S-sulfate or ^{14}C-glycerol, which established the product-precursor relationship between the glycolipids of this bacterium as follows (Deroo and Kates, unpublished)(see Figure 1):

$$\begin{array}{ccc}
\text{UDP-Glc} & & \text{UDP-Man} \\
\text{glycolipid precursor} \longrightarrow & \text{Glc-archaeol(MGA-1)} & \longrightarrow
\end{array}$$

$$\begin{array}{ccc}
\text{UDP-Gal} & & \text{PAPS} \\
\text{DGA-1} \longrightarrow & \text{TGA-1} \longrightarrow & \text{S-TGA-1}
\end{array}$$

The minor glycolipid S-TeGA could by biosynthesized by galactofuranosylation of S-TGA-1 or by galactofuranosylation of TGA-1 followed by sulfation with PAPS. The minor non-sulfated glycolipids TGA-1 and TeGA-1 could be formed by deletion of the appropriate sulfation steps or by the action of sulfatases on S-TGA-1 and S-TeGA. Such pathways would account for the presence of the glycolipids found in the *Halobacteria* (see Fig. 1 and Table 1). In *Haloferax* species, the major glycolipid S-DGA-1 (Figure 1) could be formed by deletion of the galactosylation step of DGA and insertion of a specific DGA sulfation step. In *Haloarcula*, the major glycolipid TGA-2 (Figure 1) could be formed by replacement of the galactosylation step with a glucosylation reaction using UDP-Glc.

Thus the characteristic glycolipid composition of the three genera of halobacteria could be achieved by deletion and/or insertion of the appropriate glycosylating enzymes and sulfating enzymes. Studies on the isolation of the enzymes and corresponding genes involved in glycolipid and phospholipid biosynthesis would help to establish the biosynthetic pathways of these lipids with greater certainty and to elucidate the taxonomic relationships between the genera of extreme halophiles.

ACKNOWLEDGEMENTS

The studies reported here have been supported by the Medical Research Council of Canada.

REFERENCES

[1]   M. Kates, *Prog. Chem. Fats Other Lipids*, 15: 301 (1978)

[2]   M. Kamekura and M. Kates, *in* "Halophilic Bacteria" (F. Rodriguez-Valera, ed.) Vol. II, p. 25-54, CRC Press, Boca Raton, Florida (1988)

[3]   M. DeRosa, A. Gambacorta, B. Nicolaus, H. N. M. Ross, W. D. Grant and J. D. Bulock, *J. Gen. Microbiol.* 128: 343 (1982)

[4]   M. DeRosa, A. Gambacorta, B. Nicolaus and W. D. Grant, *J. Gen. Microbiol.* 129: 2333 (1983)

[5]   M. Nishihara, H. Morii and Y. Koga, *J. Biochem.* 101: 1007 (1987)

[6]   M. Torreblanca, F. Rodriguez-Valera, G. Juez, A. Ventosa, M. Kamekura and M. Kates, *System. Appl. Microbiol.* 8: 89 (1986)

[7]   S. C. Kushwaha, G. Juez, F. Rodriguez-Valera, M. Kates and D. J. Kushner, *Can. J. Microbiol.* 28:1365 (1982)

[8]   M. Koçur and W. Hodgkiss, *Int. J. System. Bacteriol.* 23: 151 (1973)

[9]   C. G. Montero, A. Ventosa, F. Rodriguez-Valera and F. Ruiz-Berraquero, *J. Gen. Microbiol.* 134: 725 (1988)

[10]   C. G. Montero, A. Ventosa, F. Rodriguez-Valera, M. Kates, N. Moldoveanu and F. Rodriguez-Valera, *System. Appl. Microbiol.* 12: 167 (1989)

[11]   M. Kates, B. Palameta, C. N. Joo, D. J. Kushner and N. E. Gibbons, *Biochemistry*, 5: 4092 (1966)

[12]   B. J. Tindall, H. N. M. Ross and W. D. Grant, *System. Appl. Microbiol.* 5: 41 (1984)

[13]   S. Morth and B. J. Tindall, *System. Appl. Microbiol.* 6: 247 (1985)

[14]   M. DeRosa, A. Gambacorta, W. D. Grant, V. Lanzotti and B. Nicolaus, *J. Gen. Microbiol.* 134: 205 (1988)

[15]   V. Lanzotti, B. Nicolaus, A. Tricone, M. DeRosa, W. D. Grant and A. Gambacorta, *Biochem. et Biophys. Acta.* 1001: 31 (1989).

[16]   M. Kates, M. K. Wassef and D. J. Kushner, *Can. J. Microbiol.* 46: 971 (1968)

[17]   M. Kates and S. C. Kushwaha, *in* "Energetics and Structure of Halophilic Microorganisms" (S. R. Caplan and M. Ginzburg, eds.) Elsevier Biomedical Press, Amsterdam, pp. 461-489 (1978).

[18]   I. Ekiel, D. G. Sprott and I. C. P. Smith, *J. Bacteriol.* 166: 559 (1986)

[19]   N. Moldoveanu and M. Kates, *Biochem. et Biophys. Acta.* 960: 164 (1988)

[20]   M. Kates, M. K. Wassef and E. L. Pugh, *Biochem. et Biophys. Acta.* 202: 206 (1970)

[21]   M. DeRosa, A. Gambacorta and A. Gliozzi, *Microbiol. Rev.* 50: 70 (1986)

[22]   D. R. Storm and J. L. Strominger, *J. Biol. Chem.* 248: 3940 (1973)

[23]   N. Moldoveanu and M. Kates, *J. Gen. Microbiol.* 135:2503 (1989)

VARIATIONS IN THE LIPID COMPOSITION OF

AEROBIC, HALOPHILIC ARCHAEOBACTERIA

B. J. Tindall[1], Birgit Amendt[2] and Christiana Dahl[2]

[1]DSM - Deutsche Sammlung von Mikroorganismen und Zellkulturen
GmbH.
Mascheroder Weg 1b
D-3300 Braunschweig
Federal Republic of Germany

[2]Institut für Mikrobiologie der Universität Bonn
Meckenheimer Allee 168
D-5300 Bonn 1
Federal Republic of Germany

ABSTRACT

Members of the alkaliphilic genera *Natronobacterium* and *Natronococcus* have
been shown to be capable of synthesising at least two different diether lipids. It has
been possible to show, not only that the relative amounts of the two diethers vary in
different strains, but that their relative composition is influenced by the growth
conditions. Initial studies indicated that salinity plays a role in determining the relative
distribution of the diethers in the haloalkaliphiles tested. However, it has been possible
to show that other factors, such as growth phase play a significant role.

All members of the family *Halobacteriaceae* contain MK-8 and MK-8(VIII-
H_2) in varying amounts. Using representatives of the various taxonomic groupings it
has been possible to show that growth conditions reproducibly affect the relative
amounts of these two lipoquinones. Present investigations have not been able to fully
account for the factors responsible for these changes, although there seems to be a
correlation between growth phase and growth rate. The extent of the changes appears
to have some significance in the chemotaxonomy of this group of organisms.

General and Applied Aspects of Halophilic Microorganisms
Edited by F. Rodriguez-Valera, Plenum Press, New York, 1991

INTRODUCTION

Investigations on the polar lipids, fatty acids and respiratory lipoquinone composition of eubacteria have shown that they may serve as useful chemotaxonomic markers. While the presence or absence of certain compounds may be used to differentiate certain taxa, the use of such lipid material is complicated by the fact that a variety of factors may influence the relative composition of some of the components. Work on a variety of eubacteria has shown that factors such as temperature, pH and salinity may affect the distribution of both fatty acids or polar lipids. In contrast, little work has been carried out on the influence of such parameters on the lipid composition of archaeobacteria.

De Rosa and co-workers [1] have shown, using the thermoacidophilic archaeobacterium *Sulfolobus solfotaricus*, that temperature plays a significant role in the degree of cyclisation of the isopranyl side chains of the tetraether lipids, which predominate in this organism. Similarly, Langworthy and Pond [2] have reported that growth under aerobic or anaerobic conditions plays a significant role in the distribution of nonitol based tetraether lipids in *Acidianus brierleyi*. Kushwaha et al. [3] have also demonstrated the influence of salinity on the polar lipid and squalene composition of some members of the family *Halobacteriaceae*.

Considering that members of the haloalkaliphilic archaeobacteria, of the genera *Natronobacterium* and *Natronococcus* have been shown to contain a 2,3-di-*O*-diphytanyl-*sn*-glycerol diether ($C_{20}.C_{20}$) and a 2-*O*-sesterterpanyl-3-*O*-phytanyl-*sn*-glycerol diether ($C_{25}.C_{20}$) in various concentrations [4, 5], it was interesting to investigate which parameters, such as salinity, growth phase, and temperature, would affect the relative distribution of these two diether lipids. The work presented here summarises the results of work on haloalkaliphilic archaeobacteria whch demonstrates quite clearly the influence of environmental parameters on the distribution of these hydrophobic cell membrane components.

Studies on the respiratory lipoquinone composition of a wide range of microorganisms have shown that the structural diversity among this group of compounds is quite large (see [6] and [7] for a review). In many cases a single lipoquinone predominates, although it is not uncommon to find both benzoquinones and naphthoquinones, or two different types of naphthoquinone in gram-negative eubacteria. In gram-positive eubacteria, such as the actinomycetes and coryneform bacteria it is usual to find a complex pattern of menaquinones (2-methyl-3-poly-isoprenyl-1,4-naphthoquinones), in which the length of the isoprenoid chain and the degree of saturation varies. In one study, using *Streptomyces cyaneus* [8], it has been shown that the relative composition of unsaturated lipoquinones with different chain lengths changes throughout the growth curve. Members of the family *Halobacteriaceae* [9] are characterised by the presence of menaquinone-8 (MK-8; 2-methyl-3-octaprenyl-1,4-naphthoquinone) and a dihydrogenated menaquinone-8, in which saturation occurs at the end of the isoprenoid chain (MK-8[VIII-H_2]; 2-methyl-3-VIII-dihydrooctaprenyl-1,4-naphthoquinone). These are the only archaeobacteria known to date in which both a fully unsaturated and a partially saturated menaquinone occur in significant quantities. Drawing parallels from the influence of growth

conditions on the relative composition of squalenes in *Halobacterium cutirubrum* it has
been suggested that a similar effect may occur in the menaquinone composition of
members of the *Halobacteriaceae* [10]. The object of the work on the menaquinone
composition reported here was to test this hypothesis, and to establish to what degree,
if any, such effects could be observed.

MATERIALS AND METHODS

Growth Conditions

Haloalkaliphilic organisms were grown in the medium of Tindall [5], while
neutrophilic strains were grown in the medium of Tindall and Collins [11]. All strains
were grown, with shaking at 150 rpm, at 40°C. Cells were freeze dried and stored for
later analysis of the lipid components.

Diether Lipid Hydrolysis and Analysis

Diether lipids were released from 100mg of freeze dried cells by acid
hydrolysis, purified by thin layer chromatography, and analysed by gas chromatography as described previously [5].

Menaquinone Extraction and Analysis

Menaquinones were extracted from 100mg of freeze dried cells using
chloroform:methanol (2:1), purified by thin layer chromatography, and analysed by
HPLC as described by Tindall and Collins [11].

RESULTS AND DISCUSSION

Diether Lipids

A previous study [12] has shown that although all haloalkaliphilic members
of the family *Halobacteriaceae* synthesize $C_{20}.C_{20}$ and $C_{25}.C_{20}$ diethers, their relative
composition is very much strain dependant, and the $C_{25}.C_{20}$ diether may account for
anything from 0.1% to 90% of the diether lipids. Furthermore, preliminary studies on
the effect of salt concentration on the relative composition of diether lipids in a limited
number of strains indicated that there was indeed a relationship between the salinity
of the growth medium and the distribution of the two diether lipids. The influence of
salinity on the diether lipid distribution in some strains of haloalkaliphilic natronobacteria is illustrated in Figure 1.

Considering that external parameters can influence the relative diether lipid
compositon, further work was undertaken to establish what other parameters might
play a significant role. These studies included the effect of growth phase and a
combination of different salinties, temperature, and growth phase on the diether lipid
composition of two different strains. Initial work on the effect of growth phase was

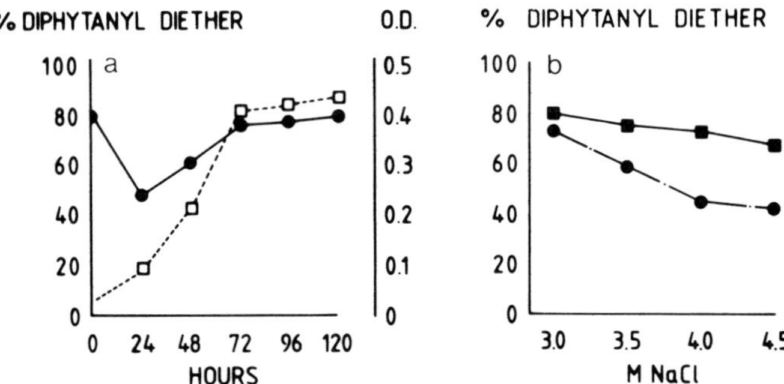

Figure 1. Influence of a) salinity on the relative diether composition of *Nb. pharaonis* Gabara ●— - —● and *Nb. magadii* MS3 ■——■ and b) growth phase □- - - -□ on the relative diether composition ●———● of *Nb. gregoryi* SP2.

limited to *Natronobacterium gregoryi*. Interestingly, growth phase appeared to have a significant effect on the relative composition of the two diethers (Figure 1). These results indicate that during the exponential phase the diether lipid composition may change quite dramatically, while the changes are significantly less in the stationary phase. This suggests that when determining the relative diether lipid composition the influence of growth phase should be taken into account. In order to minimise the effect of growth phase on the diether lipid composition, comparisons between strains are best made using cells which have reached the stationary phase.

In a more complex study the influence of temperature, salinity, and growth phase was determined using a *Natronobacterium* strain with a generation time considerably shorter than the majority of isolates. It was possible to show that the diether lipid composition in this strain was influenced by the growth phase, as well as the salinity of the medium. However, the influence of temperature was also shown to play a role in determining the diether lipid composition.

The results presented here summarise experiments which indicate that the diether lipid composition of haloalkaliphilic members of the genera *Natronobacterium* and *Natronococcus* is influenced by parameters such as growth phase, temperature, and salinity. It is, therefore, to be expected that differences between different batches of cells or between different laboratories can be attributed to effects such as medium composition and growth conditions. These results present a rather simplified view of what factors influence the diether lipid composition of these organisms. It does not attempt to answer the more complex question of how such changes affect the physical properties of the membrane, nor does it provide an insight into why such changes may be significant in the membranes of the haloalkaliphiles. In many other groups within the family *Halobacteriaceae* only a single $C_{20}.C_{20}$ diether lipid is present [4], and such changes cannot occur. Further studies are needed on the influence of such parameters on all membrane components (i.e. diether lipids, squalenes, carotenoids, and proteins) before these mechanisms can be fully understood.

202

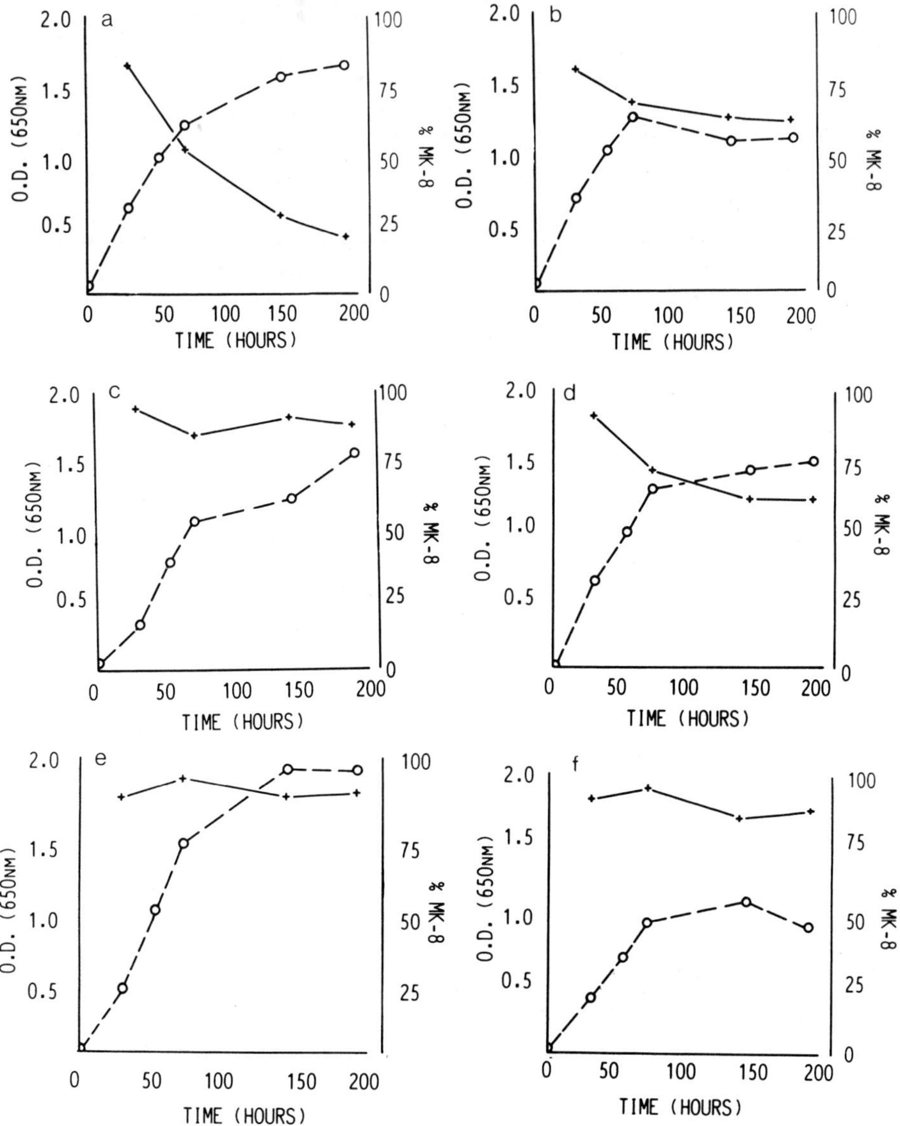

Figure 2. Changes in the relative composition of MK-8 and MK-(8-VIII-H2) in a) *H. volcani* NCMB 2012, b) *H. halobium*" NCMB 777, c) *H. vallismortis* CCM 3404, d) *H. saccharovorum* CCM 2887, e) "*H. cutirubrum*" NCMB 763, and f) *H. cutirubrum* NRC 34002. Optical density O- - - - -O ; relative MK-8 compostion +———+.

Menaquinones

Considering that a variety of factors influence the carotenoid, polar lipid, squalene and diether lipid composition of members of the family *Halobacteriaceae*, it was of interest to investigate whether changes in the relative composition of MK-8 and MK-8(VIII-H$_2$) could also be influenced by external factors.In these experiments the following strains were used: *Halobacterium cutirubrum* NRC 34001, *Haloferax volcanii* NCMB 2012, *Haloarcula vallismortis* CCM 3404, *Halobacterium saccharovorum* CCM 2887, "*Halobacterium cutirubrum*" NCMB 763, "*Halobacterium halobium*" NCMB 777, and *Natronobacterium gregoryi* SP2. In all strains examined it was possible to detect changes in the menaquinone composition during the different growth phases of the strains. These effects were most pronounced in *Haloferax volcanii* NCMB 2012 and *Halobacterium halobium* NCMB 777 (Figure 2), and indicated that, as with the diether lipid composition of the haloalkaliphiles, the changes were most pronounced in the exponential phase and became more stable in the stationary phase. That these effects were not artifacts is supported by examining the menaquinone composition of some 150 strains harvested in the stationary phase. Strains related to members of the genus *Halococcus* and non-gas vacuolate members of the genus *Haloferax* contained MK-8(VIII-H$_2$) as the major menaquinone, while strains related to *Halobacterium halobium* NCMB 777 contained 40-50% MK-8(VIII-H$_2$). In all other groups the MK-8(VIII-H$_2$) composition ranged from 10-30%, the lowest values being found in the classical members of the genus *Halobacterium*. All gas vacuolate strains of the genus *Haloferax* had MK-8 as the major menaquinone, indicating an interesting division within the genus.

A variety of factors may be responsible for these changes in the menaquinone composition, including factors such as oxygen tension, as demonstrated in the case of variations in the squalene composition. In order to test this hypothesis *Haloferax mediterranei* (R-4) and *Haloferax volcanii* (NCMB 2012) were grown aerobically and the oxygen tension in the medium recorded. Similarly, *Haloferax mediterranei* was also grown anaerobically on nitrate (attempts to grow *Haloferax volcanii* on nitrate, anaerobically were not successful). No clear relationship could be detected between oxygen tension in the medium, although some difficulty was experienced with the oxygen electrode probe, due to the adverse effect of the high salinity. That oxygen tension alone was not the sole factor influencing the menaquinone composition could, however, be demonstrated by the fact that changes in the relative levels of MK-8 and MK-8(VIII-H$_2$) could be detected at different points through anaerobic growth on nitrate of *Haloferax mediterranei*. While more complex factors may be involved in determining the menaquinone composition in members of the family *Halobacteriaceae*, it should not be forgotten that this work does not exclude the possibility that the concentration of the terminal electron acceptor (i.e. oxygen or nitrate) may play an important role. However, the reason for these changes remains unclear, since changes in the degree of unsaturation of the isoprenoid side chain of menaquinones does not appear to affect the redox potential, nor are they present in concentrations high enough to significantly affect the fluidity of the membrane.

The effects observed in the diether lipid and menaquinone composition of various members of the family *Halobacteriaceae* provide an interesting insight into the biology of these organisms. Despite the fact that changes in the diether lipid or menaquinone composition do not appear to be essential for growth, the fact that such changes do occur is worthy of further study, in order to understand more about the biology of the aerobic, extremely halophilic archaeobacteria.

ACKNOWLEDGEMENTS

This work was supported by grants from the Deutsche Forschungsgemeinschaft to H. G. Trüper (Tr 133/17-2 and Tr 133/17-3).

REFERENCES

[1]   M. De Rosa, E. Esposito, A. Gambacorta, B. Nicolaus and J. D. Bulock, *Phytochem.* 19: 827 (1980)

[2]   T. A. Langworthy and J. L. Pond, *System. Appl. Microbiol.* 7: 253 (1986)

[3]   S. C. Kushwaha, G. Juez-Pérez, F. Rodríguez-Valera, M. Kates and D. J. Kushner, *Can. J. Microbiol.* 28: 1365 (1982)

[4]   H. N. M. Ross, M. D. Collins, B. J. Tindall and W. D. Grant, *J. Gen. Microbiol.* 123: 75 (1981)

[5]   B. J. Tindall, *System. Appl. Microbiol.* 6: 243 (1985)

[6]   M. D. Collins and D. Jones, *Microbiol. Rev.*, 45: 316 (1981)

[7]   M. D. Collins, *Methods Microbiol.* 18: 329 (1985)

[8]   G. S. Saddler, M. Goodfellow, D. E. Minnikin and A. G. O'Donnell, *J. Appl. Bacteriol.* 60: 51 (1986)

[9]   B. J. Tindall and M. D. Collins, unpublished results.

[10]   B. J. Tindall, Doctoral thesis, University of Leicester, U. K. (1980)

[11]   B. J. Tindall and M. D. Collins, *FEMS Microbiol. Lett.* 37: 117 (1986)

[12]   S. Morth and B. J. Tindall, *FEMS Microbiol. Lett.* 29: 285 (1985)

BACTERIORUBERINS REINFORCE RECONSTITUTED

HALOBACTERIUM **LIPID MEMBRANES**[1]

Y. Nakatani, T. Lazrak, A. Milon, G. Wolff and G. Ourisson

Laboratoire de Chimie Organique des Substances Naturelles
Centre de Neurochimie
Université Louis Pasteur
5 rue Blaise Pascal
67084 Strasbourg
France

ABSTRACT

We have developed methodologies for the determination of the topology of α,ω-dihydroxylated carotenoids incorporated into vesicles and for the evaluation of the reinforcing effects of these carotenoids on lipid bilayers.

This development has been applied to a model of red membranes of *Halobacterium*, reconstituted from their total polar lipids and bacterioruberins. Bacterioruberins are well incorporated into the lipid vesicles and, when incorporated, they decrease the water permeability and increase the rigidity of the bilayers. Thus, at least in a system related to natural archaeobacterial membranes, carotenoids do play their postulated role of reinforcers.

INTRODUCTION

We have postulated, as part of a general theory of the molecular evolution of biomembrane constituents, that carotenoids might play an important role in bacteria

---

[1]Dedicated to Professor Morris Kates on the occasion of his 65th birthday.

General and Applied Aspects of Halophilic Microorganisms
Edited by F. Rodriguez-Valera, Plenum Press, New York, 1991

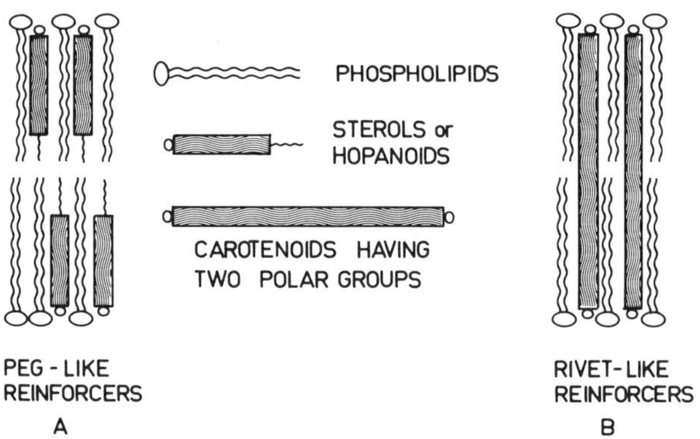

Figure 1.   Hypothetical reinforcement of lipid bilayers by sterols or hopanoids (A) or α,ω-dihydroxylated carotenoids (B).

as membrane stabilizers [1]. As schematically shown in Figure 1, the terminally hydroxylated carotenoids, typical of so many bacteria, could act as transmembrane rivets, stabilizing both halves of the bilayer. This mechanism of stabilization would be contrasting with the one-layer stabilization brought about by cholesterol in eucaryotes [2] and by hopanoids in many bacteria [3].

To confirm the hypothesis that bacterial carotenoids play a significant mechanical role in membranes, we have developed methodologies for:

(i)   the preparation and purification of unilamellar vesicles composed of dipolar carotenoids and phospholipids,

(ii)   the characterization of these systems by optical methods (UV, CD) to answer the following questions: Do the carotenoids enter the bilayer? What is the concentration of incorporated carotenoid? Are they incorporated in a transmembrane manner?,

(iii)   the evaluation of the effects of these carotenoids on the elasticity and the water permeability of lipid bilayers.

These methods have been at first tried on model systems [4, 5, 6], composed of four α,ω-dihydroxylated carotenoids (zeaxanthin $\underline{1}$, astaxanthin $\underline{2}$ and their C_{50} synthetic isoprene homologs $\underline{3}$ and $\underline{4}$) and dimyristoylphosphatidylcholine (DMPC), dipalmitoylphosphatidylcholine (DPPC) or diphytanylphosphatidylcholine (DPhPC) $\underline{5}$ (cf. Figure 2 and Figure 6). DPhPC $\underline{5}$ was prepared by modification of the polar heads of the phospholipids of *Halobacterium halobium*.

Then, we have applied the same biophysical methods to a reconstituted *Halobacterium* lipid membrane, composed of its total polar lipids and its major carotenoids, the bacterioruberins [7].

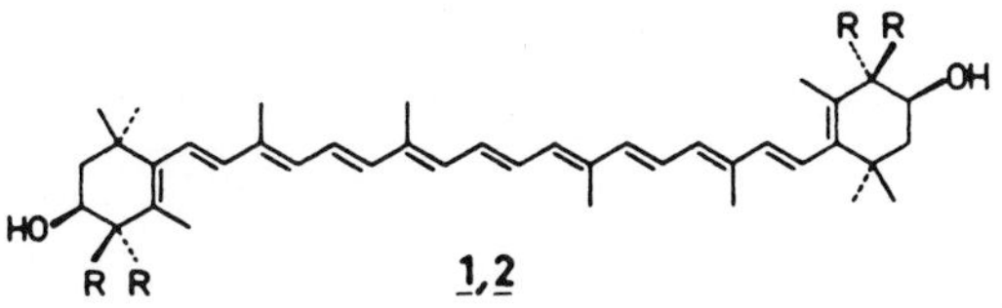

Figure 2.   Structure of (3R, 3'R)-zeaxanthin (R = H; <u>1</u>), (3S, 3'S)-astaxanthin (R,R = O; <u>2</u>), (3R. 3'R)-decapreno-zeaxanthin (R = H; <u>3</u>) and (3S,3'S)-decaprenoastaxanthin (R,R = O; <u>4</u>).

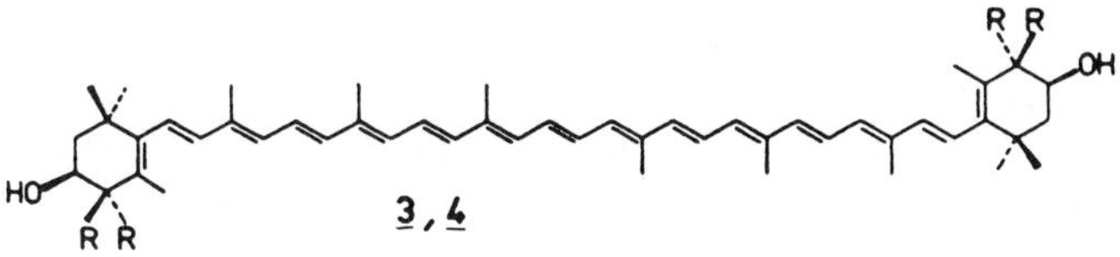

ORGANIZATION OF CAROTENOID-PHOSPHOLIPID BILAYER SYSTEMS

Incorporation of carotenoids into vesicles

Vesicles using mixtures of carotenoid and phospholipid were prepared by the sonication method, the ether injection method or the reverse phase evaporation method. The vesicles obtained were filtered through polycarbonate filters and passed over a Sepharose 4BC1 column. Analysis of the carotenoid and the phospholipid after gel filtration shows that only a small fraction of the carotenoid is really incorporated into the vesicles. Using different proportions of the constituents, we have shown that there is a limit to the solubility of the carotenoids (Table 1).

Table 1. **Maximum incorporation ratio (mol %) of zeaxanthin, decaprenozeaxanthin, astaxanthin, decaprenoastaxanthin in DMPC, DPPC and DPhPC.**

Lipids	Zeaxanthin (C_{40})	Decapreno-zeaxanthin (C_{50})	Astaxanthin (C_{40})	Decapreno-astaxanthin (C_{50})
DMPC	8.5	2.5	15	10
DPPC	0.5	5.0	–	–
DPhPC	1.5	1.5	–	–

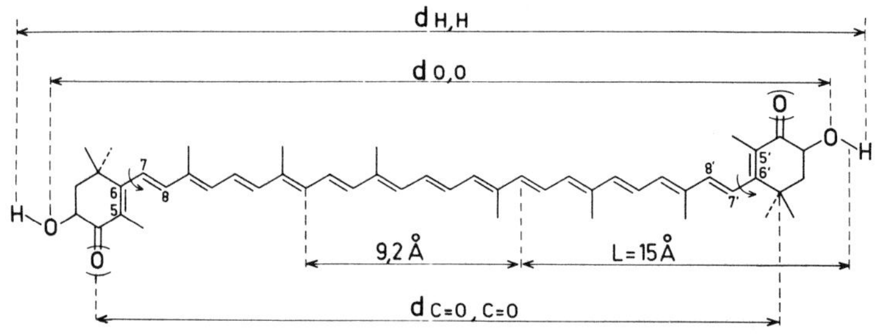

	ZEAXANTHIN 1	C_{50} ZEAXANTHIN 3	ASTAXANTHIN 2	C_{50} ASTAXANTHIN 4
$d_{H,H}$ (Å)	31.7	40.9	31.7	40.9
$d_{O,O}$ (Å)	30.2	39.4	30.2	39.4
$d_{C=O,C=O}$ (Å)			25.0	34.2

Figure 3. Structure and intramolecular dimensions of four analogs of zeaxanthin. L = 15 Å corresponds to the length of the phospholipidic segment, i.e. the average distance between the carbonyl group and the methyl group of DMPC, above Tc. This agrees much better with the half-length of the C_{40} carotenoids than with that of the C_{50} carotenoids.

It can be seen that C_{40} carotenoids are incorporated more easily into DMPC vesicles whereas C_{50} carotenoids are better incorporated into DPPC. In both cases the best incorporation was found when the length of the carotenoid molecule was close to the bilayer thickness (Figure 3). This, in itself, is in favour of a transmembrane insertion into the bilayer.

Topology of carotenoids incorporated into bilayers

The incorporation into the bilayer has been examined by UV and CD studies. The refractive index is related to the main transition absorption maximum as predicted by Bayliss' law (v_{max} proportional to $(n^2 - 1)/(2n^2 + 1)$, where n is the refractive index of the solvent [8]). The absorption of the carotenoids in the vesicles is typical of a lipidic environment having n = 1.44. So we can conclude that the carotenoids are well integrated in the membrane.

At the phase transition temperature (Tc) occurs an abrupt and reversible change in the visible spectrum of incorporated carotenoids (Figure 4): above Tc, only the spectrum of the monomer was observed, whereas below Tc the absorption maximum is shifted to a shorter wavelength, which is characteristic of molecular aggregation. This means that the phase transition of the lipids influences the behaviour of the carotenoids. This is also in favour of a good incorporation into the membranes.

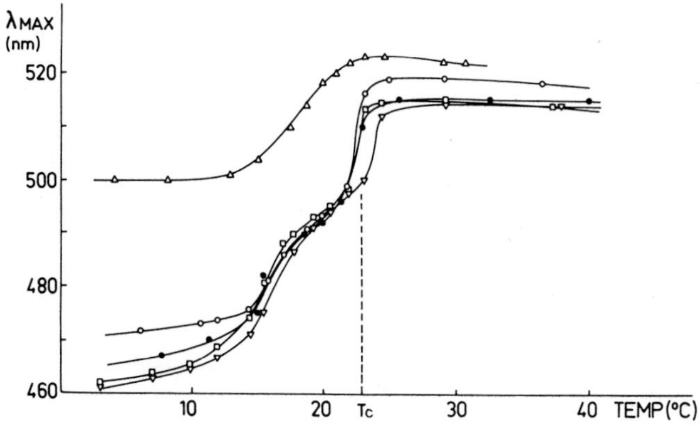

Figure 4.   Variation of the absorption maximum with temperature for several proportions of incorporated carotenoid. Vesicles were prepared from DMPC and decaprenoastaxanthin at the following proportions: 1% (Δ), 3.4% ○, 6.6% •, 8.0% □ and 9.0% ▽; Tc: phase transition temperature.

EFFECTS OF INSERTED α,ω-DIHYDROXYLATED CAROTENOIDS ON THE MECHANICAL PROPERTIES OF LIPID BILAYERS

We have employed the stopped-flow light scattering method to follow the osmotic swelling of unilamellar vesicles homogenous in size (radii varying from 45 nm to 150 nm).

Stopped-flow experiments

After rapid mixing of a suspension of vesicles (in 350 mM NaCl buffer) with the same volume of 50 mM NaCl buffer, the intensity of scattered light decreases. This is due to the swelling of vesicles by an inflow of water into the vesicles. The scattered light decreases exponentially with time; the kinetics can thus be characterized by the reaction half-time $t_{1/2}$. We have shown that the measured time constant for the variation of light scattering intensity and the time constant of vesicle swelling are identical. We have also established that the relative light scattering change ($\Delta I/Io$) is proportional to the relative radius change, ($\Delta R/Ro$), in which the value of the proportionality "constant" depends on the initial radius. With this system, we can evaluate two parameters: elasticity of the bilayer and water permeability.

Elasticity of the bilayer

We found that $\Delta I/Io$ is proportional to (Z-1) for Z<15 (Z is the dissymmetry determined by light scattering) for vesicles made of DMPC + 30 mol% cholesterol. We have confirmed this empirical relationship for DPhPC and egg PC vesicles (Figure 5). As $\Delta I/Io.(Z-1)$ is independent of the vesicle radius, it can be taken as a measure of membrane elasticity.

Table 2. Water permeability ($t_{1/2}$) and bilayer elasticity ($-\Delta I/Io.(Z-1)$) for vesicles of a given composition.

Composition	$t_{1/2}$ (ms)	$-\Delta I/Io.(Z-1)$ (%)
DMPC	20	3.7
DMPC + 30 mol% cholesterol	115	1.4
DMPC + 8.5 mol% zeaxanthin	30	2.1
DMPC + 8 mol% astaxanthin	45	2.4
DMPC + 10 mol% decaprenoastaxanthin	55	1.3

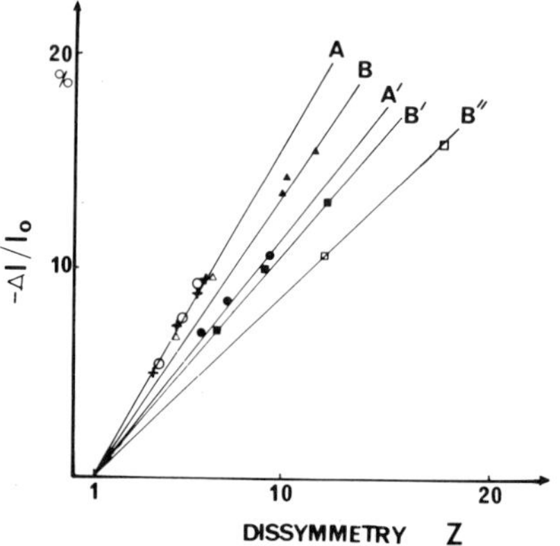

Figure 5. Linearity between the amplitude change of light scattering ($\Delta I/Io$) and dissymmetry Z. A: vesicles made of egg PC (+), egg PC + 9.5 mol% zeaxanthin (o), egg PC + 6.5 mol% decaprenozeaxanthin (Δ); A': egg PC + 30 mol% cholesterol ($\bullet$); B: DPhPC ($\blacktriangle$); B': CPhPC + 1.5 mol% decaprenozeaxanthin ($\blacksquare$); B": DPhPC + 1.5 mol% zeaxanthin ($\square$).

Water permeability

By comparing the reaction half-time $t_{1/2}$ measured by Lawaczeck's method [9] and by theosmotic shock method, we have shown that the water permeability is the limiting kinetic factor for the swelling after an osmotic shock.

Influence of cholesterol and α,ω-dihydroxylated carotenoids on mechanical properties of lipid bilayers

Table 2 summarizes our results in the study of the influence of the bilayer composition. All the carotenoids tested lower the water permeability and increase the bilayer rigidity.

212

RECONSTITUTED *HALOBACTERIUM* LIPID MEMBRANES

Kates et al. [10] have shown that *Halobacterium* species contain exclusively the phytanyl analogues of phospholipids (mainly phosphatidylglycerophosphate, with small amounts of phosphatidylglycerol and phosphatidylglycerosulphate) and glycolipids (mainly a sulfated triglycosylglycerol diether); they have also shown that these archaeobacteria produce a red membrane [11] in which the major red pigments consist of acyclic C_{50} carotenoids called bacterioruberins (Figure 6). To confirm the stabilizing role of bacterial carotenoids in natural membranes, we have studied *Halobacterium* lipid systems.

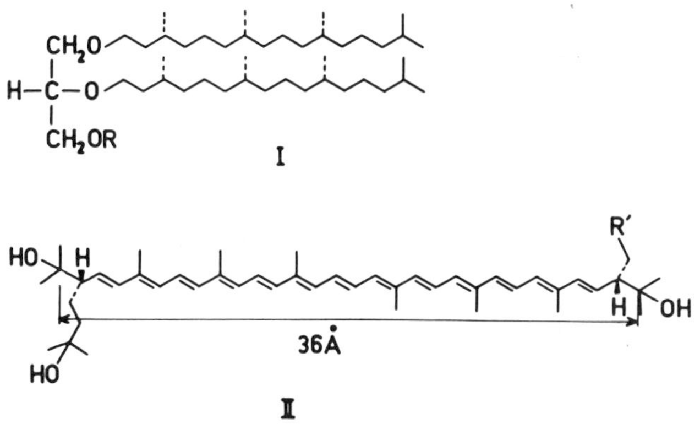

Figure 6. Structure I represents diphytanylphosphatidylcholine (DPhPC) $\underline{5}$ (R is phosphocholine) of *Halobacterium* total polar lipids $\underline{6}$ (R is mainly 1,3-glycerol bisphosphate and $3\text{-}SO_3^-\text{-Gal-Man-Glc}$). Structure II represents bacterioruberins: bacterioruberin $\underline{7}$ (R', $CH_2\text{-}C(CH_3)_2OH$) or monoanhydrobacterioruberin $\underline{8}$ (R', $CH = C(CH_3)_2$).

The extent of maximum incorporation of bacterioruberins into vesicles was 11 mol% in DPhPC and 9 mol% in total polar lipids of *Halobacterium*. These values are much higher than the incorporation value of zeaxanthin (1.5 mol% in DPhPC) or decaprenozeaxanthin (1.5 mol% in DPhPC) (cf. Table 1).

By using the stopped-flow light scattering method, we have shown that the incorporation of bacterioruberins does lower the water permeability and enhance the rigidity both of DPhPC bilayers (an effect similar to cholesterol) and of reconstituted total lipid membranes (Table 3).

213

Table 3. Water permeability ($t_{1/2}$) and elasticity of bilayers ($-\Delta I/Io.(Z-I)$) for the vesicles composed of diphytanylphosphatidylcholine or total polar lipids of *Halobacterium cutirubrum* with cholesterol or bacterioruberins.

Lipids	$t_{1/2}$ (ms)	$-\Delta I/Io.(Z-1)$ (%)
DPhPC	40	1.6
DPhPC + 11 mol% bacterioruberins	110	1.0
DPhPC + 5 mol% cholesterol	100	0.9
Total polar lipids (*Halobacterium cutirubrum*)	50	2.9
Total polar lipids + 9 mol% bacterioruberins	110	1.8

Thus, at least in this system, analogous to natural archaeobacterial membranes, carotenoids do act as membrane reinforcers, as predicted by the theory presented earlier [1].

ACKNOWLEDGEMENTS

We thank Prof. M. Kates, Ottawa, for valuable discussion, and for a gift of total polar lipids of *Halobacterium cutirubrum*, and Dr. A. Escaut, Gif-sur-Yvette, for the extraction of the acetone-insoluble lipid fraction of *Halobacterium halobium* which afforded our bacterioruberin samples. We also thank Dr. A. M. Albrecht, Prof. G. Weill and Prof. T. Tanaka for stopped-flow light scattering studies, and Mrs. M. Miehé for electron microscopy.

This work was supported by the CNRS, France.

REFERENCES

[1]   M. Rohmer, P. Bouvier and G. Ourisson, *Proc. Natl. Acad. Sci. USA* 76: 847 (1979)

[2]   R. A. Demel and B. de Kruyff, *Biochim. Biophys. Acta* 457: 109 (1976)

[3]   G. Ourisson, M. Rohmer and K. Poralla, *Ann. Rev. Microbiol.* 41: 301 (1987)

[4]   A. Milon, G. Wolff, G. Ourisson and Y. Nakatani, *Helv. Chim. Acta* 69: 12 (1986)

[5]   T. Lazrak, A. Milon, G. Wolff, A. M. Albrecht, M. Miehé, G. Ourisson and Y. Nakatani, *Biochim. Biophys. Acta* 903: 132 (1987)

[6]   A. Milon, T. Lazrak, A. M. Albrecht, G. Wolff, G. Weill, G. Ourisson and Y. Nakatani, *Biochim. Biophys. Acta* 859: 1 (1986)

[7]   T. Lazrak, G. Wolff, A. M. Albrecht, Y. Nakatani, G. Ourisson and M. Kates, *Biochim. Biophys. Acta* 939: 160 (1988)

[8]   N. S. Bayliss, *J. Chem. Phys.* 18: 292 (1950)

[9]   R. Lawaczeck, *Biophys. J.* 45: 491 (1984)

[10]   M. Kates, B. Palameta, C. C. Joo, D. J. Kushner and N. E. Gibbons, *Biochemistry* 5: 4092 (1986)

[11]   S. C. Kushwaha, M. Kates and W. G. Martin, *Can. J. Biochem.* 53: 284 (1975)

THE 'TRUE' INTRACELLULAR ENVIRONMENT OF

MODERATELY HALOPHILIC EUBACTERIA

Margot Kogut

Department of Biochemistry
University of Wales
P. O. Box 903
Cardiff CF1 1ST
Wales, U. K.

ABSTRACT

Different types of halophilic bacteria adapt to their environment in different ways. Some moderately halophilic eubacteria can grow in a wide range of NaCl concentrations. This report deals with studies on the truly intracellular concentrations of sodium ions and compatible solutes in two such organisms. It discusses the problems involved in making such measurements, and their interpretation. As a rough estimate, it is suggested that the 'true' intracellular environment may balance about 60-80% of the extracellular solute concentration. A suggestion is also made, on the basis of this and earlier work, that haloadaptation in these organisms may have a discontinuous character. There is evidence that more profound changes in cellular properties occur in adaptation to NaCl concentrations above 2 M than in those up to 2 M.

INTRODUCTION

In much of what is said and written about halophilic organisms, the work "halophilism" makes its appearance and has the effect of subtly traducing one's thought into dealing with it as if it were a unitary phenomenon. Is there such a single set of adaptive changes which can be regarded as "halophilism"? I don't believe there is. There is a lot of salt in the biosphere, in many parts of the globe and through many time spans. Many kinds of organisms came to "live with it" - in their own way, adopting their own answers. We all know now how vast is the difference between archaeobacteria and eubacteria [1] as well as the halophilic properties of a considerable array of eukaryotes.

General and Applied Aspects of Halophilic Microorganisms
Edited by F. Rodriguez-Valera, Plenum Press, New York, 1991

Halophilic eubacteria, the subject of this contribution, are very diverse, with representatives in many genera and families, Gram-positives and Gram-negatives, rods cocci and spirochaetes, as well as many metabolic types [2]. It would be nice if at least among this group we could study just one or two "representative" species; a kind of *Escherichia coli* of eubacterial halophiles. Unfortunately, however, our knowledge of the halophilic properties and behaviour among the eubacteria is very incomplete in its extent, and very fragmentary in its distribution among such a large number of different organisms. Our work, and what I am going to report, has been entirely with two Gram-negative bacteria, namely *Vibrio costicola*, which is a facultatively-anaerobic curved rod, and the organism called Ba1, which is an aerobic rod [3].

METHODS AND PROBLEMS OF MEASURING INTRACELLULAR IONS

We knew as long ago as 1977 from our own work on protein synthesis *in vitro* [4] that these organisms, grown at certain salt concentrations (i.e. when they must have synthesized proteins), gave cell-free systems that had different salt tolerances and were inhibited by the salt concentrations used for growth. Similar observations have been reported by other workers also using cell-free systems [5]. We can only assume, and this has been said many times, that the real intracellular salt or ion concentrations must be much lower than those in the external environment - or that these enzyme systems must somehow be protected *in vivo*. These inhibitory effects by salts have been observed, by and large, only in the case of enzymes which are normally intracellular. Extracellular enzymes, and some membrane-associated ones, have usually been found to be tolerant to the salt concentrations in the environment, or even to require them.

On the other hand, and again this has been pointed out by many workers [6], the osmotic pressure of the cytoplasm must balance that outside, at least approximately. And, of course, many of the "intracellular" cation concentrations recorded for eubacterial halophiles, including those we are studying, i.e. *V. costicola* [2], "appear" to balance the NaCl concentrations in the medium. Furthermore, Kushner [6] also pointed out the difficulties and uncertainties involved in determining cytoplasmic ion concentrations by the conventional methods - very much depending on correct assessment of intercellular and intracellular spaces, i.e. aqueous volumes in the samples being assayed. In addition, such determinations of intracellular ions by flame photometry or atomic absorption involve quite extensive manipulations of cultures (centrifugation and washings followed by precipitation of cell material to release the ions in soluble form). The most serious limitation of these methods, however, is their complete inability to distinguish between ions freely dissolved in the intracellular pool and those which are combined, held, bound, sequestered or occluded etc. by and within cellular structures - from membranes to compartments. The term "cell-associated ions" is therefore commonly used in reference to such quantitations [7].

The introduction of NMR measurements to the quantitative distribution of cations, such as sodium, in biological materials was reported in the sixties and seventies [8 - 12], but there were difficulties in interpretation. Although these have not yet been completely resolved, the application of this powerful analytical technique has immense advantages over the earlier methods. In the first place, the measurements are made on fresh and living cell suspensions. In the second place the method allows determination of the amounts of NMR-visible sodium ions in the intracellular and extracellular pools. Goldberg and Gilboa [13] were the first to apply such ^{23}Na-NMR investigation to a moderately-halophilic eubacterium (designated Ba1). They found that in this organism the cell-associated sodium could exist in three "states". One fraction, about 40% of the total which they termed "free" and was dissolved in the cytosol, could exchange with extracellular sodium; and a "bound" fraction, up to 60% of the total, of which part could exchange with the "free" intracellular sodium, whereas another part could not exchange and was not "seen" by the NMR [12]. Some later ^{23}Na-NMR measurements did not deal with halophilic organisms [14, 15].

'FREE' AND 'BOUND' INTRACELLULAR SODIUM

Further improvements in methods and instruments, including use of a shift reagent to separate the extracellular and intracellular signals, have now enabled us to determine the intracellular sodium ion concentrations in *V. costicola* after growth and suspension in a number of salt concentrations. The results are quite unequivocal: with cells of *V. costicola*, grown and prepared in media containing NaCl concentrations between 0.6 and 2.0 M (the latter being the highest salt concentration that could give any useful resolution of spectra) the intracellular Na$^+$ concentrations never exceeded about 15-20% of the concentration in the medium [16]. These, of course, are likely to be minimum values, because - as shown by Goldberg and Gilboa [13] - not all "intracellular" sodium ions can give appropriate signals. Thus our values might represent (or contain) 40% of the total cell-associated sodium, which is dissolved in the cytosol and able to exchange freely with extracellular sodium. But, in addition, there might be weakly-bound sodium ions, able to exchange with the dissolved ions in the cytosol, as well as tightly-bound sodium ions, which would not be seen at all. These two "bound" fractions could make up about 60% of the total sodium [12, 13].

That there must be some such "bound" fraction of sodium must follow logically from a comparison of our figures with those of Schindler *et al.* [7] and Christian and Waltho [17] for the same organism. In both studies it was found that over a similar concentration range "cell-associated" concentrations of Na$^+$ were (mostly) close to the extracellular ones. The differences between their figures and ours are very remarkable; both in the actual concentrations of cell-associated Na$^+$ given by them, compared to our figures for "free" intracellular Na$^+$ concentrations, and for the fact that their figures, but not ours, appear to vary with the external salt concentrations.

OTHER INTRACELLULAR SOLUTES

It is probably because it was believed that the intracellular cations balanced the extracellular solutes (see above) that little attention was focussed for a long time on intracellular solutes in these eubacterial halophiles: the question of "compatible solutes" [18, 19] in organisms such as *V. costicola* has only been

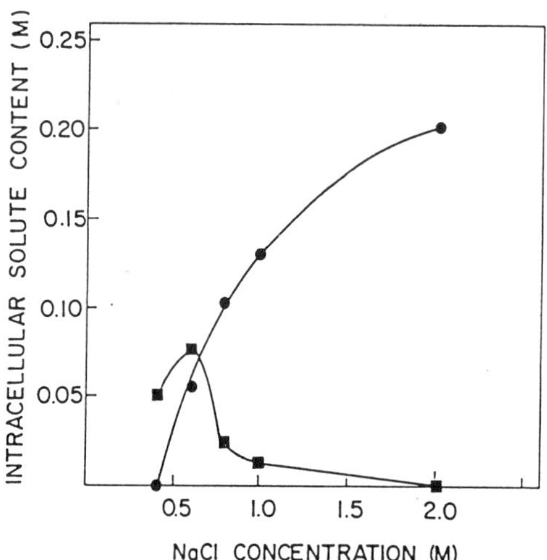

Figure 1.   The effect of extracellular NaCl concentration on the internal concentrations of compatible solutes in the halophilic eubacterium Ba1.
The concentrations of trehalose (■) and ectoine (•) were measured using ^{13}C-NMR. Data taken with permission from Regev et al. [23]

dealt with in recent years [20 - 22]. Imhoff and Rodríguez-Valera [20] suggested that betaine (glycine betaine) "is the main compatible solute in halophilic eubacteria". They showed that for *V. costicola* and a number of other halophilic eubacteria, grown with about 0.6 M, 1.8 M and 3.5 M NaCl, the intracellular concentrations of betaine varied with the salt concentration in the medium. However, even at the highest salt concentrations, the intracellular concentrations of betaine never balanced the extracellular solutes in molar terms; indeed, the highest concentrations of betaine recorded were only about 30% of the external NaCl concentration. Most recently Regev et al. [23] have found that in *V. costicola* and also in their organism Ba1, ectoine (1,4,5,6-tetrahydro-2-methyl-4-pyrimidine carboxylic acid) was the major compatible solute, provided that no betaine (or its precursor, choline) was present in the growth medium. They compared defined media with complex media, and also investigated the effect of growth phase in batch culture; they could show that for Ba1 the concentration of ectoine increased with extracellular concentrations of NaCl above 0.6 M up to a maximum of 2.0 M. They also found that the maximum concentration of ectoine occurred during the exponential phase of growth and then declined, whereas betaine continued to accumulate in the stationary phase. Although not all these findings were reported for *V. costicola* it is likely that the same applies. These variations in the conditions for accumulation of compatible solutes could explain why glycine betaine was the main compatible solute found by Imhoff and Rodríguez-Valera - since they harvested their cultures at the end of the exponential phase [20]. However, similar to the case reported for betaine [20] the molar concentrations of ectoine never approached those of the NaCl in the medium (Figure 1).

Bearing in mind all the data so far available, and the limitations and insufficiencies involved, it still seems that concentrations of compatible solutes, so far reported in these organisms, are not sufficient to balance the osmotic difference between the extracellular and intracellular milieu. Could some of the "bound" or "cell-associated" sodium ions fill the gap?

We are unable at this time to establish directly how much "bound" sodium may be associated with the cells of *V. costicola* in which the free, intracellular sodium has been determined. However, we can derive estimates of the amount of sodium that is invisible to NMR in our preparations. Such estimated amounts for "bound sodium" range from 23-35% of the external NaCl concentrations in which the cells have been grown and resuspended. They also appear to vary with the concentration of external salt, the amounts being higher at the higher salinities. The questions we now face, therefore, concern the structures and components of the cells to which these ions may be bound or attached and in what manner. "Bound" usually has the connotation of covalently bound. But here the meaning is "invisible to NMR". This leaves open the question of what roles such invisible Na^+ may play; can they be considered part of the "true internal environment"? Must such "bound" or cell-associated ions necessarily be osmotically inactive? This applies especially with reference to the weakly-bound fraction that can exchange with the ions dissolved in the cytosol [12, 13].

OSMOTIC BALANCE AND HALOADAPTATION

All halophilic cells, in common with non-halophilic ones, must osmoregulate, i.e. control their cell volume and turgor pressure [19], as well as maintain electrolyte balance. The manner in which this is achieved is likely to have important consequences for the cellular events involved in haloadaptation. This is particularly so with regard to the cell membrane, the site of ion pumps and other osmoregulatory functions. Like Janus, it must present two faces - one to the external world and another to the intracellular milieu. It is probably significant that changes in membrane lipid composition and conformation involved in haloadaptation have been described in *V. costicola* and other eubacterial halophiles [24 - 26].

There need not necessarily be just one mechanism responsible for maintaining osmotic balance. The data presented here allow one to make a rough tally of the possible balance between intracellular and extracellular solutes in *V. costicola* (Table 1).

Table 1. Estimates of the intracellular solute composition of *V. costicola*.

Solute	Percentage of total[1]
"Free" Na^+	15-20
"Bound" Na^+	25-35
Betaine/ectoine	20-30

[1]The percentage of intracellular solute is given relative to the total extracellular concentration taken as 100%.

These estimates show that the total intracellular solute can match only 60-80% of that in the extracellular medium. But there is room for other variations in cell composition and properties. Thus we have also determined the intracellular volume, i.e. water space, of *V. costicola*, in connection with our measurements of intracellular Na^+ concentrations. Using the method of Stock *et al.* [27] we found that this varied as a function of the external salt concentration, but not in a linear fashion. Over the range 1.0 - 2.0 M NaCl the intracellular space is practically constant or slightly decreased; however, when grown in 3 M NaCl, the intracellular space of *V. costicola* is decreased by almost 50% [28]. Similar changes of intracellular volume with extracellular salt concentration were reported for a marine pseudomonad by Takacs *et al.* [29].

There are, therefore, several parameters (at least in *V. costicola*, which is probably the most studied of the eubacterial halophiles) which change quite dramatically at or above external salt concentrations of 2.0 M. This would suggest that the organism is "tolerant" in the sense of showing relatively slight differences in cellular characteristics, over the range of about 0.8 - 2.0 M external salt, but then undergoes quite profound changes to enable it to grow at the higher external salinities [24 - 26].

ACKNOWLEDGEMENTS

I would like to thank, first and foremost, my colleague Dr N . J. Russell (Cardiff), together with Professors H. Gilboa and Y. Avi-Dor (Technion, Haifa), who furnished the facilities and collaboration for most of the work reported here. My thanks are also due to The Wellcome Trust for a travel grant which allowed the collaborative work on the NMR determinations to take place.

REFERENCES

[1]   C. R. Woese, Bacterial evolution, *Microbiol. Rev.* 51: 221 (1987)
[2]   D. J. Kushner and M. Kamekura, "Physiology of halophilic eubacteria", *in*: Halophilic Bacteria, Vol. 1., F. Rodriguez-Valera, ed. CRC Press, Boca Raton (1988)
[3]   D. Rafaeli-Eshkol, Studies on halotolerance in a moderately halophilic bacterium, *Biochem. J.* 109: 679 (1968)
[4]   R. M. Wydro, W. Madira, T. Hiramatsu, M. Kogut and D. J. Kushner, Salt-sensitive *in vitro* protein synthesis by a moderately halophilic bacterium, *Nature (London)* 269: 824 (1977)
[5]   M. Kamekura, Production and function of enzymes of eubacterial halophiles, *FEMS Microbiol. Rev.* 39: 145 (1986)
[6]   D. J. Kushner, What is the true internal environment of halophilic and other bacteria? *Can. J. Microbiol.* 34: 481 (1988)

[7]   D. M. Schindler, R. M. Wydro and D. J. Kushner, Cell-bound cations of the moderately halophilic bacterium *Vibrio costicola*, *J. Bacteriol.* 130: 698 (1977)

[8]   F. W. Cope, Nuclear magnetic resonance evidence for complexing of sodium ions in muscle, *Proc. Natl. Acad. Sci., U.S.A.* 54: 225 (1965)

[9]   T. Ogino, J. A. Den Hollander and R. G. Shulman, ^{39}K, ^{23}Na and ^{31}P NMR studies of ion transport in *Saccharomyces cerevisiae*, *Proc. Natl. Acad. Sci., U.S.A.* 80: 5185 (1983)

[10]   M. Shporer and M. M. Civan, Effects of temperature and field strength on the NMR relaxation times of ^{23}Na in frog striated muscle, *Biochim. Biophys. Acta* 354: 291 (1974)

[11]   M. M. Civan, H. Degani, Y. Margalit and M. Shporer, Observations of ^{23}Na in frog skin by NMR, *Am. J. Physiol.* 245: C213 (1983)

[12]   P. G. Morris, NMR spectroscopy in living systems, *Ann. Rep. NMR Spectr.* 20:1 (1988)

[13]   M. Goldberg and H. Gilboa, Sodium exchange between two sites, the binding of sodium to halotolerant bacteria, *Biochim. Biophys. Acta* 538: 268 (1978)

[14]   B. A. Wittenberg and R. K. Gupta, NMR studies of intracellular sodium ions in mammalian cardiac myocytes, *J. Biol. Chem.* 260: 2031 (1985)

[15]   A. M. Castle, R. M. MacNab and R. G. Shulman, Measurement of intracellular sodium transport in *Escherichia coli* by ^{23}Na nuclear magnetic resonance, *J. Biol. Chem.* 261: 5288 (1986)

[16]   H. Gilboa, R. Regev, M. Kogut, N. J. Russell and Y. Avi-Dor, Nuclear magnetic resonance to measure intracellular sodium in a halophilic bacterium, submitted for publication.

[17]   J. H. B. Christian and J. Waltho, Solute concentrations within cells of halophilic bacteria, *Biochim. Biophys. Acta* 65: 506 (1962)

[18]   A. D. Brown, Microbial water stress, *Bacteriol. Rev.* 40: 803 (1976)

[19]   A. D. Brown, K. F. Mackenzie and K. K. Singh, Selected aspects of microbial osmoregulation, *FEMS Microbiol. Rev.* 39: 31 (1986)

[20]   J. F. Imhoff and F. Rodriguez-Valera, Betaine is the main compatible solute of halophilic eubacteria, *J. Bacteriol.* 160: 478 (1984)

[21]   J. F. Imhoff, Osmoregulation and compatible solutes in eubacteria, *FEMS Microbiol. Rev.* 39: 57 (1986)

[22]   E. Galinski and H. G. Trüper, Betaine, a compatible solute in the extremely halophilic phototrophic bacteria *Ectothiorhodospira halochloris*, *FEMS Microbiol. Lett.* 13: 357 (1982)

[23]   R. Regev, I. Pery, H. Gilboa and Y. Avi-Dor, ^{13}C-NMR study of the interrelation between synthesis and uptake of compatible solutes in moderately halophilic eubacteria, *Arch. Biochem. Biophys.* 278: 106 (1990)

[24]   N. J. Russell and M. Kogut, Haloadaptation: salt-sensing and cell envelope changes, *Microbiol. Sci.* 2: 345 (1985)

[25]   M. Kogut and N. J. Russell, Life at the limits: considerations on how bacteria grow at extremes of temperature and pressure or with high concentrations of salts and solutes, *Sci. Progr. Oxford* 71: 381 (1987)

[26]   N. J. Russell, Adaptive modifications in membranes of halotolerant and halophilic microorganisms, *J. Bioenerg. Biomembr.* 21: 93 (1989)

[27]   J. B. Stock, B. Rauch and S. Roseman, Periplasmic space in *Salmonella typhimurium* and *Escherichia coli*, *J. Biol. Chem.* 252: 7850 (1977)

[28]   M. Kogut and N. J. Russell, unpublished results.

[29]   F. P. Takacs, T. I. Matula and R. A. MacLeod, Nutrition and metabolism of marine bacteria XIII, Intracellular concentrations of sodium and potassium ions in a marine Pseudomonad, *J. Bacteriol.* 87: 50 (1964)

EFFECTS OF SALINITY ON MEMBRANE LIPIDS AND

MEMBRANE-DERIVED OLIGOSACCHARIDES

Nicholas J. Russell and Rachel L. Adams

Department of Biochemistry
University of Wales
P.O. Box 903
Cardiff CF1 1ST
Wales, U. K.

ABSTRACT

A common feature of the adaptive response to elevated external salinity in microorganisms is an increase in the proportion of negatively-charged lipids in their membrane(s). In Gram-negative bacteria the change is usually a rise in phosphatidylglycerol content relative to phosphatidylethanolamine, a modification that is believed to be part of the mechanism for ensuring correct membrane function under conditions of high osmolarity. The periplasmic space in Gram-negative bacteria contains membrane-derived oligosaccharides (MDO), comprised of glucose residues substituted with phosphoglycerol and phosphoethanolamine residues that are derived from phosphatidylglycerol and phosphatidylethanolamine respectively. In *Escherichia coli* the concentration of MDO depends on the external osmotic pressure, and it has been postulated that they are part of the osmoregulatory response. This paper describes the metabolic interrelationships between MDO and membrane phospholipids in the context of the known salt-dependent changes in lipid composition of halotolerant and halophilic eubacteria and considers whether this supports the proposed role of MDO in osmoregulation of these bacteria.

INTRODUCTION

It is likely that a microorganism first senses the effects of a change in external salinity at the membrane [1], a supposition that is supported by the common observation that microbial membrane lipid composition depends on NaCl (and other solute) concentration of the culture medium [2, 3]. It is believed that the changes in lipid composition preserve appropriate membrane function [4], an important feature of which is the maintenance of the internal ionic and solute environment. It is significant, therefore, that a major aspect of osmoregulation is the accumulation of compatible solutes, which may be taken up from the external medium and/or synthesised *de novo* [5]; whichever route is taken the appropriate level of compatible solutes must be retained inside the cell; although in some microorganisms there may be considerable leakage of compatible solutes from the cell [6], growth in high solute concentrations will depend on proper membrane function.

The critical membrane in this respect is the plasma (cytoplasmic) membrane, but the situation in Gram-negative bacteria is complicated by the fact that, in addition to having a plasma membrane, they are surrounded by the so-called outer membrane, the space between the plasma and outer membranes being known as the periplasm [7, 8]. The plasma membrane has no great mechanical strength and is osmotic whereas the outer membrane is mechanically stronger but is not osmotic. Thus, if a Gram-negative bacterium was placed in medium of low osmolarity, the plasma membrane would swell unless the periplasm contained substances to compensate for the difference in osmotic pressure between the inside and outside of the cell. It has been postulated that this is a function of MDO, which generally comprise about 1% of the cellular dry weight and are retained by the outer membrane unless its permeability is impaired [9, 10]. It is argued that the MDO not only balance the osmolarity between cytoplasm and periplasm but also provide fixed anions in the periplasm, thereby creating a Donnan membrane potential across the outer membrane that in turn leads to the concentration of protons and cations in the periplasm [11]. When the external osmolarity is raised, compatible solutes are accumulated in the cytoplasm in order to balance the osmotic pressure difference across the plasma membrane [12, 13]; thus MDO are no longer required in the periplasm and their level drops by 16-fold [9]. Therefore, the role of MDO in osmoregulation proposed for *E. coli* is to protect the bacteria at low osmolarities, rather than as an adaptation for growth in high solute concentrations.

E. coli is not particularly halotolerant and most strains do not tolerate > 0.5 M NaCl [14]. The problem for these bacteria is not so much to protect themselves against low solute concentrations but rather to adapt for conditions of high osmolarity. It is of interest, therefore, to know whether halotolerant and halophilic bacteria contain MDO and if so whether their amounts are regulated in response to external osmolarity, particularly in view of the known salt-dependent changes in composition of phospholipids whose metabolism is linked to MDO synthesis (see below).

STRUCTURE AND METABOLISM OF MDO IN *E. COLI*

The MDO are a family of anionic oligosaccharides containing on average 10-12 glucose residues joined by the ß-1,2 and ß-1,6 glycosidic bonds to give compact, highly-branched molecules with M_r values around 2400 [15]. The oligosaccharide chains, possibly linked to a lipid carrier, are built up from UDP-glucose. It is one of the early stages of polymerisation or initiation of synthesis of the oligosaccharide chains that is osmoregulated, but the enzyme has not been identified. The anionic

226

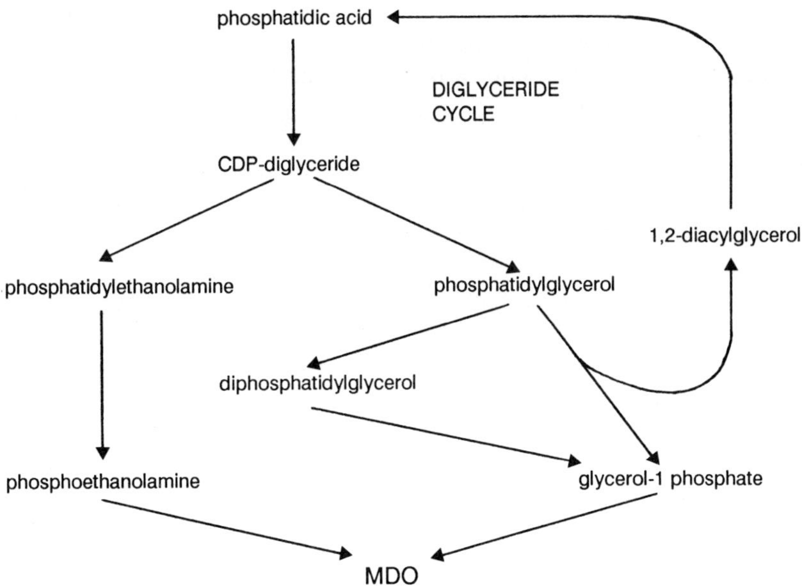

Figure 1. The interrelationships of phospholipid metabolism and MDO synthesis showing the diglyceride cycle.

character of MDO is due to substitution mainly with sn-1-glycerol phosphate residues ($\approx$3 per molecule), together with smaller amounts of phosphoethanolamine (<1 per molecule) and sometimes O-succinyl ester residues [15]. The sn-1-glycerol phosphate and phosphoethanolamine residues are derived from phosphatidylglycerol and phosphatidylethanolamine respectively [1516 16]. It is this fact which links the metabolism of MDO with that of membrane phospholipids (summarised in Figure 1).

In *E. coli* the major phospholipid is phosphatidylethanolamine (70-80%), which is metabolically stable, in contrast with phosphatidylglycerol (15-20%), which is metabolically active [17]. In *E. coli* [2-³H] glycerol specifically labels cell products such as MDO and phospholipids which contain glycerol residues, because the ³H label is lost during metabolism via dihydroxyacetone phosphate to other compounds. Thus it was possible to show in pulse-chase labelling experiments that during the synthesis of MDO there is transfer of glycerophosphate from phosphatidylglycerol [18], which is catalysed by a plasma-membrane-bound phosphoglycerol transferase I that has been partially characterised [19]; a second transferase enzyme in the periplasm shuffles phosphoglycerate residues between MDO molecules [15]. A similar reaction transfers phosphoethanolamine residues from phosphatidylethanolamine to MDO but at a much lower frequency [16], which reflects the metabolic stability of this phospholipid in *E. coli*. In mutants unable to transfer phosphoglycerol there is a corresponding increase in phosphoethanolamine transfer, which maintains the anionic nature of MDO [20].

When phosphatidylglycerol and phosphatidylethanolamine act as donors of sn-1-glycerol phosphate and phosphoethanolamine to MDO, sn-1,2-diacylglycerol is released during the transfer reaction (Figure 1). This is normally phosphorylated by diglyceride kinase in the plasma membrane giving phosphatidic acid which is a precursor of cellular phospholipids, to complete the so-called "diglyceride cycle" [21]. Although this is only a minor route of phospholipid formation in *E. coli*, experiments with mutants show that production of diacylglycerol is obligatorily linked to MDO formation. Moreover, the turnover of phospholipids, phosphatidylglycerol in particular, in *E. coli* results largely from MDO synthesis (Figure 1).

OCCURRENCE OF MDO IN OTHER GRAM-NEGATIVE BACTERIA

Nearly all studies of MDO have used *E. coli* as the test organism. Miller et al. [22] have demonstrated that the synthesis of cyclic (1→2)-ß-D-glucans in the related Gram-negative bacteria *Rhizobium trifolii* and *Agrobacterium tumefaciens* is also osmoregulated. The structure and size of these cyclic glucans is similar to that of MDO and they are also located in the periplasm. This provides further, albeit circumstantial, evidence for the proposed role of MDO in osmoregulation.

We have detected MDO in the moderate halophile *Vibrio costicola* and in four halotolerant food-spoilage bacteria, two *Pseudomonas* spp. and two *Vibrio* spp. [23]. The total amounts of MDO in all species are up to 10-fold less than in *E. coli* and there is no consistent pattern in their response to changes in external osmolarity; moreover the labelling of MDO from radioactive precursors in response to alterations in the concentration of different solutes is not uniform and varies between species. Although we do not know the basis of these differences, they do not seem to be related to culture medium osmolarity; instead they could reflect solute-concentration-dependent changes in the metabolism of phospholipids which varies between species and solutes, and so could cause the observed differences in MDO metabolism.

SALT-DEPENDENT CHANGES IN LIPID COMPOSITION OF HALO-TOLERANT AND HALOPHILIC EUBACTERIA

The effect of NaCl concentration on lipid composition has been investigated in a variety of halotolerant and halophilic bacteria, both Gram-positive and Gram-negative; despite the fact that the different species display a wide range of lipid compositions, the common response to elevated salinity is an increase in the relative proportion of anionic membrane lipids [2, 4]. The Gram-negative species which have been investigated all belong to the genera *Pseudomonas* and *Vibrio*, which have relatively simple lipid compositions comprising phosphatidylethanolamine and phosphatidylglycerol with small amounts of diphosphatidylglycerol [25]. These bacteria respond to a rise in external osmolarity by increasing the proportion of phosphatidyl-glycerol, usually at the expense of phosphatidylethanolamine, thus increasing the amount of anionic compared with zwitterionic membrane lipid [2, 3]. It was originally believed that the reason for this change was to give charge balance at the membrane surface when the Na^+ concentration was elevated, but it has been reasoned that the amount of phosphatidylglycerol involved could make only a small contribution [1].

228

Instead, it is believed that the reason for the changes is to preserve the lipid bilayer conformation at high salinities. Such a mechanism is necessary because a lipid such as phosphatidylethanolamine, when dispersed in solution on its own, will form non-bilayer phases in high solute concentrations, but this can be prevented by mixing it with a bilayer-forming lipid such as phosphatidylglycerol [26]. In a biophysical study of purified phospholipids extracted from *V. costicola*, we have shown that the increased proportion of phosphatidylglycerol at higher salinities does prevent the formation of non-bilayer phases in lipid mixtures which mimic the ratio of phosphatidylglycerol to phosphatidylethanolamine in native membranes [27]. Thus it does appear that the common observation that halotolerant and halophilic bacteria modify their membrane lipid compositions at high salinities is a mechanism for ensuring that their membranes retain the appropriate properties, including selective permeability which is destroyed by the formation of non-bilayer phases.

IS THE METABOLISM OF MDO AND LIPIDS PART OF THE OSMOREGULATORY RESPONSE?

It appears that the formation of MDO in bacteria such as *E. coli* is a mechanism to protect them against low osmolarity media and that they are not required for growth at higher osmolarities (e.g. in 0.4 M NaCl). Since halotolerant/halophilic bacteria can grow at much higher salinities, and some halophiles not at all in 0.4 M NaCl, it could be argued that MDO would be present in only small amounts or not at all. This argument is supported by our finding of MDO in four halotolerant Gram-negative bacteria in amounts considerably less than in *E. coli*. However, the moderate halophile *V. costicola* also contains MDO and, although the amounts are again relatively small compared with *E. coli*, they do not vary significantly with external osmolarity. The data indicate that MDO do not have a positive role in osmoregulation in halophilic bacteria. It should be remembered that the osmoregulatory role of MDO in enteric bacteria has not been proven categorically and indeed has been questioned [20]. It has not been explained why most of the enzymes of MDO synthesis are constitutive, even during growth at high osmolarity, and *mdoA* mutants of *E. coli* which lack MDO do not have any obvious phenotype [15]. It has been suggested that there might be other roles for MDO including the stabilisation of membranes [20].

The presence of MDO in halotolerant and halophilic Gram-negative bacteria may reflect their evolutionary origins from non-halophilic ancestors. The outer membrane is selectively permeable and allows passage of small molecules such as NaCl and other low molecular weight solutes encountered by such bacteria [7]. Thus, the osmoregulatory problem of life in high solute concentrations is to prevent water loss from the cytoplasm and the consequent cell shrinkage, which halotolerant/halophilic bacteria accomplish by accumulating compatible solutes. They also modify their membrane lipid composition in order to preserve the lipid bilayer conformation and correct membrane function. Since this response generally involves an increase in phosphatidylglycerol synthesis, one might predict that this would stimulate MDO synthesis but this does not appear to occur in *V. costicola*. It remains to be determined whether this is because there are different metabolic pathways in halophiles, but this should be worthwhile investigating since it could shed further light on the mechanism(s) of osmoregulation.

REFERENCES

[1]   N. J. Russell and M. Kogut, Haloadaptation: salt-sensing and cell-envelope changes, *Microbiol Sci.*, 2: 345 (1985)

[2]   N. J. Russell, Environmentally-induced changes in lipid composition and membrane function, *in* "Homeostatic Mechanisms in Micro-organisms," R. Whittenbury, G. W. Gould, J. G. Banks and R. G. Board, eds, Bath University Press, Bath, England (1988)

[3]   J. A. Bygraves and N. J. Russell, Solute tolerance and membrane lipid composition in some halotolerant food-spoilage bacteria, *Food Microbiol.* 5: 109 (1988)

[4]   N. J. Russell, Adaptive modifications in membranes of halotolerant and halophilic microorganisms, *J. Bioenerg. Biomembr.* 21: 93 (1987)

[5]   J. F. Imhoff, Osmoregulation and compatible solutes in eubacteria, *FEMS Microbiol. Rev.*, 39: 57 (1986)

[6]   A. D. Brown, K. F. Mackenzie and K. K. Singh, Selected aspects of microbial osmoregulation, *FEMS Microbiol. Rev.*, 39: 31 (1986)

[7]   B. Lugtenberg and L. Alphen, Molecular architecture and functioning of the outer membrane of *Escherichia coli* and other Gram-negative bacteria, *Biochim. Biophys. Acta*, 737: 51 (1983)

[8]   H. Nikaido and M. Vaara, Molecular basis of outer membrane permeability, *Microbiol. Rev.*, 49: 1 (1985)

[9]   E. P. Kennedy, Osmotic regulation and the biosynthesis of membrane-derived oligosaccharides in *Escherichia coli*, *Proc. Natl. Acad. Sci. USA* 79: 1092 (1982)

[10]   E. P. Kennedy and M. K. Rumley, Osmotic regulation of biosynthesis of membrane-derived oligosaccharides in *Escherichia coli*, *J. Bacteriol.* 170: 2457 (1988)

[11]   J. B. Stock, B. Rauch and S. Roseman, Periplasmic space in *Salmonella typhimurium* and *Escherichia coli*, *J. Biol. Chem.*, 252: 7850 (1977)

[12]   W. Epstein, Osmoregulation by potassium transport in *Escherichia coli*, *FEMS Microbiol. Rev.*, 39: 73 (1986)

[13]   C. F. Higgins and I. R. Booth, Molecular mechanisms of osmotic regulation: an integrated homeostatic response, *in* "Homeostatic Mechanisms in Micro-organisms," R. Whittenbury, G. W. Gould, J. G. Banks and R. G. Board, eds, Bath University Press, Bath, England, (1988)

[14]   F. Rodriguez-Valera, The ecology and taxonomy of aerobic chemo-organotrophic halophilic eubacteria, *FEMS Microbiol. Lett.* 39: 17 (1986)

[15]   E. P. Kennedy, Membrane-derived oligosaccharides, *in:* "*Escherichia coli* and *Salmonella typhimurium*: Cellular and Molecular Biology, Vol. 1", F. C. Neidhardt, J. L. Ingraham, K. B. Low, B. Magasanik, M. Schaechter and H. E. Umbarger, eds, American Society for Microbiology, Washington, D.C. (1987)

[16]   K. Miller and E. P. Kennedy, Transfer of phosphoethanolamine residues from phosphatidylethanolamine to the membrane-derived oligosaccharides of *Escherichia coli*, *J. Bacteriol.*, 169: 682 (1987)

[17]   C. R. H. Raetz, Enzymology, genetics and regulation of membrane phospholipid synthesis in *Escherichia coli*, *Microbiol. Rev.*, 42: 614 (1978)

[18]   L. M. G. van Golde, H. Schulman and E. P. Kennedy, Metabolism of membrane phospholipids and its relation to a novel class of oligosaccharides in *Escherichia coli*, *Proc. Natl. Acad. Sci. USA* 70: 1368 (1973)

[19]   B. J. Jackson and dpk, The biosynthesis of membrane-derived oligosaccharides, *J. Biol. Chem.*, 258: 2394 (1982)

[20]   W. Fiedler and H. Rotering, Characterization of an *Escherichia coli mdoB* mutant strain unable to transfer *sn*-1-phosphoglycerol to membrane-derived oligosaccharides, *J. Biol. Chem.*, 260: 4799 (1985)

[21]   C. R. H. Raetz and K. F. Newman, Diglyceride kinase mutants of *Escherichia coli*: inner membrane association of 1,2-diglyceride and its relation to synthesis of membrane-derived oligosaccharides, *J. Bacteriol.*, 137: 860 (1979)

[22]   K. J. Miller, E. P. Kennedy and V. N. Reinhold, Osmotic adaptation by Gram-negative bacteria: possible role for periplasmic oligosaccharides, *Science*, 231: 48 (1986)

[23]   R. L. Adams, J. A. Bygraves and N. J. Russell, submitted for publication (1990)

[24]   N. J. Russell, Adaptive modificatons in membranes of halotolerant and halophilic microorganisms, *J. Bioenerg. Biomembr.*, 21: 93 (1989)

[25]   J. L. Harwood and N. J. Russell, "Lipids in Plants and Microbes," George Allen and Unwin, London (1984)

[26]   P. R. Cullis and B. de Kruijff, Lipid polymorphism and the functional role of lipids in biological membranes *Biochim. Biophys. Acta*, 559: 399 (1979)

[27]   G. S. Sutton, P. J. Quinn and N. J. Russell, *Biochem. Soc. (U.K.) Trans.*, (in press)

PHYSIOLOGY OF *HALOMONAS ELONGATA*

IN DIFFERENT NaCl CONCENTRATIONS

R. H. Vreeland[1], S. L. Daigle[1], S. T. Fields[2], D. J. Hart
and E. L. Martin[2]

[1]Dept. of Biology
West Chester University
West Chester, Pennsylvania 19383
United States of America

[2]School of Life Sciences
University of Nebraska
Lincoln
Nebraska 68588
United States of America

ABSTRACT

Several recent physiological studies of *Halomonas elongata* have shown that
this bacterium makes a variety of physiological changes in response to the NaCl
concentration of its growth medium. This previous research has indicated that *H.
elongata* responds to NaCl by tightening its cell wall structure and shifting from a
hydrophobic cell surface to one with a more hydrophilic character. In addition,
experiments on membrane transport and uptake have shown that the cell permeases
adapt to the specific NaCl concentration in which the organisms are grown.

Further examination of this bacterium has centered on specific fatty acid and
protein changes. The data produced from these studies show a variety of salt related
variations. These changes include an increase in the total fatty acid content of the cells
as well as differences in the type and proportion of fatty acids present. Cells grown in
high salts possess increased saturated and hydroxy fatty acids. The types of changes
noted are consistent with previous data indicating an overall tightening of the cell walls
in response to NaCl.

General and Applied Aspects of Halophilic Microorganisms
Edited by F. Rodriguez-Valera, Plenum Press, New York, 1991

A concomittant study of the protein profiles of this bacterium has shown that the organism also produces various salt related proteins. Whole cell profiles have shown that low salt grown cells possess several unique protein groups which appear to be either absent or modified in high salt grown cells. Preliminary immunological examination indicates that at least some of these altered proteins are present on the outer surface of the cells. Altogether the data indicate that *H. elongata* maintains its salt tolerance through the interaction of a number of cell systems.

INTRODUCTION

The halotolerant bacterium *Halomonas elongata* [1] has been the subject of a number of physiological investigations. These previous studies concentrated on the physiology of the cell wall and transport mechanisms of this organism and how these systems changed in response to NaCl. Vreeland et al [2] and Hart and Vreeland [3] have shown that the cell wall becomes significantly tighter and more hydrophilic as the cells adapt to increasing NaCl. Compared to low salt grown cultures cells that have adapted to high salts possess increased levels of cardiolipin and phosphatidyl glycerol and less phosphatidylethanolamine. As would be predicted from this phospholipid profile high salt grown cells are extremely water loving and maintain this hydrophilicity throughout their growth cycle [3].

Similar changes have been noted for the transport processes of this organism [4]. These authors showed that the transport system of *H. elongata* specifically required the Na^+ cation for activity. This same study also demonstrated that the L-alanine transport system of *H. elongata* was similar to that of *Alteromonas haloplanktis* [5] in that it was competetively inhibited by glycine, D- or L-serine, D-alanine, and L-homoserine.

The data produced to date indicate that *H. elongata* responds to salt in a variety of ways. The present paper describes an additional series of experiments on the fascinating series of physiological changes made by this euryhaline bacterium. The text is actually divided into two parts, the first part concentrates on a series of physiological studies while the second part concentrates primarily on more detailed characterization of the cells' transport mechanisms.

RESULTS AND DISCUSSION

Figure 1 presents a typical protein profile from cells grown on complex medium in three different NaCl concentrations. Examination of this profile shows a variety of proteins that are specific to the various salt concentrations. Low salt grown cells produce at least two unique proteins with molecular weights of 45,000 and 60,000 Daltons (arrows A & B, Figure 1). In addition the profiles from cells grown in 1.37 (1.3 in Figure 1) and 3.4 M show that these cells are each missing several groups of proteins with high molecular weights (arrows C & D respectively, Figure 1). The exact function and locations of all of these various proteins is still being investigated. There are however indications that at least some of them are located on the surface of the cells.

234

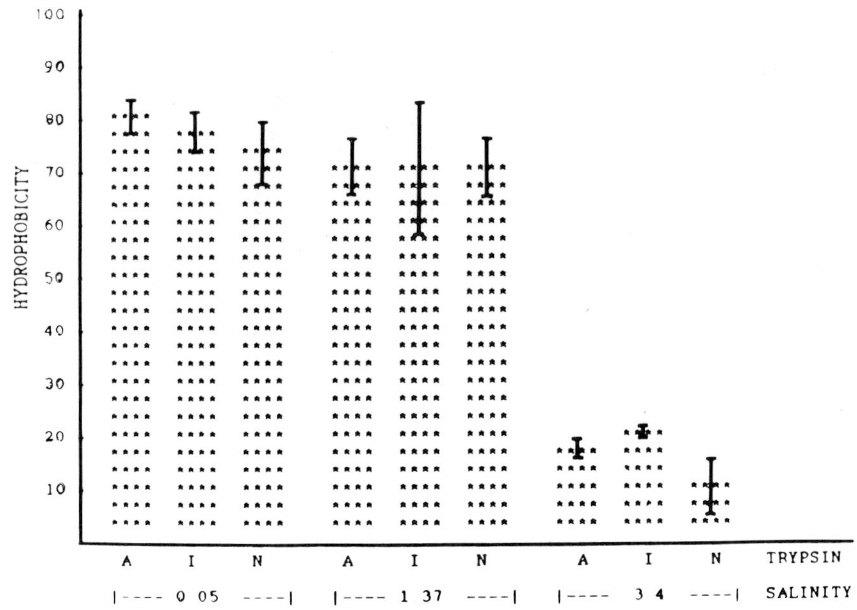

Figure 1. Whole cell protein profiles of *Halomonas elongata* following growth in complex media containing different concentrations of NaCl.
A = Active Trypsin.
I = Heat-inactivated trypsin.
N = No trypsin added.

Table 1. Serum Cross Reactivity of *Halomonas elongata* grown in various NaCl concentrations[a]

NaCl in Growth Media (M)	Anti-0.05M Serum	Anti-1.37M Serum	Anti-3.4M Serum
0.05	1:16	0	0
0.375	1:32	1:32	0
1.37	1.32	1:32	0
2.5	1:32	1:64	1:128
3.4	1:32	1:64	1:12

[a]The cells used to produce the anti-serum were grown in complex medium at the indicated salt concentration then injected without adjuvant into New Zealand white rabbits. The cross reactivity of each anti-serum was then determined using agglutination of whole cells grown in the same complex media having the NaCl concentrations indicated in the first column.

235

Table 1 shows the results of a preliminary series of immunization experiments in which rabbits were immunized (without adjuvants) to live cells of *H. elongata* grown in different NaCl concentrations. When these anti-sera were cross titered the data showed that while anti-serum prepared against low salt grown cells reacted with cells grown in all salt concentrations the high salt anti-sera was specific for high salt grown cells. These data can be interpreted in two ways, 1) either the low salt cells posessed a unique antigen, or 2) some type of conformational change in the surface antigens had caused these rabbits to respond to a determinant that was not exposed in low salt grown cells. Interestingly, these antigenic and protein differences correlate well with the observed hydrophilicity (Figure 2).

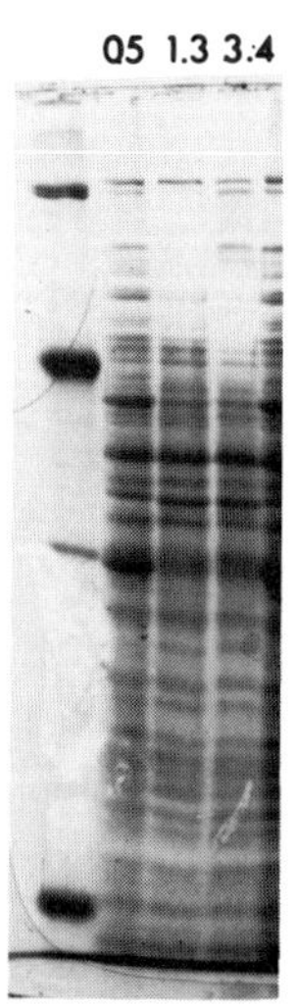

Figure 2.   Surface hydrophobicity of *Halomonas elongata* following exposure to Trypsin.

The data in Figure 2 show the hydrophobicity of *H. elongata* cells from various NaCl concentrations following exposure to trypsin. These data indicate that the cell surface hydrophobicity of this organism arises from either non-trypsin sensitive proteins or from other cell wall constituents. This type of response differs from that seen with many other microorganisms where trypsin treatment converted hydrophobic cells to hydrophilic forms [6, 7, 8, 9, 10, 11]. However the pattern of increasing hydrophilicty in cells that are deficient in several proteins (such as 3.4 M *H. elongata* Figure 1) has been seen in non-halophilic bacteria [7, 8]. This increased hydrophilicty is also consistent with the data on wall tightening and phospholipid changes presented by Vreeland et al [2] and is also indicated by changes in the fatty acid profiles of this bacterium (Table 2).

Table 2 shows the effect of medium changes on some of the fatty acids found in *H. elongata*. There are several significant changes apparent in these data, most notable of which are the increased levels of cyclic fatty acids and of the 18:1a fatty

Table 2. Effect of Medium NaCl on the concentration of various fatty acids in *Halomonas elongata.*

All values in mg/g dry weight			
0.05M NaCl	1.37M NaCl	3.4M NaCl	Shorthand Designation
0.13	N.D.	0.79	2-0H-12:0
N.D.	N.D.	1.15	2-0H-14:O
0.09	0.16	0.40	3-OH-14:0
0.11	0.27	0.25	a-15:0
0.13	0.32	0.54	a-17:0
4.37	2.29	1.98	16:1
1.04	2.39	2.31	18.1a
34.80	14.60	5.59	18:1b
4.49	0.45	6.55	cy 17:0
45.70	32.80	82.00	cy 19:0

N.D. = Not Detected

acid. This latter fatty acid (trans-9 octadecenoic) represents 0.75% of the total fatty acid produced by low salt grown cells while in high salt adapted cells it constitutes 1.54% of the total. This particular increase would be sufficient to allow the cell membrane to pack more efficiently and form a tighter structure. The other significant increase is that seen with the cyclo-propanoic (cy-17:0 and cy-19:0) fatty acids (Table 1). These two species, which would arise via conversion of the corresponding 16:1 and 18:1b species (M. Kates personal communication), would serve to maintain membrane fluidity and prevent salting out problems in high salts.

On a more functional level, *H. elongata* also makes several changes in its membrane transport systems. Table 3 shows data on the efficiency of the cell permeases for a variety of amino acids. These data show that the membrane transport systems of this organism increase their transport efficiency as the salt concentrations increase.

In this experiment transport efficiency for proline, alanine and lysine increased while the efficiency for glutamic acid declined. Transport of threonine was not significantly affected. This efficiency arises primarily from an increase in Vmax which compensates for a salt related decrease in the Km value. While there is as yet no data showing that the efficiently transported amino acids are directly involved in osmoregulation in this bacterium their increased levels of transport may reflect the cells' need for metabolizable substrates for use during adaptation.

Table 3. Efficiency of amino acid transport systems of *Halomonas elongata* in different NaCl concentrations.

Efficiency of transport expressed as Vmax/Km.		

Amino Acid	NaCl (M)		
	0.05	1.37	3.4
Proline	1.40	1.65	2.05
Glutamic Acid	1.61	0.31	0.25
Alanine	1.16	1.38	1.70
Lysine	0.46	0.68	0.83
Threonine	0.02	0.09	0.05

Table 4. Competitive inhibition of transport by other Amino acids.

Competitor (100x)	% Inhibition of labelled substrate		
	L-Proline	L-Glutamic Acid	L-Alanine
L-Proline	95	0	0
L-Glutamic acid	0	96	0
L-Alanine	0	0	91
D-Proline	10	0	0
D-Glutamic acid	0	95	0
D-Alanine	0	0	71
L-Serine	0	0	5
D-Serine	0	0	57
Glycine	0	0	90
L-Aspartic acid	0	0	0
L-Glutamine	0	0	0
L-Phenylalanine	0	0	0
L-Leucine	0	0	0
L-Lysine	0	0	0
L-Threonine	0	0	0
L-Hydroxyproline	56	ND	ND

ND = Not Done

Table 5.   Effect of Potential Electron Donors on Transport of L-Alanine by *Halomonas elongata*

Donor	Concentration	% Stimulation of Uptake
Ethanol	0.021 M	20
	0.21 M	30
	2.1 M	- 10[a]
Ascorbate/TMPD	20mM/150μM	37
Ascorbate/PMS	20mM/100μM	- 5
Succinate	25 mM	43
Malate	25 mM	- 28
NADH	10 mM	49

[a] = Negative values indicate inhibition of transport

Finally the highly efficient proline transport (Table 3) may also be a reflection of the cells' need for the newly discovered amino acid Ectoine which has been found in high levels in this bacterium (Trüper et al., this volume).

Apparently, recent studies on the halophilic photosynthetic bacterium *Ectothiorhodospira halophila* have shown that the proline transport systems of this bacterium will preferentially transport ectoine in high salt concentrations (H. Trüper and J. Imhoff, personal communication).

Table 4 shows the amino acid competition pattern for the proline, glutamic acid and alanine transport systems of *H. elongata* grown in 1.37M NaCl. The fact that most of these amino acids failed to competitively inhibit transport suggests that the bacterium posesses several distinct and highly specific transport systems. In this organism only the alanine system showed a rather broad specificity. In this respect *H. elongata* appears to be similar to *Alteromonas haloplanktis* [5].

The transport systems of *H. elongata* are also similar to other bacterial systems in their response to various electron donors (Table 5). Compounds such as ethanol, ascorbate–TMPD, succinate and NADH all stimulated transport of L-alanine into the cells. These data correlate well with similar information from *E. coli, Mycobacterium phlei, Bacillus subtilis* and *A. haloplanktis* [12, 13, 14, 15, 16, 17]. In view of the different levels of salt tolerance shown by these five bacteria the similarity in their transport systems is rather interesting. It is tempting to speculate that the salt tolerance of *H. elongata*'s transport system arises more from the physiological changes of the membrane environment than from any inherent tolerance of the transport enzymes.

The similarity in the membrane transport systems is also evident from the data in Table 6. This table shows the results of a series of experiments using various

Table 6. Effect of Metabolic Inhibitors on L-Alanine Transport by *halomonas elongata*

Inhibitor tested	Concentration (mM)	% Inhibition
KCN	1	61
	10	82
	100	95
$Na_2HASO_4.7h_2O$	1	10
	10	33
	100	50
HOQNO[a]	0.05	$- 23$[b]
2,4 Dintrophenol[a]	1	5
N-ehtylmaleimide[a]	1	68

[a] = compounds suspended in ethanol
[b] = negative values indicate stimulation of transport

transport inhibiting agents. The very strong inhibition demonstrated by the sulfhydral reagents NEM and IAA indicate that the transport enzymes require at least one functional SH group. Further inhibition of transport by cyanide provides strong evidence that alanine transport in *H. elongata* is energized by the cells' respiratory chain. This possibility is further supported by the fact that addition of exogenous electron acceptors to reaction mixtures containing inhibitors will at least partially alleviate inhibition (Data not shown).

SUMMARY

The data available to date show that *Halomonas elongata* is constitutively salt tolerant. Simply put this bacterium has evolved in ways that make its natural lifestyle one geared toward surviving in different NaCl concentrations. This salt tolerance arises from a fascinating interplay of salt stable systems which exhibit few changes (i.e. Peptidoglycan) and salt affected systems which show extensive salt mediated alterations (i.e. Proteins, Fatty acids). Further, the data presented here show that this bacterium has been very adept at using common "normal" biological systems (i.e. Transport enzymes) in ways that allow these systems to function under a variety of different conditions. Whether this extra tolerance involves incorporating minor structural modifications, the use of isozymes, or some type of modification of the molecular environment allowing non-tolerant systems to function in extreme conditions remains to be elucidated.

REFERENCES

[1]   R. H. Vreeland, C. D. Litchfield, E. L. Martin and E. Elliot. *Int. J. Syst. Bacteriol.* 30:485 (1980)

[2]   R. H. Vreeland, R. Anderson and R. G. E. Murray, *J. Bacteriol.* 160:879 (1984)

[3]   D. J. Hart and R. H. Vreeland, *J. Bacteriol.* 170:132 (1988)

[4]   E. L. Martin, T. Duryea, R. H. Vreeland, L. A. Hilsabeck and C. Davis. *Can. J. Microbiol.* 29:407 (1983)

[5]   J. E. Fein and R. A. MacLeod, *J. Bacteriol.* 170:132 (1988)

[6]   L. M. Hesketh, J. E. Wyatt and P. S. Handley, *Microbios* 50:131 (1987)

[7]   B. C. McBride, M. Song, B. Krasse and J. Olsson, *Infect. Immun.* 44:68 (1984)

[8]   H. Miorner, P. Albertsson and G. Kronvall. *Infect. Immun.* 36:336 (1983)

[9]   E. J. Morris, N. Ganeshkumar and B. C. McBride. *J. Bacteriol.* 164:255 (1985)

[10]   I. Ofek, E. Whitnack and E. H. Beachey, *J. Bacteriol.* 154:139 (1983)

[11]   F. Reifsteck, S. Wee and B. J. Wilkinson, *J. Med. Microbiol.* 24:65 (1987)

[12]   H. Hirata, A. Asano and A. Brodie. *Biochem. Biophys. Res. Commun.* 44:368 (1971)

[13]   H. R. Kaback and E. M. Barnes, *J. Biol. Chem.* 17: 5523 (1971)

[14]   W. Konings and E. Freese *FEBS Lett.* 14:65 (1971)

[15]   G. D. Sprott and R. A. MacLeod, *Biochim. Biophys. Res. Commun.* 47:838 (1972)

[16]   G. D. Sprott and R. A. MacLeod, *J. Bacteriol.* 117:1043 (1974)

[17]   J. Thompson and R. A. MacLeod, *J. Bacteriol.* 120:598 (1974)

DO PERIPLASMIC OLIGOSACCHARIDES PROVIDE A ROLE IN THE

OSMOTIC ADAPTATION OF GRAM-NEGATIVE BACTERIA?

Karen J. Miller

Department of Food Science
The Pennsylvania State University
111 Borland Laboratory
University Park, PA 16802
USA

ABSTRACT

A wide diversity of Gram-negative bacteria has been shown to strictly osmoregulate the biosynthesis of anionic periplasmic oligosaccharides. Specifically, the concentrations of these oligosaccharides and their counterions increase when cells are grown in media of low osmolarity. Based on the amounts of these oligosaccharides, their presence should substantially influence periplasmic osmotic strength and volume. In addition, the development of a Donnan potential across the outer membrane of Gram-negative bacteria should also derive largely from the concentrations of these anionic periplasmic oligosaccharides. This paper provides a review of several studies which indicate that the cell-envelope structure of Gram-negative bacteria is influenced greatly by the presence of periplasmic oligosaccharides. The possible role for these compounds in the osmotic adaptation of Gram-negative bacteria is also considered.

INTRODUCTION

The study of the microbial response to osmotic stress historically has focused upon adaptation to extracellular environments of high osmolarity. Under such conditions, a variety of low molecular weight solutes has been found to accumulate within the cytoplasmic compartment. These accumulated solutes have been termed "compatible" solutes [1] because of their innocuous effects on cytoplasmic metabolic activities, even at high concentrations. The accumulation of these compatible solutes is presumed to provide two functions for the organism: (1) restoration of cytoplasmic volume and (2) restoration of an appropriate level of turgor pressure across the cytoplasmic membrane.

General and Applied Aspects of Halophilic Microorganisms
Edited by F. Rodriguez-Valera, Plenum Press, New York, 1991

At the other extreme of osmotic stress, is the adaptation of microorganisms to extracellular conditions of low osmolarity. Upon exposure to a very dilute environment, there is an immediate tendency for water to flow into the microbial cell. Although the presence of a cell wall will prevent lysis of the microorganism under such conditions, it is likely that both cytoplasmic volume as well as turgor pressure across the cytoplasmic membrane are increased significantly. Growth under such conditions may not be optimal. Thus, it is possible that microbial adaptation to extracellular conditions of low osmolarity may also involve the maintenance of appropriate cytoplasmic volume and cytoplasmic membrane turgor pressure values.

In 1982 Kennedy proposed that periplasmic oligosaccharides synthesized by *Escherichia coli* may provide a role in the adaptation of this Gram-negative bacterium to extracellular environments of low osmolarity [2]. More recently, Miller and Kennedy have proposed that periplasmic oligosaccharides may provide a general role for the osmotic adaptation of a wide variety of Gram-negative bacteria [3]. The present paper reviews accumulating evidence which supports the role for periplasmic oligosaccharides in the osmotic adaptation of Gram-negative bacteria.

PERIPLASMIC OLIGOSACCHARIDES OF *ESCHERICHIA COLI*

Periplasmic oligosaccharides of gram-negative bacteria were first discovered by Van Golde and Kennedy during their studies of the phospholipid metabolism of *E. coli* [4]. During these studies, it was shown that the synthesis of periplasmic oligosaccharides by *E. coli* was linked to the turnover of membrane phospholipids. Because of this metabolic link, these oligosaccharides were termed the membrane-derived oligosaccharides (MDO).

The MDO of *E. coli* are composed of 8-10 glucose residues in a branched structure linked by ß-1,2 and ß-1,6 glycosidic bonds [5]. These compounds have a net negative charge of 5 due to the presence of anionic substituents. These substituents include phosphoglycerol, phosphoethanolamine and succinic acid. Phosphoglycerol and phosphoethanolamine substituents are respectively derived from the head groups of the phospholipids, phosphatidylglycerol and phosphatidylethanolamine [6, 7].

The biosynthesis of MDO is strictly osmoregulated [2]. Specifically, these oligosaccharides are synthesized only when cells are grown in low osmolarity, dilute media. Under these conditions, these oligosaccharides may amount to 7% of the total cellular dry weight [2]. Because the biosynthesis of MDO was found to be strictly osmoregulated, Kennedy proposed a possible role for these compounds in the osmotic adaptation of *E. coli* (see below).

POSSIBLE FUNCTIONS FOR THE PERIPLASMIC OLIGOSACCHARIDES IN THE OSMOTIC ADAPTATION OF *ESCHERICHIA COLI*

Based on the amounts of MDO synthesized by *E. coli* at low osmolarity, Kennedy proposed that these anionic oligosaccharides and their counterions would represent the major source of osmotically active solute within the periplasmic compartment [2]. In fact, concentrations were found to be sufficient to provide the periplasmic compartment with an osmotic strength similar to that of the cytoplasmic compartment. Furthermore, Kennedy's calculations have been found to be consistent with measurements performed by Stock and co-workers [8] and Sen et al. [9] that have indicated that the periplasmic compartment of *E. coli* is, in fact, isoosmotic with respect to the cytoplasm.

As a major anionic solute within the periplasmic compartment, it is likely that MDO concentrations will strongly influence the cell-envelope architecture of the cell. For example, MDO concentrations should impact periplasmic osmotic strength, periplasmic volume, as well as the development of a Donnan potential across the outer membrane of the cell [2]. The modulation of all of the above may represent important aspects of the adaptation of Gram-negative bacteria to environments of low osmolarity.

PERIPLASMIC OLIGOSACCHARIDES OF GRAM-NEGATIVE BACTERIA OTHER THAN *ESCHERICHIA COLI*

Structural similarities between the MDO of *E. coli* and a class of glucans produced by the plant-infective bacteria of the family *Rhizobiaceae*, prompted Miller and Kennedy to examine various genera of this family for the presence of periplasmic oligosaccharides [3]. Previous studies had reported the presence of cyclic ß-1,2-glucans within cultures of bacteria within this family. Specifically, these glucans were shown to be composed of 17 to 24 glucose residues in a cyclic structure, linked solely by ß-1,2 glycosidic bonds [10-15]. As reported in 1986 by Miller et al. [3], the cyclic ß-1,2 glucans of *Agrobacterium tumefaciens* were found to be localized within the periplasmic compartment of this organism [3]. Furthermore, the biosynthesis of these compounds was shown to be strictly osmoregulated, in the same manner as that for MDO biosynthesis. Additional studies have demonstrated that the cyclic ß-1,2-glucans of the *Rhizobiaceae* also are substituted with phosphoglycerol moieties derived from the head group of phosphatidylglycerol [16, 17]. Thus, the cyclic ß-1,2-glucans of the *Rhizobiaceae* share many similarities with the MDO of *E. coli*. As a result of these studies, it has been concluded that periplasmic oligosaccharides may provide a role in the osmotic adaptation of a wide variety of Gram-negative species, including enteric and soil bacteria [3, 16].

ANALYSIS OF MUTANTS DEFECTIVE FOR THE BIOSYNTHESIS OF PERI-PLASMIC OLIGOSACCHARIDES

A variety of mutants defective for the biosynthesis of periplasmic oligo-saccharides has been identified for both *E. coli* [18-20] and the *Rhizobiaceae* [21-23]. It is of interest, however, that these mutants are still able to grow in media of low osmolarity [18]. It would, thus, appear that the periplasmic oligosaccharides are not essential for the growth of Gram-negative bacteria in low osmolarity environments. It should be noted, however, that alternative adaptive responses to environments of low osmolarity may occur in such mutants.

245

The further analysis of the periplasmic oligosaccharide-deficient mutants of *E. coli* and the *Rhizobiaceae* has revealed that these compounds may provide a role in osmotic adaptation, perhaps through their influence on the cell-envelope architecture of the Gram-negative cell. For example, Holtje and co-workers [19] have demonstrated that MDO-deficient mutants of *E. coli* have a more pronounced separation between their cytoplasmic and outer membranes with fewer zones of adhesion as determined by electron microscopy. These mutants have also been shown to contain fewer flagella and are impaired for motility [20]. An additional characteristic of these MDO mutants noted by Fiedler and Rotering [20] was the apparent increased production of slime by these cells at 27ºC.

A phenotypic trait of the MDO-deficient mutants of particular interest was also reported by Fiedler and Rotering [20]. Specifically, the outer membrane porin composition of these mutants was found to be altered such that it resembled that of wild-type cells grown in a medium of high osmolarity. As a result of this observation, it was suggested that MDO itself or its biosynthesis may affect the osmotic regulation of porin biosynthesis (i.e. through an interaction with the inner membrane protein, EnvZ) [20]. Thus, it is possible that the presence of MDO within the periplasm may influence the structure and/or activity of certain inner membrane proteins.

The examination of cyclic ß-1,2-glucan deficient mutants of the *Rhizobiaceae* has revealed some striking parallels with the MDO mutants of *E. coli*. For example, cyclic glucan-deficient mutants of *Agrobacterium tumefaciens* and *Rhizobium meliloti* have been shown to possess a lower number of flagella and to be impaired for motility [21, 23]. In addition, these mutants have also been shown to produce higher amounts of extracellular polysaccharide [23]. Additional characteristics of these mutants have been revealed through the study of their interactions with higher plants. For example, cyclic ß-1,2-glucan deficient mutants of *Agrobacterium* are avirulent and unable to attach to plant cells [21], while mutants of *Rhizobium* are unable to induce the formation of effective nitrogen-fixing nodules on the roots of leguminous plants [22, 23].

CONCLUSIONS AND SUGGESTIONS FOR FUTURE STUDIES

As described above, several lines of evidence indicate that the presence of oligosaccharides within the periplasmic compartment of Gram-negative bacteria may greatly influence the cell envelope structure of these organisms. However, whether or not these compounds provide an adaptive advantage for the growth of Gram-negative bacteria in low osmolarity environments still remains unclear. The fact that mutants lacking periplasmic oligosaccharides are able to grow in low osmolarity environments indicates that these compounds are not essential for growth under these conditions. However, as noted above, it is possible that alternative adaptive mechanisms are operating in such mutants. Thus, it would appear worthwhile to examine such mutants for the presence of alternative periplasmic anions. Likewise, additional studies should examine the periplasmic volume as well as the magnitude of the Donnan potential across the outer membranes of such mutants. To date, such measurements have only been performed using wild-type cells.

An additional possible factor to be considered is that the presence of periplasmic oligosaccharides may provide a competitive advantage for Gram-negative bacteria in low osmolarity, natural environments. Such an adaptive advantage may not be clearly revealed through the study of single bacterial strains in low osmolarity laboratory media. Thus, it should be of interest to perform growth experiments in which competition between wild-type cells and periplasmic oligosaccharide-deficient mutants may be established and examined under low osmolarity conditions.

ACKNOWLEDGEMENTS

Research in my laboratory on periplasmic oligosaccharides is funded by grant DCB-8803247 from the National Science Foundation and from a grant from the College of Agriculture at The Pennsylvania State University.

REFERENCES

[1]   A. D. Brown, *Bacteriol. Rev.* 40: 803 (1976)

[2]   E. P. Kennedy, *Proc. Natl. Acad. Sci. USA* 79: 1092 (1982)

[3]   K. J. Miller, E. P. Kennedy and V. N. Reinhold, *Science* 231:48 (1986)

[4]   L. M. G. Van Golde, H. Schulman and E. P. Kennedy, *Proc. Natl. Acad. Sci. USA* 70: 1368 (1973)

[5]   J. E. Schneider, V. Reinhold, M. K. Rumley and E. P. Kennedy, *J. Biol. Chem.* 254: 10135 (1979)

[6]   B. J. Jackson and E. P. Kennedy, *J. Biol. Chem.* 258: 2394 (1983)

[7]   K. J. Miller and E. P. Kennedy, *J. Bacteriol.* 169: 682 (1987)

[8]   J. B. Stock, B. Rauch and S. Roseman, *J. Biol. Chem.* 252: 7850 (1977)

[9]   K. Sen, J. Hellman and H. Nikaido, *J. Biol. Chem.* 263: 1182 (1988)

[10]   A. Dell, W. S. York, M. McNeil, A. G. Darvill and P. Albersheim, *Carbohyd. Res.* 117: 185 (1983)

[11]   M. Hisamatsu, A. Amemura, T Matsuo, H. Matsuda and T. Harada, *J. Gen. Microbiol.* 128: 1873 (1982)

[12]   M. Hisamatsu, A. Amemura, K. Koizumi, T. Utamura and Y. Okada, *Carbohyd. Res.* 121: 31 (1983)

[13]   K. Koizumi, Y. Okada, S. Horiyama and T. Utamura, *J. Chromatography* 265: 89 (1983)

[14]   W. S. York, M. McNeil, A. G. Darvill and P. Albersheim, *J. Bacteriol.* 142: 243 (1980)

[15]   L. P. T. M. Zevenhuizen and H. J. Scholten-Koerselman, *Ant. van Leeuwenhoek* 45: 165 (1979)

[16]   K. J. Miller, V. N. Reinhold, A. C. Weissborn and E. P. Kennedy, *Biochim. Biophys. Acta* 901: 112 (1987)

[17]   K. J. Miller, R. S. Gore and A. J. Benesi, *J. Bacteriol.* 170: 4569 (1988)

[18]   J.-P. Bohin and E. P. Kennedy, *J. Bacteriol.* 157: 956 (1984)

[19]   J.-V. Holtje, W. Fiedler, H. Rotering, B. Walderich and J. van Duin, *J. Biol. Chem.* 263: 3539 (1988)

[20]   W. Fiedler and H. Rotering, *J. Biol. Chem.*263: 14684 (1988)
[21]   V. Puvanesarajah, F. M. Schell, G. Stacey, C. J. Douglas and E. W. Nester, *J. Bacteriol.* 164: 102 (1985)
[22]   T. Dylan, L. Ielpi, S. Stanfield, L. Kashyap, C. Douglas, M. Yanofsky, E. Nester, D. R. Helinski and G. Ditta, *Proc. Natl. Acad. Sci. USA* 83: 4403 (1986)
[23]   R. A. Geremia, S. Cavaignac, A. Zorreguieta, N. Toro, J. Olivares and R. A. Ugalde, *J. Bacteriol.* 169: 880 (1987)

PART IV

MOLECULAR BIOLOGY AND GENETICS

THE RIBOSOMAL RNA OPERONS OF HALOPHILIC ARCHAEBACTERIA

Patrick P. Dennis

Department of Biochemistry
The University of British Columbia
Vancouver, B.C., Canada, V6T 1W5

The comparative analysis of both 5S rRNA and small subunit rRNA sequences provided the first compelling evidence that archaeobacteria were a unique and coherent phylogenetic group, separate and distinct from both the eubacteria and the eucaryotes [1]. Within the archaebacteria, the extreme halophiles were shown to be a closely related phylogenetic clade and descendant from the anaerobic methanogens. As an alternative to using sequence comparison of structural RNAs and as part of an overall effort to characterize some of the unique features of halophilic archaebacteria at the molecular level, we have been examining the genetic organization and transcription of rRNA operons and the processing and assembly of the transcripts into ribosomal particles.

The operons encoding rRNAs have been cloned and partially or completely sequenced from a number of halophilic archaebacteria [2 - 7]. Most species appear to have only one or two rRNA operons per genome; when multiple, these operons are unlinked and usually preceded by three or more tandemly arranged transcription start sites. The operons all share a common 16S-23S-5S gene organization that is reminiscent of the rRNA gene organization found in eubacteria. The primary rRNA transcripts also contain an alanine tRNA sequence in the 16S-23S intercistronic space and a cysteine tRNA sequence distal to the 5S cistron. These tRNA sequences are inefficiently processed from the primary transcript; as a consequence, their stoichiometry relative to ribosomes is considerably less than one [8]. When there are two operons, the cysteine tRNA gene is not always present in both and the 16S-23S intergenic spaces are probably also different [6, 9].

In *Halobacterium cutirubrum* (halobium)[1] there is a single rRNA operon that

---

[1]*Halobacterium cutirubrum*, H. halobium and H. salinarium are believed to be separate isolates of a single species. *Haloferax volcanii* and *Haloarcula marismortui* were previously called *Halobacterium volcanii* and*Halobacterium marismortui* respectively.

General and Applied Aspects of Halophilic Microorganisms
Edited by F. Rodriguez-Valera, Plenum Press, New York, 1991

is growth rate regulated [10]. The 5' ends of transcripts derived from the operon have been extensively characterized by S1 nuclease protection and primer extension analysis [11, 8, 5, and our unpublished results]. Transcripts are initiated at any one of eight tandemly arranged promoters that are spaced at intervals of 120 to 367 nucleotides within the 1300 nucleotides of 5' flanking sequence. The eight promoters are aligned at their start sites with each other and with the promoters from other halophilic rRNA operons (Figure 1). Three regions of high nucleotide sequence similarity are apparent. The first is centered around the transcription start site; the major start site is almost always a G residue; when the start nucleotide is not G, the promoter is generally low in efficiency. The second and third regions of similarity are centered about 30 and 40 nucleotides respectively upstream from the transcription start site. The -30 sequence represents the halophilic version of the general sequence TTTAAA which is present in most archaebacterial promoters. This sequence may be the equivalent of the eucaryotic TATA box that is also located about 30 nucleotides upstream of the transcription start site in polymerase II type promoters [12]. The -40 sequence appears to be unique to halobacteria. It is important to recognize that these conserved sequence blocks may define the minimal features of an rRNA promoter but they do not explain the clearly observable differences in the relative strengths of the eight tandem rRNA promoters of *H. cutirubrum*; the sequences that are involved in the modulation of these promoters have not yet been identified.

For purposes of illustration, the promoters of halophilic protein encoding genes are also aligned in Figure 1. They differ from rRNA promoters in several ways. First, there is little if any sequence conservation around the major start nucleotide although this start nucleotide is almost always A or G. In many cases, the 5' untranslated leaders of these transcripts are short and the ATG translation initiation codons appear at or adjacent to the transcription start sites. The -30 and -40 sequences although present exhibit more variability than seen in rRNA promoters. This is especially true of the -40 sequence which is sometimes either absent or not easily recognized in the promoter region of protein encoding genes.

Transcripts initiated at the first rRNA promoter POA of the *H. cutirubrum* rRNA operons are relatively infrequent. However, these transcripts contain near their 5' end the start of a long open reading frame that potentially encodes a protein of 175 amino acids [13]. This putative protein is unusual compared to most halophilic proteins; it is overall basic, containing 26 Arginine and 5 Lysine and only 13 Aspartic and 8 Glutamic acid residues. The significance of this open reading frame and its role in regulation of ribosome synthesis is unclear; the rRNA operons from other species apparently do not contain a related open reading frame.

The 16S and 23S rRNA coding regions in the *H. cutirubrum* operons are surrounded by long nearly perfect inverted repeat sequences that are capable of forming an interrupted helical structure within the primary transcript (Figure 2). Nuclease protection analysis has demonstrated that the interruptions in the putative helices are early cleavage sites used to remove precursor 16S and 23S rRNAs from the long primary transcript [3, 8]. The sequence and structure of these rRNA processing

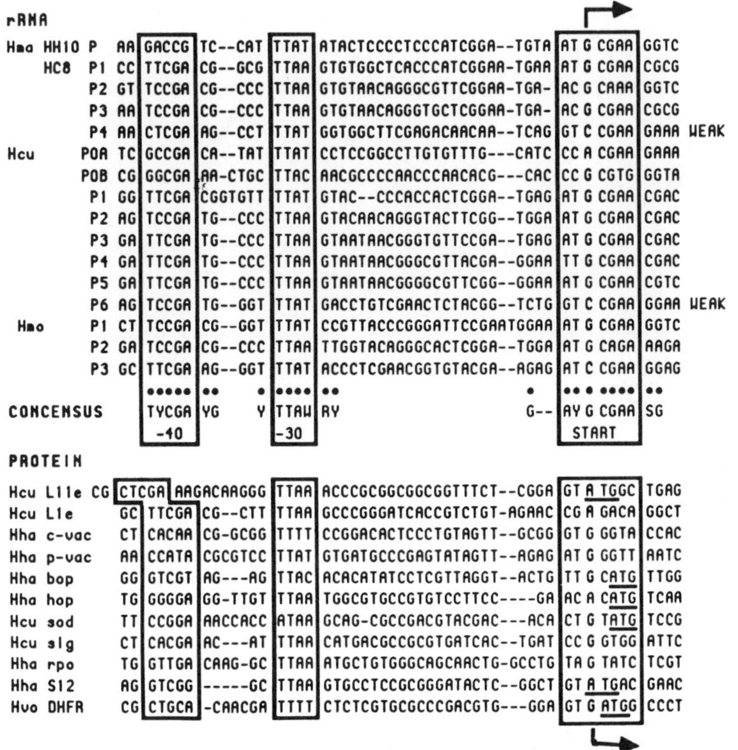

Figure 1. Halobacterial promotors: A summary of the conserved sequence blocks within halobacterial promoter regions. Sixteen rRNA and eleven protein promoter sequences from halobacterial species are summarized. *Haloarcula marismortui* (Hma) contains two rRNA operons, designated HH10 and HC8; the first has a single promoter and the second has four tandem promoters. *Halobacterium cutirubrum* (Hcu)* has a single rRNA operon with eight tandem promoters. *Halococcus morrhuae* (Hmo) has a single rRNA operon with three apparent tandem promoters; the primary transcripts from this rRNA operon have not been mapped. The most prominent 5' start sites (bold characters) are aligned and located at position 3 within the highly conserved heptanucleotide start sequence (boxed). Two other conserved segments designated -30 and -40 (boxed) are also aligned. Consensus nucleotides within the rRNA promoters are present in at least 75% of the promoter sequences: Y, pyrimidine; R, purine; W, A or T; S, G or C. The promoters of protein encoding genes from *H. cutirubrum* (Hcu), *H. halobium* (Hha), and *Haloferax volcanii* (Hva) are aligned below (the hop gene transcript has not been characterized and its alignment is based only on the position of the TTAA sequence and matches to the rRNA promoter consensus sequences). The site of transcription initiation is often at or immediately adjacent to the ATG translation initiation codon. When this occurs the ATG codon is underlined. The protein promoters exhibit less sequence conservation near the transcript start site and only sometimes conserve the -40 conserved pentanucleotide.

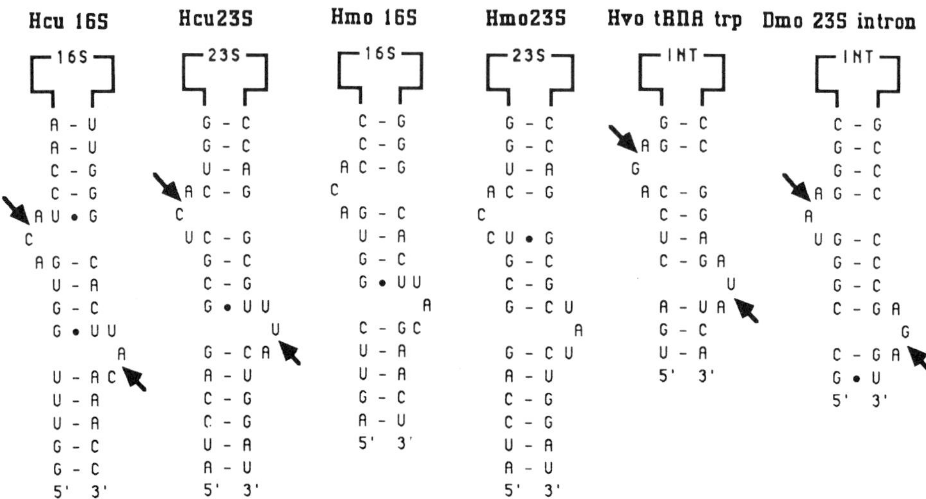

Figure 2. Sequence and structure of RNA processing sites. The intramolecular secondary structure associated with excision of precursor 16S and 23S rRNA from the primary transcripts of the *H. cutirubrum* (Hcu) and *Halococcus morrhuae* (Hmo) are illustrated. The (→) indicates the positions of incision and the sequence above the arrows represents precursor 16S or 23S rRNA that is released. The endonuclease incision sites have not been determined in the *Halococcus morrhuae* transcript. The sequence and secondary structure of the exon-intron boundaries in the *H. volcanii* (Hvo) tryptophan tRNA gene transcript and the *Desulfurococcus mobilis* (Dmo) 23S rRNA gene transcript and the sites of endonuclease incision for intron removal are illustrated. The sequence above the arrows represents the excised introns; the two exon pairs (below the arrows) are ligated together to produce respectively the anticodon stem of mature tryptophan tRNA and intact 23S rRNA.

sites in *H. cutirubrum* and other halophiles are remarkably conserved and resemble the exon-intron junctions in archaebacterial tRNA and rRNA intron containing genes [7, 14, 15, 16, 17] (see Figure 2).

The endonuclease from *Haloferax volcanii* involved in excision of the tryptophan tRNA intron has been partially purified and characterized with respect to its substrate specificity and the structure of its cleavage products [14]. Unlike most endonucleases, the enzyme cleaves to produce a 5' hydroxyl and a 2', 3' cyclic phosphate. Analysis of synthetic substrates produced in vitro by T7 RNA polymerase transcription of altered or deleted intron containing tryptophan tRNA genes indicate that neither the tRNA structure nor the internal region of the intron sequence are required for excision. Rather, the enzyme appears to recognize the limited sequence and secondary structure at the intron-exon boundaries. A length of four nucleotide base pairs between the two loops situated on opposite sides of the helix is essential for substrate cleavage by the enzyme. Surprisingly, the intron excision enzyme is

254

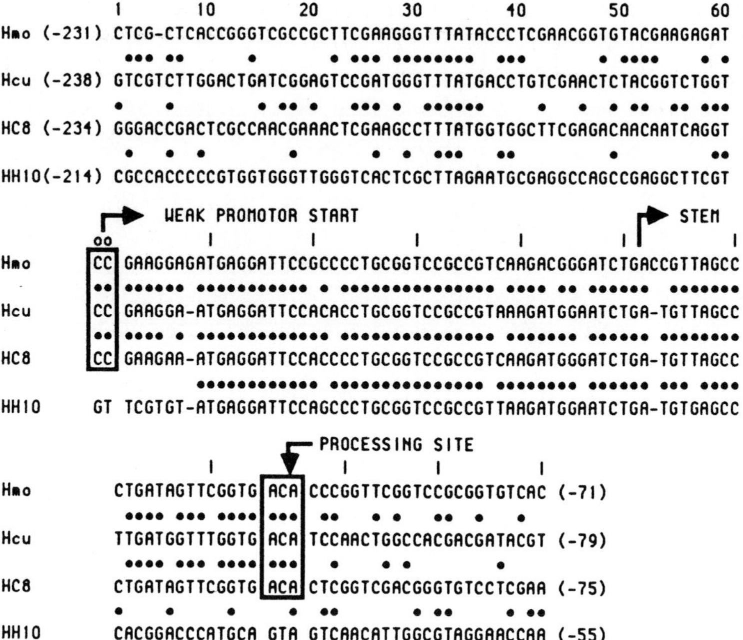

Figure 3. Sequence similarity in the 5' leader regions of halophilic rRNA operons. The 5' leader sequences of the rRNA operons from *Halococcus morrhuae* (Hmo), *Halobacterium cutirubrum* (Hcu) and the two operons from *Haloarcula marismortui* (HC8 and HH10) are aligned at the processing site (the HH10 operon lacks the conserved processing sequence). The nucleotide position at the beginning and end of the leader sequences are given relative to the first nucleotide in the 16S rRNA gene (position +1). Identical nucleotides in pairwise comparisons are indicated (•). The beginning of the inverted repeat sequence that forms the processing stem for excision of the precursor 16S rRNA sequence from the primary transcript is indicated. The 5' end sites of the most proximal (and very weak) promoter for the Hmo, Hcu and HC8 operons is also indicated. The -40 and -30 elements of these three promoters are not highlighted but are easily identified at positions 22-26 and 32-35 respectively.

capable of cleaving short synthetic hairpin substrates in vitro that are identical in sequence and structure to the 16S and 23S rRNA processing sites in *H. cutirubrum* rRNA primary transcripts (Thompson, Daniels, Dennis and Joshi, unpublished results). These observations lead one to speculate that the intron excision machinery and the rRNA processing machinery share a common excision endonuclease. Furthermore, it raises the intriguing possibility that during rRNA processing the leader and trailer sequences liberated by excision of p16S and p23S from the primary transcript might be recognized as exons and subject to subsequent ligation.

Another intriguing observation that bears on this point comes from inter-species comparison of the sequences of the 5' leaders liberated by excision of p16S rRNA from the rRNA primary transcript (Figure 3). In three distantly related halophiles, *Halobacterium cutirubrum*, *Haloarcula marismortui* and *Halococcus morrhuae*, the 75-80 nucleotides preceding the cleavage site are about 90% identical; remarkably, this level of nucleotide identity is at least as high as between the corresponding mature rRNA sequences. Surprisingly, the HH10 operon of *H. marismortui* appears to lack the highly conserved endonuclease processing site and retains only about 50 nucleotides from the conserved region [6]. The significance of this difference in the HC8 and HH10 operons of *H. marismortui* is currently being investigated. Immediately 3' to the endonuclease cleavage site (and extending toward the 16S coding region) there is no conservation in the compared sequences. The rRNA operons of eubacteria contain in their 5' leaders much shorter and less highly conserved sequences that function as transcriptional antitermination signals [18]. We are currently attempting to ascertain the function of this conserved sequence.

In summary, the intron excision and the rRNA processing machinery may share a common endonuclease component. This enzyme has a simple and well defined substrate specificity and cleaves to produce a 5' hydroxy and a 2', 3' cyclic phosphate. A highly conserved sequence in the leader region of the rRNA primary transcript that immediately precedes the endonuclease cleavage site is a potential substrate for a splicing reaction and might represent part or all of a yet to be identified cellular RNA.

ACKNOWLEDGEMENTS

I wish to thank Chuck Daniels, Moshe Mevarech, Lawrence Shimmin and Phalgun Joshi for their comments and suggestions. This work was supported by grants from the MRC (Canada) and the ONR (USA). Patrick P. Dennis is a fellow of the Canadian Institute for Advanced Research.

REFERENCES

[1]   C. R. Woese and G. E. Fox, Phylogenetic structure of the procaryotic domain: the primary kingdoms. *Proc. Nat. Acad. Sci. USA*, 74:5088 (1977).

[2]   R. Gupta, A. Broccoli and C. R. Woese, Sequence of the 16S ribosomal RNA from *Halobacterium volcanii*, an archaebacterium. *Science* 221:656 (1983)

[3]   I. Hui and P. P. Dennis, Characterization of the ribosomal RNA gene clusters in *Halobacterium cutirubrum. J. Biol. Chem.* 260:899 (1985)

[4]   N. Larsen, H. Leffers, J. Kjems and R. Garrett, Evolutionary divergence between ribosomal RNA operons of *Halococcus morrhuae* and *Desulfurococcus mobilis. System. Appl. Microbiol.* 7:49 (1986).

[5]   A. S. Mankin and V. K. Kagramanova, Complete nucleotide sequence of the single ribosomal RNA operon of *Halobacterium halobium*: secondary structure of the archaebacterial 23S rRNA. *Mol. Gen. Genet.*, 202:152 (1986).

[6]   M. Mevarech, S. Hirsch-Twizer, S. Goldman, E. Yakobson, H. Eisenberg and P. P. Dennis, Isolation and characterization of the rRNA gene clusters of *Halobacterium marismortui. J. Bacteriol.* 171:3479 (1989).

[7]   P. P. Dennis, Molecular biology of archaebacteria. *J. Bacteriol.* 186:471 (1986).

[8]   J. Chant and P. P. Dennis, Archaebacteria: Transcription and processing of ribosomal RNA sequences in *Halobacterium cutirubrum. EMBO J.* 5:1091 (1986).

[9]   C. J. Daniels, J. D. Hofman, J. G. McWilliams, W. F. Doolittle, C. R. Woese, K. R. Lenhresen and G. E. Fox, Sequence of 5S ribosomal RNA gene regions and their products in the archaebacteria *H. volcanii. Mol. Gen. Genet.* 198:270 (1985).

[10]   J. Chant, I. Hui, D. de Jong-Wong, L. Shimmin and P.P. Dennis, The protein synthesis machinery of the archaebacterium *Halobacterium cutirubrum*: molecular characterization. *Syst. Appl. Microbiol.* 7:106 (1986).

[11]   P. P. Dennis, Multiple promoters for the transcription of the ribosomal RNA gene cluster in *Halobacterium cutirubrum. J. Mol. Biol.* 186:457 (1985).

[12]   W. A. Reiter, P. Palm and W. Zillig, Analysis of transcription in the archaebacterium *Sulfolobus* indicates that archaebacterial promoters are homologous to eucaryotic pol II promoters. *Nucl. Acids Res.* 16:1 (1988).

[13]   A. S. Mankin and V. K. Kagramanova, Complex promoter pattern of a single ribosomal RNA operon of an archaebacteria, *Halobacterium halobium. Nucl. Acids Res.* 16:4679 (1988).

[14]   L. Thompson and C. Daniels, A tRNA trp intron endonuclease from *H. volcanii*: Unique substrate recognition properties. *J. Biol. Chem.* 263:17951 (1988).

[15]   L. Thompson, L. Brandon, D. Neuewlandt and C. Daniels, Transfer RNA intron processing in halophilic archaebacteria. *Can. J. Microbiol.* 35:36 (1989).

[16]   J. Kjems and R. Garrett, Novel splicing mechanism for the ribosomal RNA intron in the archaebacterium *Desulfurococcus mobilis. Cell*, 54:693 (1988).

[17]   J. Kjems, J. Jensen, T. Oleson and R. Garrett, Comparison of transfer RNA and ribosomal RNA intron splicing in the extreme thermophile and archaebacteria *Desulfurococcus mobilis. Can. J. Microbiol.* 35:210 (1989.

[18]   S. Li, C. Squires and C. L. Squires, Antitermination of *E. coli* rRNA transcription is caused by a control region segment containing lambda nut-like sequences. *Cell*, 38:851 (1984).

BACTERIO-OPSIN GENE EXPRESSION IN *HALOBACTERIUM HALOBIUM*

Mary C. Betlach and Richard F. Shand

Department of Biochemistry and Biophysics
University of California at San Francisco
San Francisco, California 94143

ABSTRACT

The retinal binding protein, bacterio-opsin, functions as a proton pump in the purple membrane of the extremely halophilic archaebacterium *Halobacterium halobium*. Under growth conditions of low oxygen tension and high light intensity, purple membrane synthesis is induced and *H. halobium* will grow phototrophically depending on the proton pump as the sole energy source. The bacterio-opsin gene (*bop*) is located within a cluster of genes which affect its expression. The bacterio-opsin activator gene (*bat*) appears to activate expression of the *bop* gene and the nearby bacterio-opsin related protein gene (*brp*). Transcription from these three genes is inducible by conditions of high light intensity and low oxygen tension.

INTRODUCTION

Bacterio-opsin is the sole protein found in the purple membrane of the extreme halophile, *Halobacterium halobium* [1]. Bacterio-opsin complexed with the chromophore retinal, constitutes bacteriorhodopsin (bR) which functions as a light driven proton pump [2]. The resultant electro-chemical gradient is used to drive energy-requiring metabolic processes and provides sufficient energy to sustain phototrophic growth [3].

Purple membrane synthesis involved processing of a precursor [4] and some level of coordination between bacterio-opsin synthesis and retinal synthesis [5,6]. In addition, growth conditions of high light intensity and low oxygen tension result in a 5-fold increase in purple membrane which then constitutes as much as 50% of the cell membrane surface area [1].

General and Applied Aspects of Halophilic Microorganisms
Edited by F. Rodriguez-Valera, Plenum Press, New York, 1991

THE BACTERIO-OPSIN GENE

Since DNA transformation systems for the extreme halophiles are just becoming available, analyses of the gene encoding bacterio-opsin (*bop*) thus far have relied on recombinant DNA approaches. The *bop* gene has been cloned, the nucleotide sequence determined and the transcript analyzed [4,7]. The *bop* gene is 786 bp in size and encodes a 26.000 Da (248 amino acids) protein. The *bop* mRNA has a 2 nucleotide leader and a heterogeneous 3' terminus, giving rise to a major species of 0.83 kb and a minor species of 1.0 kb [7]. Spontaneous *bop* mutants arise at a high frequency (10-4)[8] and in the majority of cases the mutational defect responsible for the phenotype is the integration of an insertion element (ISH) within or up to 3800 bp upstream of the *bop* gene [9, 10, 11]. Approximately 50 such mutants have been

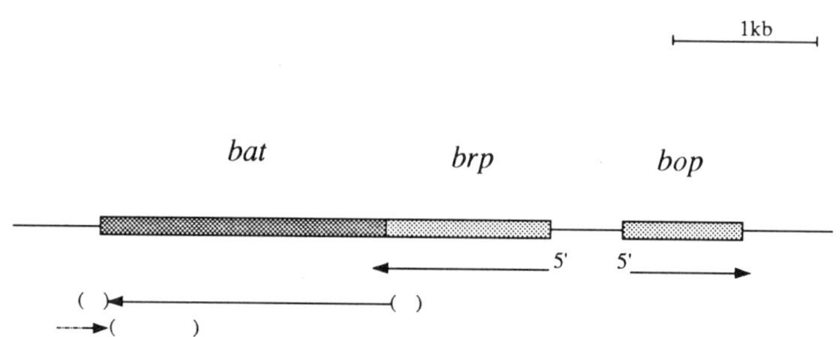

Figure 1.   Map of the *bop* gene cluster. The *bat*, *brp* and *bop* genes are indicated by shaded bars. Beneath the map, the array of horizontal arrows depict the direction and length of transcripts arising from this region in the genome. Transcription initiation start points are indicated where known. Termini for the *bat* transcript and the 0. 57 kb transcript near the 3' terminus of the *bat* gene have not been determined, but are probably located within the bracketed region. The dotted line at the 3' terminus of the 0. 57 kb transcript indicates that a start docon for the corresponding ORF has not yet been determined.

analyzed so far. These analyses revealed that the *bop* gene is embedded within a cluster of genes which are involved in its expression [12] (Figure 1). At least two additional genes have been identified in the "*bop* gene cluster" [10, 11]. These two genes were the targets of alterations (insertion elements and one deletion) in the spontaneous *bop* mutants.

THE BACTERIO-OPSIN RELATED PROTEIN GENE

The 1077 bp "bacterio-opsin related protein" gene (*brp*) is located 526 bp upstream of the *bop* gene and is transcribed in the opposite orientation [10]. The 5' terminus of the *brp* transcript is coincident with the ATG start codon of the *brp* gene and the 3' terminus is heterogeneous. The largest *brp* mRNA specie is approximately 1.2 kp in size and is present at 2% of *bop* mRNA levels in wild-type halobacterial cells grown under aerobic conditions [13]. The *brp* gene could encode a protein of 36.500 Da (359 amino acids). A secondary structure prediction of the putative *brp* protein extrapolated from the nucleotide sequence reveals a membrane protein-like structure with 6-7 hydrophobic alpha-helices of sufficient size to span the membrane [12]. An immunological approach has been used to detect the *brp* protein. A synthetic *brp* peptide was made and used as an antigen to generate a polyclonal antiserum. This antiserum was able to detect a portion of the *brp* protein expressed in an E. coli expression vector but was unable to detect a specific protein expressed in wild-type halobacterial cells grown under aerobic conditions. Attempts to obtain a different and perhaps more sensitive antiserum have focussed on using the entire *brp* protein expressed in E. coli as an antigen. Several expression vectors for the *brp* protein have been constucted. One of these encodes 11 heterologous amino acid residues at the 5' terminus of the *brp* gene. A second encodes a *bop/brp* protein fusion. Attempts to detect the *brp* protein expressed from these two vectors in E. coli using the antiserum described above and antiserum to bR are currently under way.

THE BACTERIO-OPSIN ACTIVATOR GENE

The 2022 bp "bacterio-opsin activator" gene (*bat*) is located 1602 bp upstream of the *bop* gene and is transcribed in the same orientation as the *brp* gene [11]. The start codon of the *bat* gene overlaps the stop codon of the *brp* gene. Preliminary RNAse mapping experiments indicate that the 5' terminus of the 2.2 kb *bat* transcript is located approximately 100 bases upstream of the *bat* gene start codon. Thus, the 5' terminus of the *bat* transcript is present at 4% of *bop* mRNA levels in halobacterial cells grown under aerobic conditions [13]. The *bat* gene could encode an acidic protein of 73.000 Da 674 amino acids) with a predicted secondary structure typical of a soluble alpha-beta type protein. This type of secondary structure is significantly different than the hydrophobic structure predicted for the putative *brp* protein. The *bat* protein has not yet been detected in wild-type halobacterial cells.

ROLES OF THE GENES IN THE *bop* GENE CLUSTER

To determine the roles of the *brp* and *bat* genes in the regulation of *bop* gene expression, transcript levels arising from all three genes were quantitated in a series of spontaneous *bop* mutants grown under aerobic conditions [13]. These mutants contain either insertion or deletion mutations in one of these three genes.

Bop mutants with mutations in the *bop* gene had normal levels of *brp* and *bat* transcripts suggesting that *bop* gene expression is not required for *brp* or *bat* gene expression. *Bop* mutants with mutations in the *brp* gene had drastically decreased levels of *bat* transcript (0-9%) whereas the levels of *bop* mRNA were decreased somewhat less drastically (0-23%). This data suggested that *bat* gene expression is affected by *brp* gene expression and that some *bop* gene expression can occur in the absence of *brp* and *bat* transcription. *Bop* mutants with mutations in the *bat* gene had virtually undetectable levels of transcripts from all three genes suggesting that the *bat* gene is involved in activating *bop* and *brp* gene expression. *Bop* mutants with mutations in the 526 bp intergenic region between the *bop* and *brp* genes had normal levels of *bat* transcripts but very low levels of *brp* transcripts and no *bop* mRNA. This data indicated that normal levels of *bat* gene expression are not sufficient to activate *bop* and *brp* gene expression when the region between these two genes is disrupted.

Although the interactions between these three genes are complex, the data supports a model (Figure 2) in which a trans-acting *bat* gene product activates *bop* and *brp* gene expression. Since *bat* gene expression is affected by *brp* gene expression, while the *brp* gene in turn requires the *bat* gene product in order to be expressed, the *bat* gene would be autoregulated. The role of the *brp* gene is less clear. A few feasible possibilities are: 1) the *brp* gene affects *bop* gene expression via an indirect effect of

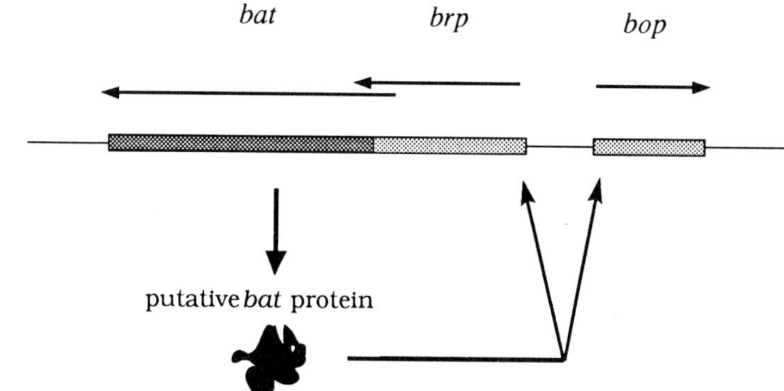

Figure 2. Model of relationships between the genes in the *bop* gene cluster. The *bat*, *brp* and *bop* genes are indicated by shaded bars. Transcripts are depicted by horizontal arrows above the corresponding genes. The putative *bat* protein is shown in its proposed role of activating *brp* and *bop* gene expression.

brp transcription on *bat* gene expression, 2) the putative *brp* protein plays a role in purple membrane assembly, and/or 3) the putative *brp* and *bat* proteins may constitute a two-component regulatory system in which the *brp* protein functions as a membrane bound sensor and the *bat* protein as a soluble regulatory protein. These roles would be consistent with the secondary structures predicted for the two proteins.

Preliminary data suggest that a third genetic locus exists which may affect *bop* gene expression. An open reading frame (ORF) of 381+ bp has been recently discovered for which the start codon has not yet been determined. ORF 381+ is located distal to the 3' terminus of the *bat* gene, is transcribed in the opposite orientation and generates a transcript of 0.57 kb (see Figure 1). It is uncertain whether this ORF and its transcript play a role in the regulation of *bop* gene expression, although levels of this transcript are affected in the few *bop* mutants analyzed thus far.

262

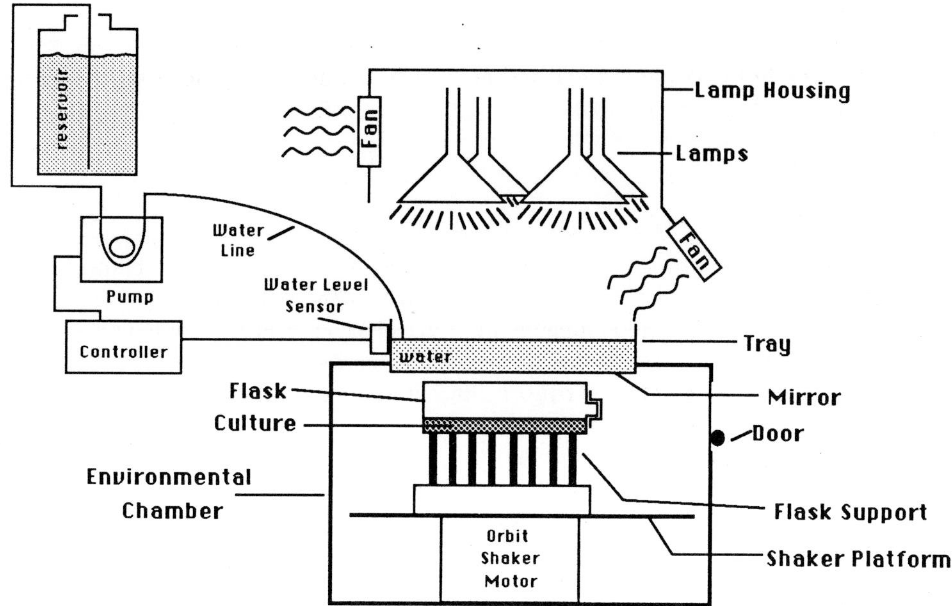

Figure 3. Schematic of a phototrophic growth chamber. The cultures are illuminated at 30,000 lux [100 mW/cm2] using 4, 150 W tungsten spot lights [General Electric 150R/SP]. The lid of the environmental chamber [Labline model 3528] has been modified to accept a 25 cm2 plexiglas tray capable of holding H2O to a depth of 3cm. The bottom of the tray is fitted with a wide band hot mirror   [Optical Coating Laboratory, Inc.]. The mirror and the environmental chamber are cooled by evaporation of H2O from the tray. Evaporated H2O is replaced automatically using a water level sensor [Instruments for Research and Industry [12R]model RM-1].

EFFECT OF HIGH LIGHT INTENSITY AND LOW OXYGEN TENSION ON EXPRESSION FROM THE *bop* GENE CLUSTER

Purple membrane synthesis is inducible five-fold by environmental factors such as high light intensity and low oxygen tension. Under such conditions *H. halobium* will grow phototrophically and depends on the activity of the proton pump in the purple membrane for a cellular energy source [3]. We aim to elucidate the role of light and oxygen in the regulation of the *bop* gene cluster and determine if any of these genes are involved in mediating these environmental signals. We have correlated bacterio-opsin protein and mRNA levels, and *brp* and *bat* transcript levels in halobacterial cells grown under steady state conditions and in shift experiments involving high light intensity, low oxygen tension or both. A schematic of the phototrophic growth chamber used for these experiments is shown in Figure 3. Preliminary date indicates that transcription from these three genes in the *bop* gene cluster is inducible by conditions of high light intensity and low oxygen tension. This data suggests that the increase in purple membrane synthesis is the result of increase in transcription of the genes in the *bop* gene cluster.

ACKNOWLEDGEMENTS

This work was supported by Public Health Service grant GM31785.

REFERENCES

[1]   D.Oesterhelt and W. Steockenius, Functions of a new Photoreceptor membrane. *Proc. Natl. Acad. Sci. USA* 70:2853 (1973).

[2]   W. Stoeckenius and R. Bogomolni, Bacteriorhodopsin and related pigments of halobacteria. *Ann. Rev. Biochem.* 52:587 (1982).

[3]   D. Oesterhelt and G. Krippahl, Phototrophic growth of halobacteria and its use for isolation of photosynthetically-deficient mutants. *Ann. Rev. Microbiol.* (Inst. Pasteur) 134B:137 (1983).

[4]   R. Dunn, J. McCoy, M. Simsek, A. Majumdar, S. Chang, U. RajBhandary and H. Khorana, The bacterio-opsin gene. *Proc. Natl. Acad. Sci. USA* 78:6744 (1981).

[5]   M. Sumper and G. Herrmann, Biogenesis of purple membrane: regulation of bacterio-opsin synthesis. *FEBS Lett.* 69:149 (1976).

[6]   M. Sumper and G. Herrmann, Biogenesis of purple membrane: control of retinal synthesis by bacterio-opsin. *FEBS Lett.* 71:333 (1976).

[7]   S. DasSarma, U. RajBhandary and H. Khorana. Bacterio-opsin mRNA in wild-type and bacterio-opsin deficient *Halobacterium halobium* strains. *Proc. Natl. Acad. Sci. USA* 81:125 (19844).

[8]   F. Pfeifer, G. Weidinger and W. Goebel, Genetic variability of *Halobacterium halobium*. *J. Bacteriol.* 145:375 (1981).

[9]   F. Pfeifer, J. Friedman, H. Boyer, and M. Betlach, Characterization of insertions affecting the expression of the bacterio-opsin gene in *Halobacterium halobium*. *Nucleic Acids Res.* 12:2489 (1984).

[10]   M. Betlach, J. Friedmand, H. Boyer and F. Pfeifer, Characterization of a halobacterial gene affecting bacterio-opsin gene expression. *Nucleic Acids Res.* 12:7949 (1984).

[11]   D. Leong, F. Pfeifer, H. Boyer and M. Betlach, Characterization of a second gene involved in bacterio-opsin gene expression in a halophilic archaebacterium. *J. Bacteriol.* 170:4903 (1988a).

[12]   M. Betlach, R. Shand and D. Leong, Regulation of the bacterio-opsin gene of a halophilic archaebacterium. *Can. J. Microbiol.* 35:134 (1989).

[13]   D. Leong, H. Boyer and M. Betlach, Transcription of genes involved in bacterio-opsin gene expression in mutants of a halophilic archaebacterium. *J. Bacteriol.* 170:4910 (1988b).

THE KINETIC OF THE GENETIC EXCHANGE PROCESS

IN *HALOBACTERIUM VOLCANII* MATING

Ilan Rosenshine and Moshe Mevarech

Department of Microbiology
George S. Wise Faculty of Life Sciences
Tel Aviv University
Tel Aviv, 69978
Israel

ABSTRACT

When cells of *Halobacterium volcanii* grow in mixed cultures on solid surfaces, they exchange genetic information. This exchange can be detected immediately after mixing the cells and the number of recombinant cells reaches a maximum after 7 hours. The genetic exchange process creates diploid or merodiploid cells ("heteroclones"). These cells can either segregate back to their parental types or recombination can occur between the two chromosomes resulting in recombinant genotypes. The stability of the diploid state was studied and found to be a function of the genetic markers that were used for the selection.

INTRODUCTION

Halobacterium volcanii is the only archaebacterium in which a genetic exchange process was detected [1]. A detailed characterization of the mechanism of this process [2] showed that this system is unique and differs from the conjugation systems of eubacteria as well as from the sexual system of eukaryotes. The mating system of *H. volcanii* is characterized by the following features: 1) each parental cell can be a donor or a recipient; 2) the genetic exchange process requires contact between cells; 3) cytoplasmic bridges are produced which enable the transfer of the chromosomal DNA but restrict the transfer of cytoplasmic markers.

The following communication describes the kinetic of the system and presents evidence that a relatively stable diploid or merodiploid state is established prior to the recombination event that leads to the production of recombinant cells.

General and Applied Aspects of Halophilic Microorganisms
Edited by F. Rodriguez-Valera, Plenum Press, New York, 1991

Table 1. Mutants of *H. volcanii* used in this study.

(WR No.)	Genotype
201	*his1*
243	*ser1,white1*
244	*gua1 , white 2*
256	*his1,arg1*
252	*his1,phe3*
266	*ade18,cys2*
268	*ade18,arg2*

MATERIAL AND METHODS

Bacterial strains

Halobacterium volcanii DS2 mutants that were used in this study are listed in Table 1.

Media and growth conditions

Halobacteria were grown in the described minimal (MM) or complete (H) media [1], except that the medium was buffered using Tris-HCl pH 7.2 at a final concentration of 50 mM, and as a carbon source (in MM media) 0.45% sodium succinate and 0.05% glycerol were used. When needed, amino acids (final concentration 50 μg/ml) or nucleotides (final concentration of 25 μg/ml) were added. Plates were incubated at 42ºC and liquid cultures at 37ºC in a shaker. Bacteria were crossed as described in [1] and [2].

Mutagenesis

Stationary culture of *H. volcanii* was diluted 1:10 in complete medium, and grown for 16-20h at 37ºC. Three ml were washed twice with SMT (3.5 M NaCl, 0.15 M $MgSO_4$ and 10 mM Tris HCl pH 7.2) and resuspended in 8 ml SMT. Then 120 μl of ethylmethansulfonate (Merck) were added followed by vigorous shaking for 30 sec. The culture was then incubated with gentle shaking at 37ºC. Aliquots (2 ml) were collected at various time intervals (typically at 0, 10, 20 and 30 min), washed twice with SMT and resuspended in complete medium. A 1:10 dilution of the culture was grown overnight for phenotypic expression. This culture was used for screening or selection. Usually survival frequencies of 1%-50% were used for mutant isolation.

Isolation of mutants

The mutagenized culture was washed with SMT and dilutions were plated on MMYE plates (minimal media supplemented with 0.02% yeast extract, Difco). These plates contained limited amounts of growth factors and therefore the auxotrophic colonies were small. Small colonies were picked from the MMYE plates and tested for growth on minimal and complete media. Typically 30%-50% of these small colonies

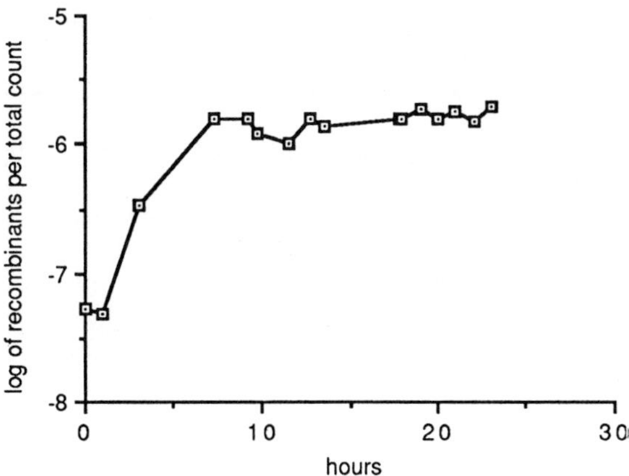

Figure 1. The kinetic of the mating process.

were found to be auxotrophs. The auxotrophic mutants were further analyzed for their specific growth requirement as described by Davis et al. [3]. White colony mutants were isolated by visualizing colonies of mutated culture.

RESULTS

The kinetic of the mating process

The kinetic of mating was measured by following the ratio of recombinants in the population at different times after plating the mixed parental strains.

Strains WR201 (*His*1) and WR243 (*Ser*1, *White*1) were mixed, filtered, placed on rich plates and incubated. The filters were removed from the plates at different times and the cells were suspended and washed twice with SMT. Aliquots were plated for viable count on rich plates and for recombinant selection on minimal plates. The results of this experiment are summarized in Figure 1.

As can be seen recombinants were obtained also at time – 0. These recombinant colonies were readily distinguishable from revertants (that appear at a similar frequency) by their red-white sectorial appearance. Spontaneous revertant prototrophs, on the other hand, give rise to uniform red or white colonies. The mating frequency reaches its maximum after 7 hours, and remains at the same level for at least 20 hours.

Some recombinants are diploids (or merodiploids) for few generations

The white-colony mutants WR243 or WR244 were crossed with a red strain WR201 and recombinants were selected. Fifteen percent of the recombinant colonies showed red-white sectorial phenotype. It seems therefore that the first cell in the colony had the two colour allels. The diploid state can also be shown using auxotrophic

Table 2. Analysis of growth requirements of isolated colonies derived from "heteroclone" colonies.

"heteroclone" colony *	Mumber of colonies which were grown on the indicated media:				
	MM+ Ade,Arg	MM+ His,Phe	complete medium	MM+ Ade	MM+ or Ade His,Phe
A	3		7		
B	2	6	2		
C	2		5	3	
D			10		
E		10			
F				10	
G	1	1		7	1
H	3	2	3		2
I				8	2
J				8	2
TOTAL	11	19	27	26	7

* 10 colonies were analysed for each "heteroclone"

markers. The strains WR252 (*His*1, *Phe*3) and WR268 (*Ade*18, *Arg*2) were crossed, and recombinants were selected on minimal plates supplemented with adenine and phenylalanine. Only 50% of the recombinant colonies were able to grow when restreaked on the same plates (minimal plates supplemented with adenine and phenylalanine). Those colonies which could not grow when restreaked on the selective plates were usually smaller and contained a mixture of bacteria of different phenotypes. The analyses of the phenotypes of these colonies, termed "heteroclones" (analogous to a similar situation which exists in Streptomyces [4]), was performed as follows. Each colony was streaked onto a rich plate and the phenotypes of the resulting offspring colonies were determined. Table 2 summarizes the phenotype analyses of ten colonies derived from ten heteroclones (A-J). As can be seen, some of the heteroclones segregated to their parental phenotypes, indicating that the heteroclone colony contains diploid or merodiploid bacteria.

Under different selection conditions (such as minimal plates supplemented with histidine and argenine, or histidine and adenine, or phenylalanine and arginine), the ratio of the heteroclones in the recombinant population amounts to 1% - <0.5%. These heteroclones did not show the small colony phenotype. Thus, the ratio of the heteroclones in the population of the recombinants is a characteristic of specific strains and specific selection conditions.

These results suggest that after mating, recombinants can stay as diploids or merodiploids for several generations, and by using specific conditions it is possible to select against segregation. Similar results were obtained when the strains WR256 (*His*1, *Arg*1) and WR266 (*Ade*18, *Cys*2) were crossed in a similar experiment.

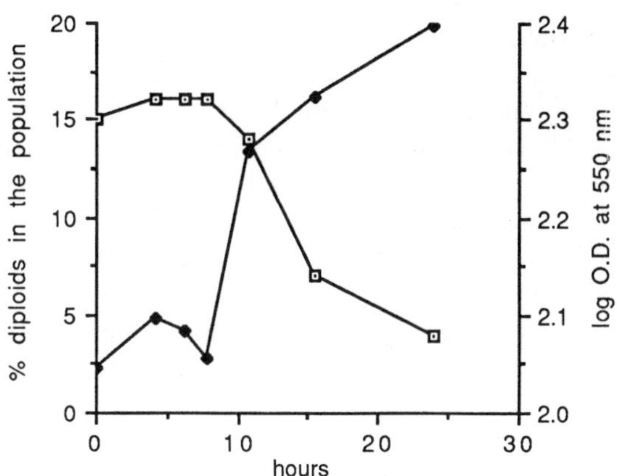

Figure 2. The kinetic of segregation of the red (w.t.) and white (*white*1) allels after mating.

The kinetic of allel segregation after mating

The red and white markers were used to measure the kinetic of segregation of unselected markers. Diploid cells give rise to sectorial red-white colonies while haploid cells form uniform red or white colonies. The strains WR201 and WR243 were crossed, as described, and the filters were incubated for 23h at 42ºC on rich plates. Then, the culture which grew on the filter was suspended in rich liquid medium and grown at 37ºC with vigorous shaking. Under these conditions the mating process was stopped and only segregation continued.

Aliquots were then taken from the culture at time intervals, plated on minimal selective plates and incubated at 42ºC. In addition the optical density (at 550 nm) of the aliquots was measured. After 10 days the colour of the colonies was examined, and the ratio of diploids (sectorial colonies) in the population (red, white and sectorial colonies) was determined (Figure 2).

DISCUSSION

From the original description of the genetic exchange system of *H. volcanii* [1] it was clear that the frequency of the exchange is a function of the length of time of contact between the parental cells. When cells are grown in liquid cultures the mechanical separation interferes with the establishment of the close contact required for the exchange and no recombinants could be detected. It was therefore surprising to find that a very brief contact between two mutants on the filter was sufficient to produce recombinants, though at a very low frequency. The frequency of the genetic exchange increases fifty fold with increase in the time of contact and reaches a saturation after 7 hours. During this time the number of cells increases only about two fold. It seems, therefore, that although a small number of cells will establish strong

269

enough connections at the instance of contact, most of the cells require time to establish these connections.

Once DNA of one cell enters the recipient cell a rather stable diploid or merodiploid cell is established. The existence of the diploid state was demonstrated in two ways. The first way was by crossing red and white auxotrophic mutants that resulted in sectorial colonies. These sectorial colonies derived from cells that contain the chromosomes of both parental cels. The second way of showing the diploid state was to analyze the phenotype of "heteroclone" colonies. These colonies probably arise from crosses between mutants having closely linked mutations. The close linkage between the mutations prevents recombination between them and the selective pressure forces the maintenance of the two chromosomes. The small size of the colonies is probably due to the fact that daughter cells which do not contain the two chromosomes cannot grow anymore on the selective medium. The genetic heterogeneity of the "heteroclone" colony, as demonstrated in Table 2, is the result of segregation of the chromosomes as well as the result of recombination between genetic markers that are not closely linked. This explanation fits the observation that the frequency of heteroclones in the population of recombinants is a function of the genetic markers used for the selection. When the selection is made for two markers that are far away, the frequency of recombination is high and most of the resulting colonies are true recombinants.

The third experiment was designed to estimate the rate of segregation of unselected markers (or chromosomes). Unlike the heteroclone situation, in which the medium exerts selective pressure on the cells to maintain the two parental chromosomes, when the observed markers are the colour of the colony, no selection pressure is applied and the chromosomes are free to segregate. As seen from Figure 2, after about one doubling only 4% of the population still maintain the diploid state.

ACKNOWLEDGEMENT

This work was supported by The Endowment Fund for Basic Research in Life Sciences: Charles H. Revson Foundation.

REFERENCES

[1]   M. Mevarech and R. Wervzberger, Genetic transfer in *Halobacterium volcanii*, *J. Bacteriol.* 162: 461 (1985)

[2]   I. Rosenshine, R. Tchelet and M. Mevarech, The mechanism of DNA transfer in the mating system of an archaebacterium, *Science* 245: 1387 (1989)

[3]   R. W. Davis, D. Botstein and J. R. Roth "Advanced Bacterial Genetics; A manual for genetic engineering", Cold Spring Harbor Laboratory Press, New York (1980)

[4]   D. A. Hopwood, K. F. Chater, J. F. Dowding and A. Vivian, *Streptomyces coelicolor* genetics, *Bacteriol. Rev.* 31: 373 (1973)

PHYSICAL MAPPING AND GENE TRANSFER METHODS FOR

HALOBACTERIUM (HALOFERAX) VOLCANII

W. Ford Doolittle, Robert L. Charlebois, Leo C. Schalkwyk, Wan. L. Lam, R. Keith Conover, Dina Tsouluhas, Steven W. Cline, Jason D. Hofman and Annalee Cohen

Department of Biochemistry
Dalhousie University
Halifax, Nova Scotia B3H 4H7
Canada

ABSTRACT

Molecular archaebacteriology is now about to enter a manipulative phase. Many laboratories have contributed to the development of methods for proper genetic analysis. Here we describe our efforts with *Halobacterium volcanii*. We have nearly completed a physical map of the genome of this species, have optimized procedures for its transformation by exogenous DNA, and developed shuttle vectors alternately selectable and maintainable in this species and *Escherichia coli*.

INTRODUCTION

A dozen years ago, Woese and his collaborators proposed that the contemporary biological world is sundered not by a single profound evolutionary discontinuity - that separating procaryotes from eucaryotes - but by two, which define three modern "kingdoms", the eubacteria, eucaryotes and archaebaceria (for review, see [1]). Archaebacteria had of course been known before, with halophiles, methanogens and thermophiles separately enjoying attention because of their quite different adaptations to quite different extreme environments. Woese's claim gave us a reason to try to see beneath these differences to an underlying deep biochemical and molecular genetic unity, and a reason to begin analyses of basic features of archaebacterial biology, so that these might inform a truly comparative approach to understanding the origin and early evolution of the genetic apparatus.

General and Applied Aspects of Halophilic Microorganisms
Edited by F. Rodriguez-Valera, Plenum Press, New York, 1991

Progress has been well summarized in the proceedings of three international symposia on molecular archaebiology, held in 1981, 1985 and 1988 [2-4]. Even at the time of the most recent of these we were still clearly wedded to an approach that John Reeve at the earliest symposium called "stare and compare". Genes - cloned by indirect techniques often requiring considerable ingenuity - are sequenced, and sequences are scanned and aligned against each other and against presumed eubacterial and eukaryotic homologs, for clues to function. Although conclusions of surprising generality and power have indeed been extracted from the sequence data in this way (see papers from the laboratories of Dennis, Betlach, Oesterhelt, DasSarma, Kohiyama and Zillig in this volume, and many other groups in reference 4), the approach remains limited. Deeper understanding requires the development of faithful *in vitro* systems for transcription and translation (an area in which there is at last real promise [4]) and efficient and flexible means for genetic exchange, linkage analysis and functional studies of introduced exogenous DNA.

In 1985, Mevarech and Werczberger [5] described a conjugation-like system of bidirectional genetic exchange for *Halobacterium (Haloferax) volcanii* and further refinement of this extremely useful "natural" process is described by Mevarech in this volume. In 1986, we decided to attempt to develop methods for surrogate genetics in this same species by concentrating in three areas; physical mapping, genetic transformation and shuttle vector construction. Here we summarize progress to date, and describe a few instances of application of these methods.

MAPPING

Approaches to chromosomal mapping can be described as either "top-down" (chromosomes are first cut into a minimum number of large fragments resolvable by, for instance, pulsed-field gel electrophoresis and these are then successively subdivided into smaller and smaller mapped units) or "bottom-up" (overlaps between individual clones from a random library are identified by comparisons of restriction site distribution and used to link them into ever longer and longer "contigs"). We have used both in looking at the genome of *Halobacterium volcanii* [6].

Pulsed-field gel electrophoretic separations of *Bam*HI digests of genomic DNA yield fragments with a total length of about 3.8 million base pairs. Resolution of partial digests obtained with the rarely-cutting enzymes *Xba*I and *Spe*I allow the mapping of stretches of more than 500 kilo-base-pairs (kbp). In several cases, cloned and sequenced genes have been localized within such large fragments by hybridization to Southern blots made from pulsed-field gels.

Bottom-up mapping, using libraries made in cosmid vectors has, however, proved of more use. We have screened more than 500 clones of *Mlu*I partially-digested *H. volcanii* DNA in Lorist vectors, by what we call a "landmark" strategy [6]. For this, *Mlu*I single digests are resolved in parallel with double digests obtained with *Mlu*I and one of ten second enzymes, chosen because they cut the genome about one-tenth as frequently as *Mlu*I. Clones have on average ten *Mlu*I sites and only about one site for each of these second enzymes, so a typical double digest reveals that a single *Mlu*I fragment (of known size) is cleaved by the second enzyme into two fragments, of known sizes. Very likely there is within the entire genome only one *Mlu*I fragment of this size with a site for the second enzyme in such a position - the information is unique and uniquely identifies a "landmark". Clones showing the same landmark thus must overlap, and (because ten second enzymes are used) overlapping clones often share several landmarks.

We now have linked cosmid clones into 25 contigs, representing about 4 million base pairs. The longest contig comprises 24 cosmids. We are closing the gaps between contigs by using probes derived from the terminal cosmids of each contig in hybridizations to total genomic DNA digested with several infrequently-cutting enzymes. Probes giving similar fingerprints are likely to mark and orient contigs which will lie adjacent to each other on the final map.

Top-down mapping yields only a map of restriction sites. Bottom-up methods provide also an ordered set of cosmid clones. This resource allows direct localization of known cloned genes, by hybridization of those clones to ordered dot blots of cosmid DNAs representing a minimal overlapping set. By this means we have localized to clones (and thus ordered within contigs) nearly two dozen sequenced or otherwise identifiable genes, 29 tRNA-coding genes of as yet unknown specific identity, 32 copies of the previously characterized insertion sequence ISH51 [7], and eleven copies of an as yet uncharacterized repeat element. We have located a few more protein-coding genes (the homolog of the *E. coli hisC* locus, an arginine pathway gene and a leucine pathway gene) by a more nearly genetic approach. Cosmid DNA will transform *H. volcanii* auxotrophs to prototrophy. Even though cosmid DNA prepared from *E. coli* is subject to severe restriction in this halobacterium, transformation is sufficiently efficient to allow not only unambiguous identification of cosmids bearing the wild-type versions of genes for which we have auxotrophs, but also localization of those genes on forty kbp cosmid inserts to within a kbp or so, by transformation with restriction fragments eluted from gels.

More recently, we have assembled and classified a collection of auxotrophs obtained after ethylmethane sulfonate mutagenesis. This collection includes strains requiring tryptophan, histidine, proline, lysine, leucine, arginine, phenylalanine, valine/isoleucine, glutamine, threonine, asparagine, methionine, serine, uracil, adenine, guanosine and biotin. These are being mapped by cosmid transformation or by shotgun cloning of the wild type allele into shuttle vectors (see below) and localization to cosmids by probing of dot blots.

TRANSFORMATION

Methods for getting endogenous DNA into cells are essential for genetic linkage analyses, clone selection and assessment of the effects of structural alterations on gene function. Development of such methods requires improving simultaneously techniques for introducng DNA and ways to score for its introduction - sometimes a difficult feat of double optimization. We chose to look at transfection of *Halobacterium halobium* by double-stranded DNA purified from the naturally occurring phage ϕH [8]. Initially we were lucky to detect transfection at all, because it requires conditions (removal of divalent cations, low molecular weight polyethylene glycol) quite different from those used in many eubacterial transformation protocols, as well as care in preventing cell lysis. Now we can obtain more than 10^7 plaque-forming units per microgram of phage DNA with *H. halobium* and readily score, on lawns of *H. halobium*, virus produced after transfection of *H. volcanii* (which does not adsorb phage particles). Using in part this heterologous transfection assay, we optimized conditions for transformation of *H. volcanii* [9, 10]. We can demonstrate transformation of auxotrophs with uncloned linear fragments of wild-type DNA, and wild-type DNA cloned either into (i) the naturally-occurring 6 kbp sequenced [11] *H. volcanii* plasmid pHV2, (ii) shuttle vectors obtained by insertion into pHV2 of a drug-resistance determinant and bits of *E. coli* plasmid vectors, or (iii) cosmids, as described above. Even in the worst cases (unfractionated or unmodified DNA), hundreds of transformants can be detected under conditions where control (no DNA, wrong DNA) plates show no colonies.

SHUTTLE VECTORS

The plasmid pHV2 has provided very useful starting material for the development of several shuttle vectors. Its 6, 354 bp sequence contains several open reading frames, but much of the sequence can be eliminated without affecting plasmid maintenance. Plasmid maintenance has no obvious effect on cell growth, and we obtained a cured derivative of *H. volcanii*, a strain we call WFD11, after growth in ethidium bromide. We then demonstrated that pHV2 can be reintroduced into WFD11, scoring for success in transformation by colony hybridization with labeled plasmid as probe.

Useful vectors must be selectable, not simply scorable. In 1986, Cabrera et al. [12] reported that mevinolin, an inhibitor of mammalian HMGCoA reductase, strongly inhibits this enzyme in halobacterial extracts, and also abolishes cell growth. We find that mutants resistant to this drug (at levels up to 100 μM - 50 fold higher than levels inhibiting wild-type cells) arise at convenient low frequency, about 1 in 10^9 cells plated. We used a derivative of pHV2 which had suffered a spontaneous ISH51 insertion near position 5,500 of the pHV2 sequence as a vector in shot-gun cloning of *Mlu*I-digested DNA from mevinolin resistant cells. (A unique *Mlu*I site was introduced into the plasmid by the inserted ISH51.) We obtained several recombinant plasmids which conferred resistance on transformed sensitive cells. All contained a common fragment which we assume carries the halobacterial HMGCoA reductase gene; sequencing under way should bear this out. We have also noted that some mevinolin-resistant mutants seem to result from amplification of this gene - these are not useful in shuttle vector construction, but the amplification phenomenon itself is of great interest.

Resistance-conferring plasmids were made into shuttle vectors by addition of sequences from pAT153, a pBR322 derivative selectable in *E. coli* by virtue of ampicillin or tetracycline resistance determinants. Our smallest useful construct to date is pWL102. Its 10.5 kbp include the ampicillin resistance determinant, the mevinolin resistance determinant and unique cloning sites for *Kpn*I, *Xba*I, *Sph*I, *Bam*HI, *Nco*I, *Cla*I and *Eco*RI. Although replication in *E. coli* reduces *H. volcanii* transformation efficiencies some 10^4-fold, we still obtain thousands of resistant colonies in transformation experiments where control plates show no colonies. This vector, and other bulkier ones with additional cloning sites, will allow us to bring the full armamentarium of modern *E. coli* genetics to bear on questions about the molecular biology of the archaebacteria [13].

Experiments in which portions of the pHV2 moiety of shuttle vectors were excised have allowed us to identify regions necessary for maintenance in the halobacterial host. Their removal does not, however, reduce transformation frequencies to zero. We can show that in transformants obtained with such reduced vectors, pBR322 sequences have become stably integrated into *H. volcanii* DNA. Thus we will likely be able to create targeted mutations, by transformation with appropriately modified DNAs which cannot be maintained as plasmids.

APPLICATIONS

We have begun to apply the methods described here to the detailed study of the molecular biology of *H. volcanii*. For instance, we now have a dozen or more mutants which require tryptophan. Some will grow also on indole and others on indole or anthranilic acid, while some grow only on tryptophan - we must have at least three of the genes of the tryptophan biosynthetic pathway cloned. We have used several of these mutants in shot-gun cloning experiments with wild-type DNA, and are mapping the genes thus cloned to the minimal overlapping set of cosmids, by hybridization. There appear to be at least two, not-closely-linked, groups of tryptophan biosynthetic genes. Sequencing of one of these clones showed that it bears the halobacterial homolog of the *E. coli trp*B gene, and that, from raw sequence similarity, this gene is equally distant from its eubacterial and yeast homologs.

We have begun similar analyses of genes of histidine and leucine biosynthesis. At the same time, we are constructing new vectors, and vectors which bear cloned genes for tryptophan and serine tRNA. These are being altered at their anticodons by site specific mutation, to produce nonsense suppressors, which should be useful in the characterization of mutant strains and other more sophisticated analyses.

DISCUSSION

This laboratory is not the only one attempting to develop systems for genetic analyses of archaebacteria. The reader should also consult chapters from the groups of Dennis, Betlach, Mevarech, Oesterhelt, Rdest, Pfeifer, Amils, DasSarma, Kohiyama, Zillig and Forterre elsewhere in this volume. We are probably almost through with the "stare and compare" phase in the development of molecular archaebacteriology, ready to move into a period of modern molecular manipulation. Soon we can hope to approach halobacteria with the experimental confidence we had acquired with *Escherichia coli* by the mid-1970's. In one sense, this facility has been a long time coming. In another, the fact that only 35 years after the discovery of the nature of the genetic material we are ready to rearrange genes which took their current shape 3.5 billion years ago, beggars the imagination.

ACKNOWLEDGEMENTS

This work was supported by grants from the Medical Research Council of Canada and the U. S. Office of Naval Research. W. F. D. is a Fellow of the Canadian Institute for Advanced Research. R. L. C. and L. C. S. held Studentships from the Medical Research Council and Natural Sciences and Engineering Research Council of Canada.

REFERENCES

[1]   C. R. Woese, *Microbiol. Revs.* 51:221 (1987).

[2]   O. Kandler, "Archaebacteria", Gustav Fischer Verlag, Stuttgart (1982).

[3]   O. Kandler and W. Zillig, eds., "Archaebacteria '85", Gustav Fischer Verlag, Stuttgart (1986).

[4]   P. P. Dennis and A. T. Matheson, eds., Molecular Biology of Archaebacteria. *Can. J. Microbiol.* 35, No. 1. (1989).

[5]   M. Mevarech and R. Werczberger, *J. Bacteriol.* 162:461 (1985).

[6]   R. L. Charlebois, J. D. Hofman, L. C. Schalkwyk, W. L. Lam and W. F. Doolittle, *Can. J. Microbiol.* 35:21 (1987).

[7]   J. D. Hofman, L. C. Schalkwyk and W. F. Doolittle, *Nucleic Acids Res.* 14:6983 (1986).

[8]   S. W. Cline and W. F. Doolittle, *J. Bacteriol.* 169:1341 (1987)

[9]   S. W. Cline, W. L. Lam, R. L. Charlebois, L. C. Schalkwyk and W. F. Doolittle. *Can. J. Microbiol.* 35:148 (1989).

[10]   S. W. Cline, L.C. Schalkwyk and W. F. Doolittle, *J. Bacteriol.*, 171: 4987 (1989).

[11]   R. L. Charlebois, W. L. Lam, S. W. Cline and W. F. Doolittle, *Proc. Natl. Acad. Sci. USA* 84:8530 (1987)

[12]   J. A. Cabrera, J. Bolds, P. E. Shields, C. M. Havel and J. A. Watson. *J. Biol. Chem.* 261:3578 (1986).

[13]   W. L. Lam and W. F. Doolittle, *Proc. Natl. Acad. Sci. USA.* 86:5478 (1989).

GAS VACUOLE GENES IN HALOBACTERIA

Balakrishna Pillay[2], Ursula Rdest and Werner Goebel

Institut für Genetik und Mikrobiologie
Röntgenring 11
D-8700 Würzburg
Federal Republic of Germany

ABSTRACT

As previously shown by us there are two, highly homologous genes in *Halobacterium halobium* encoding gas vacuole proteins (GVP). These genes are located on the plasmid pHH1 (gene for GVP-A) and on the chromosome (gene for GVP-B). Recently we found that copies of both genes are also located on "minor circular DNA" (MCD), a heterogeneous collection of extrachromosomal DNA molecules which occur in very low concentrations in *H. halobium*. The origin of MCD is discussed using the *gvp* genes as markers.

In cyanobacteria there is an additional protein (GVP-c) involved in the formation of gas vesicles. Using a 25mer oligonucleotide deduced from a conserved region of the cyanobacterial *gvpC* gene, we could show hybridization to chromosomal DNA from *H. halobium*. Other halobacterial strains, natural isolates, rod shaped and purple membrane forming but not closely related to *H. halobium* also synthesize gas vacuoles. Examination of the strain *Halobacterium* sp. GRA revealed the presence of the *H. halobium gvpB* type gene and the *gvpC* gene in the chromosomal DNA and in the 160 kbp MCD. These genes were cloned from the plasmid DNA and sequenced. *gvpB* from GRA was shown to be almost identical to *gvpB* in *H. halobium*. For the *gvpC* gene our sequence data from both strains indicate that this gene is located at the same distance downstream from *gvpB*, the sequence between the two genes being identical.

---

[2]Present address of Balakrishna Pillay:
University of Durban-Westville, Private Bag X 54001, Durban, South Africa.

General and Applied Aspects of Halophilic Microorganisms
Edited by F. Rodriguez-Valera, Plenum Press, New York, 1991

INTRODUCTION

Some halobacterial strains express a phenotype which may serve as a suitable
genetic marker, viz. the formation of gas vesicles (GV). These are gas-filled structures
composed of a monolayer of protein molecules which enable the bacteria to float in the
medium. Among aquatic bacteria GV-formation is common [1], the best studied GV
are those from cyanobacteria [2, 3]. Simon [4] described in *H. salinarium* strain 5 two
morphologically different GV envelopes each of which consists of one protein type.
Recently we reported [5] on two types of gas vacuoles in *H. halobium*, spindle-shaped
and cylindrical. The former type migrates in a PAU-gel-electrophoresis at 19 kDa
(GVP-A) and the latter at 20 kDa (GVP-B). The N-terminal sequence of 40 amino
acids of the two proteins was determined and we found amino acid exchanges at two
defined positions Gly-7 and Val-28 in GVP-A are replaced by Ser-7 and Ile-28 in
GVP-B.

The halobacteria strains described here contain extrachromosomal DNA [6].
H. halobium carries two species of ccc-DNA, the plasmid pHH1, where most of the
known halobacterial insertion elements (ISH) are located [7] and a low-copy number
heterogeneous DNA population of large ccc-molecules which we described earlier as
minor circular DNA (MCD) [6]. The ccc DNA of *Halobacterium* sp. GRA is not related
to pHH1 and homology to ISH elements is not shown [6], but we have demonstrated
homology between MCD clones from *H. halobium* and cccDNA from the strain
Halobacterium sp. GRA. In our previus work [5] we studied the localization of the *gvp*
genes. For *H. halobium* we could demonstrate that GVP-A is encoded by a plasmid
pHH1-borne gene and GVP-B by a chromosomal *gvp*-gene.

RESULTS AND DISCUSSION

Characterization of the plasmid- and chromosomally encoded GVP genes in *H. halobium.*

Purified cccDNA of *H. halobium* was restricted with *Pst*I and *Eco*RI and
separated by agarose gel electrophoresis. The gene probes used for Southern
hybridization were synthetic oligonucleotides originally derived from the N-terminal
amino acid sequence of GVP-A using amino acids 14 to 22. Cyanobacterial and
halobacterial gas vacuole proteins have been shown to share a high degree of homology
[8]. This information suggested that a cyanobacterial *gvp* gene could be used in
heterologous probing to identify the related halobacterial *gvp* gene. The result obtained
with the synthetic oligonucleotides was confirmed by using a probe from the cloned
Calothrix *gvpA* gene, most of which is located on a 237 bp *Hinc*II-*Hind*III fragment
(kindly provided by N. Tandeau de Marsac). Both probes hybridized to a 34 kb *Pst*I-
and a 7 kb *Eco*RI fragment of digested pHH1 DNA. The *gvp* gene carried by pHH1
was subcloned in pTZ18 and pTZ19 (supplied by Pharmacia) as a 900 bp
*Eco*RI/*Hind*III fragment and as a 144 bp *Sau*3A fragment and sequenced according to
the method of Sanger.

Using the 144 bp *Sau*3A fragment of pHH1 as a probe the restricted chromosomal DNA showed hybridization of a 7,5 kbp *Pst*I fragment and a 3,65 kbp *Bam*HI fragment. The latter was subcloned as a 2,76 kbp and a 0,84 kbp *Bam*H/*Hind*III fragment and used for sequence determination. The comparison of the plasmid- and the chromosomally encoded *gvp* gene sequences (*gvpA* and *gvpB* respectively) (Figure 1) shows a high degree of homology at the amino acid level with two amino acid

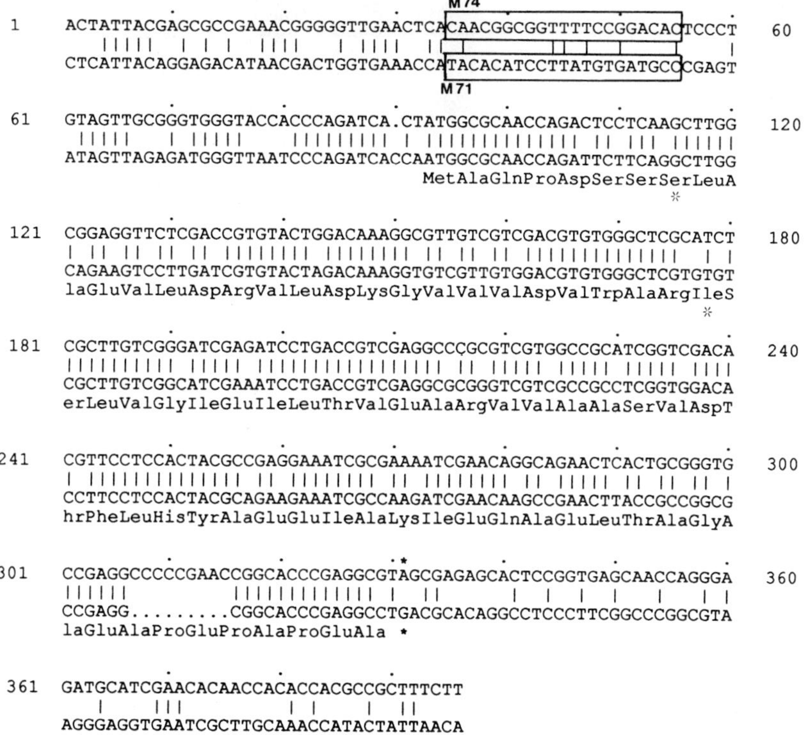

Figure 1. Nucleotide sequence of (above) gene *gvpB*, chromosomally encoded, (below) gene *gvpA*, plasmid encoded. The amino acid exchanges occur at positions 7 (S* → G) and 28 (I* → V). The sequences of the chromosome-specific (M 74) and the plasmid-specific (M 71) oligonucleotide probes are boxed.

exchanges determined previously by protein sequencing [5]. At the nucleotide level there is a homology of 85%, i.e. there are 35 basepair exchanges in the structural gene. A significant difference is that the coding sequence of the *gvpB* gene (240 bp, 79 amino acids) is nine nucleotides longer than *gvpA* (231 bp, 76 amino acids) between nucleotides 316 and 326. Upstream of the ATG start codon a significant degree of homology is observed between both genes, but very little homology can be seen for the sequence downstream of the TAG and TGA stop codons.

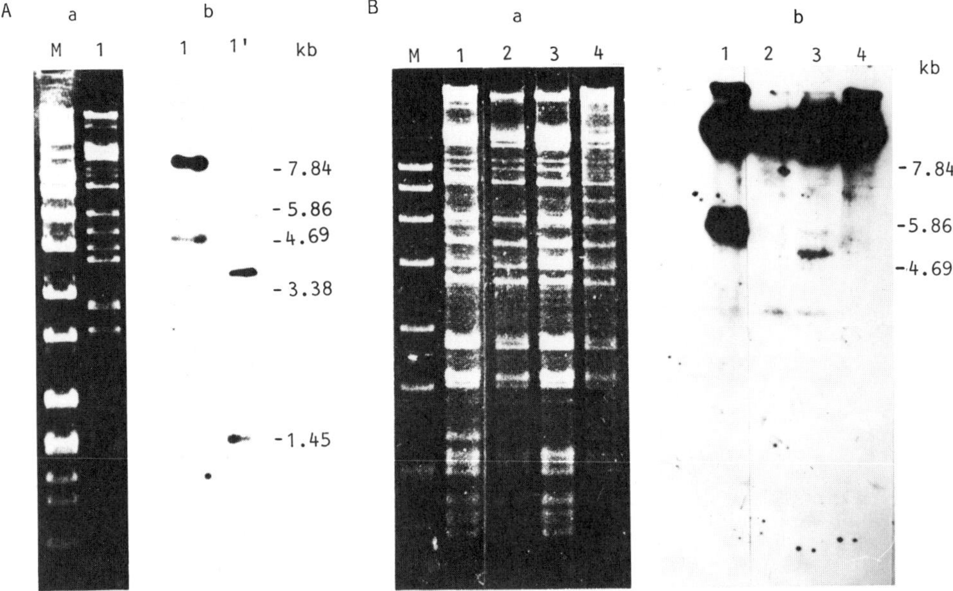

Figure 2. A. *H. halobium* cccDNA isolated from cultures of different single colonies (1-4), cleaved with *Pst*I. M71 was used as a *gvpA* specific probe. a) Ethidium bromide stained gel. b) Autoradiogram.

B. Hybridization of *H. halobium* MCD (*Pst*I/*Bam*HI digest) to both chromosome- and plasmid-specific gvp probes. a) Ethidium bromide stained gel. b) Autoradiogram with probe M71 (gvpA-specific)(1) and probe M74 (gvpB-specific)(2).

Presence of the *gvp* gene on the minor circular DNA (MCD)

It was speculated that MCD originates from (or is part of) the chromosome. Southern hybridization analysis of cccDNA of *H. halobium* with *gvp* specific probes revealed the presence of weaker hybridizing bands in addition to that produced by the strong hybridization signals of pHH1 (Figure 2a). These bands were not due to contamination by chromosomal DNA since there were several, and which also varied in size. Comparison of the autoradiograms with the ethidium bromide stained agarose gels showed that these weaker hybridizing bands correspond to the bands of the MCD.

Two *Pst*I/*Bam*HI fragments of the MCD 5,7 kb and 3,38 kb in size were cloned into pTZ19 and sequenced. From the nucleotide sequence it was observed that the clone carrying the 5,7 kb fragment has an almost identical sequence to that of the *gvpB* structural gene and its flanking sequences, with the exception of two base exchanges within the structural gene (nucleotide 221 C → T and nucleotide 278 A → G). In addition, six base-pair exchanges occur upstream of the structural *gvp* gene. A *gvpA* type gene is found in the clone containing the 3,38 kb of one alanine (nucleotide

280

306-308). Apart from this difference, identical nucleotide sequences are found both upstream and downstream of the structural gene.

Since the MCD *gvpA* gene shares, with exception of a deleted alanine, the exact same nucleotide sequence with the pHH1 *gvpA* gene it seems likely that the MCD *gvpA* and the part of the MCD in which it is found originates from the pHH1 plasmid. MCD isolated from *H. halobium* cultures inoculated with different single colonies shows a variable restriction pattern and also the *gvp* gene hybridizing to the specific *gvpA* probe M71 (Figure 1) is located on fragments of variable size (Figure 2b). This could be related to high variability found in pHH1 [7] due to insertion elements. In

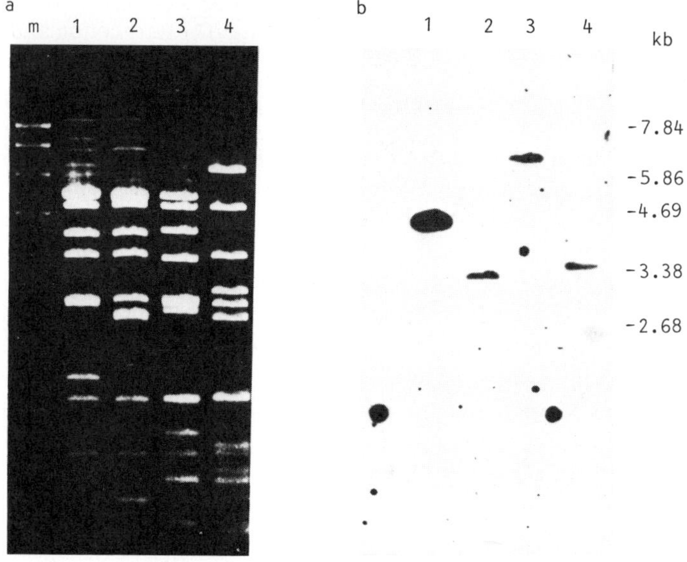

Figure 3.   *Halobacterium* sp. GRA cccDNA digests with *Pst*I, and *Hin*dIII (1), *Sph*I (2), *Eco*RI (3) and BamHI (4). Ethidium bromide stained gel. b) Autoradiogram, using the 144 bp *gvp* specific probe.

contrast the MCD *gvpB* gene appears more conserved and is always found on the same restriction fragment for different DNA preparations of *H. halobium* when hybridized to the *gvpB* specific probe M74 (Figure 1).

Characterization of the *gvp* gene of *Halobacterium* sp. GRA

The natural isolate *Halobacterium* sp. GRA contains three species of cccDNA with sizes of 65, 38 and 160 kbp. The largest is a MCD which Ebert et al. [6] demonstrated to hybridize to cloned MCD from *H. halobium*. Purified plasmid DNA of *Halobacterium* sp. GRA was double digested with *Pst*I and one of the following enzymes, *Hin*dIII, *Sph*I, *Eco*RI and *Bam*HI. Subsequent Southern hybridization analysis using the 144 bp *gvp* specific probe revealed a single hybridizing band for each of the restriction digests (Figure 3). Using the chromosomal- (M74) and plasmid- (M71) specific hybridization probes, only hybridization to M74 was obtained. The *Pst*I/*Sph*I fragments were cloned into pTZ19 and sequenced.

The extrachromosomal *gvp* structural gene of *Halobacterium* sp. GRA is located on the 160 MCD, it encodes a 76 amino acid protein and apart from four basepair exchanges (pos 118 A → T; 121 A → T; 231 T → C; 288 G → C) it is identical to that of the chromosomal *gvpB* gene of *H. halobium*. As a result of two of the base-pair exchanges the amino acid threonine instead of serine is found at positions 6 and 7. In addition six base-pair exchanges were obtained within the first 200 nucleotides 5' to the ATG start codon. The homology to the MCD *gvpB* gene of *H. halobium* is even closer. Within the first 200 nucleotides 5' to the ATG start codon and the first 230 nucleotides 3' to the TAG stop codon no base-pair exchanges occur between these two genes.

Purified chromosomal DNA from *Halobacterium* sp. GRA was subjected to the same cleavages as the plasmid DNA and hybridized to the *gvpA* (M71) and *gvpB* (M74) specific probes. The hybridization pattern was the same as shown for the plasmid DNA and also only visible with M74 as a probe in Southern analysis. For purity check of the plasmid DNA of *Halobacterium* sp. GRA the cloned 7S RNA gene which is chromosomally encoded [9] was probed against *Pst*I cleaved chromosomal and plasmid DNA. No hybridization could be detected in the plasmid fraction while a chromosomal fragment hybridized to the 7S RNA gene specific probe (data not shown). These investigations strongly suggest that a plasmid encoded *gvp* gene, in addition to a chromosomally encoded one is found in *Halobacterium* sp. GRA.

Characterization of *gvpC*.

Previous investigations of the gas vesicle composition suggested that there was only one type of protein (GVP) which formed this structure. However in *Calothrix* strain PCC7601 a gene *gvpC* which coded for a second component of 162 amino acids has been found [10]. More recently, Hayes et al. [11] sequenced the *gvpC* gene from *Anabaena flosaquae*. Using the sequence from highly homologous regions between the *gvpC* genes of the two organisms for a synthetic oligonucleotide probe, we did Southern analysis of the cloned *gvp* genes described above for detection of a *gvpC* gene.

The clones E6 (*H. halobium gvpA*), F5 (*H. halobium gvpB*), B4 (*H. halobium* MCD *gvpB*) and P/S44 (*Halobacterium* sp. GRA *gvpB*) were cleaved with *Eco*RI, *Bam*HI, *Bam*HI and *Pst*I/*Sph*I, respectively. Southern hybridization analysis of this DNA produced hybridization signals for all clones except for E6. This result indicated that the region downstream of the chromosomal type *gvp* genes were highly homologous while the plasmid type *gvp* genes did not share any homology in this region to the chromosomal *gvp* genes. Nucleotide sequence analysis revealed an open reading frame of 519 nucleotides which encoded a hydrophilic protein of 172 residues (data not shown). The characterization of the *gvp* genes may provide useful markers in vectors used in the development of a halobacterial transformation system. Further analysis of the gas vesicle genes and their transcripts in mutants of *Halobacterium* sp. GRA should reveal the mechanism of gas vacuole formation in halobacteria.

ACKNOWLEDGEMENTS

We thank W. Schmitt for typing the manuscript and B. Yoon for helpful discussions. This work was supported by grants from the Deutsche Forschungsgemeinschaft (Rd1/4-4) and the DAAD (to B.P.).

REFERENCES

[1]   A. E. Walsby, Structure and function of gas vacuoles, *Bacteriol. Rev.* 36: 1 (1972)

[2]   N. Tandeau de Marsac, D. Mazel, D. A. Bryant and J. Houmard, Molecular cloning and nucleotide sequence of a developmentally regulated gene from the Cyanobacterium *Calothrix* PCC7601: a gas vesicle protein gene, *Nucl. Acids Res.* 13: 7223 (1985)

[3]   P. K. Hayes, A. E. Walsby and J. E. Walker, Complete amino acid sequence of cyanobacterial gas-vesicle protein indicates a 70-residue molecule that corresponds in size to the crystallographic unit cell, *Biochem. J.* 236: 31 (1986)

[4]   R. D. Simon, Morphology and protein composition of gas vesicles from wild type and gas vacuole defective strains of *Halobacterium salinarium* strain 5, *J. Gen. Microbiol.* 125: 103 (1981)

[5]   B. Surek, B. Pillay, U. Rdest, K. Beyreuther and W. Goebel, Evidence for two different gas vesicle proteins and genes in *Halobacterium halobium*, *J. Bacteriol.* 70: 1746 (1988)

[6]   K. Ebert, W. Goebel and F. Pfeifer, Homologies between heterogeneous extrachromosomal DNA populations of *Halobacterium halobium* and four new halobacterial isolates, *Mol. Gen. Genet.* 194: 91 (1984)

[7]   F. Pfeifer, M. Betlach, R. Martienssen, J. Griedman and H. W. Boyer, Transposable elements of *Halobacterium halobium*, *Mol. Gen. Genet.* 191: 1982 (1983)

[8]   J. E. Walker, P. K. Hayes and E. A. Walsby, Homology of gas vesicle proteins in Cyanobacteria and Halobacteria, *J. Gen. Microbiol.* 130: 2709 (1984)

[9]   A. Moritz and W. Goebel, Characterization of the 7S RNA and its gene from halobacteria, *Nucl. Acids Res.* 13: 6969 (1985)

[10]   T. Damerval, J. Houmard, G. Guglielmi, K. Csiszar and N. Tandeau de Marsac, A developmentally regulated *gvpABC* operon is involved in the formation of gas vesicles in the cyanobacterium *Calothrix* 7601, *Gene* 54: 83 (1987)

[11]   P. K. Hayes, C. M. Lazarus, A. Bees, J. E. Walker and A. E. Walsby, The protein encoded by *gvpC* is a minor component of gas vesicles isolated from the cyanobacteria *Anabaena flosaquae* and *Microcystis* sp., *Mol. Microbiol.* 2: 545 (1988)

INSERTION ELEMENTS AFFECTING GAS VACUOLE GENE

EXPRESSION IN *HALOBACTERIUM HALOBIUM*

Felicitas Pfeifer, Mary Horne, Christoph Englert, and Ulrike Blaseio

Max-Planck-Institut für Biochemie
D-8033 Martinsried
Federal Republic of Germany

ABSTRACT

In *Halobacterium halobium* gas vacuoles are synthesized constitutively due to the expression of a plasmid borne gene (*p-vac*) whereas a related chromosomal *c-vac* gene copy is not active. In contrast, strains and species lacking the *p-vac* gene region synthesize gas vesicles in stationary phase of growth. The high mutation frequency of 1% affecting the *p-vac* gene expression in *H. halobium* is caused by the action of insertion elements either integrating up or downstream of the *p-vac* gene or inducing deletions encompassing the whole *p-vac* gene region. A burst of ISH27 transpositions was analyzed and led to the characterization of three types, one of which is more closely related to the *Haloferax volcanii* ISH51 family than to the other family members present in the same strain.

INTRODUCTION

The genetic instability of *Halobacterium halobium* has been known for a long time. Phenotypic mutants in gas vacuole (Vac), bacterioruberin or purple membrane synthesis occur at high frequencies [1]. Almost all mutants in purple membrane or gas vacuole synthesis are due to the transposition of an insertion element (ISH-element)-([2-6] and unpublished results). Nine of these elements are summarized in Table 1 [2-6]. While all ISH1 and ISH2 copies characterized thus far are identical at the nucleotide level, ISH23, ISH26 as well as ISH27 appear to be insertion element families consisting of different, though highly similar members. Two types of ISH23 and ISH26 have been characterized [3,5], and three classes of ISH27 [7]. The gas vacuole production of *H. halobium* results in a pinkish-opaque appearance of the colonies on

General and Applied Aspects of Halophilic Microorganisms
Edited by F. Rodriguez-Valera, Plenum Press, New York, 1991

Table 1. Insertion elements of *Halobacterium halobium*. The copy number determined in FI-DNA (high G+C content) and FII-DNA is given. The last value given for FII-DNA is the number of ISH-elements present in pHH1.

ISH-element	size (bp)	inverted repeat	copy-number	distribution in FI-DNA	FII-DNA
ISH1	1118	TGCCTtGT	2	1	1
ISH2	520	CATTCGTCTTTAGTTAAGA	8–10	1	>4 + 4
ISH26	1384	CATCCGTCTTTAGTTA	>10	–	>8 + 2
ISH26–1	1705	"			1
ISH23	996	CGCTCTTGtGgcGgaGATTGTTGGTgAGT	2	1	1
ISH24	3000	ctgaTGTTGCGAAG	2	–	2
ISH27–1	1398	CAGTACcTCACAAAGC			
ISH27–2	1389	"	>10	–	>8 + 2
ISH27–3	1389	"			
ISH28	1000	GGCTCTGGTGAATCGC	>10	–	>8 + 1
ISHS1	1449	GGTCGcgtCGCAAAGCGGTCTAGAGTT	?	?	?

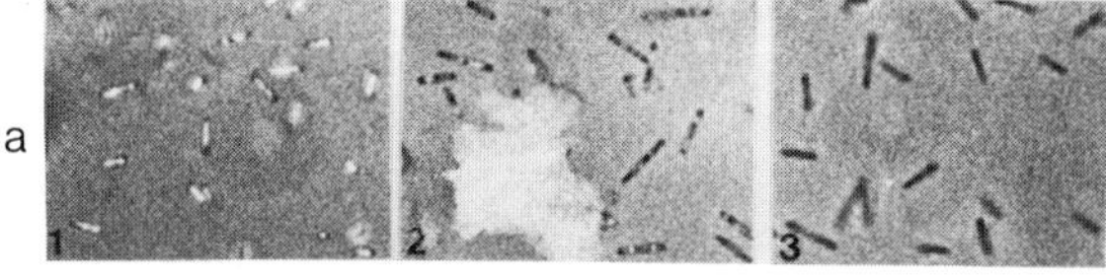

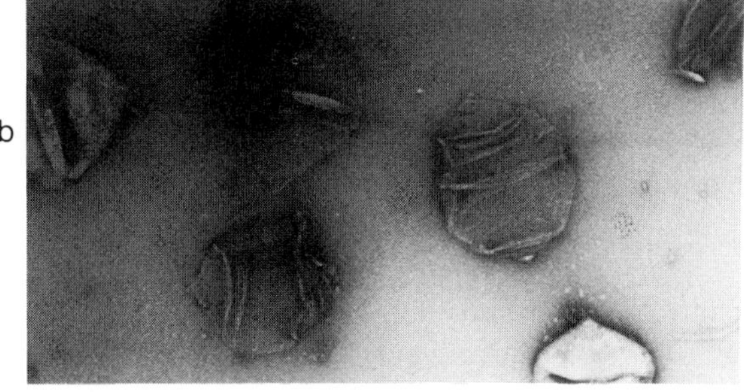

Figure 1. a) Phase contrast microscopy of H. halobium. Gas vacuoles appear as white particles within the rod-shaped cell. 1: wild-type, 2: *p-vac* deletion mutant; gas vacuoles are produced by the *c-vac* gene, 3: *vac* negative mutant.

b) Electron microscopic picture of isolated gas vesicles of *H. halobium* wild-type. The size of a single gas vesicle is about 0.3x0.2 *μ*m.

agar plates. In the phase contrast microscope these gas vesicles are visible as light refractile bodies (Figure 1a). On the average, there are about 70 gas vesicles found in a wild-type cell [8]. The electron microscopic picture of isolated gas vesicles indicates a rib-like structure (Figure 1b). The gas vesicle "membrane" contains no lipids, but consists of a hydrophobic major protein with a molecular mass of about 8 kilo Dalton (kD). *H. halobium* synthesizes gas vacuoles throughout the growth cycle, whereas the *Halobacterium* species GN101, YC819-9 and *Haloferax mediterranei* start to synthesize gas vesicles in stationary phase of growth ([10] and unpublished data). In this paper we summarize our current knowledge on the expression of genes encoding gas vacuole proteins in *H. halobium* and other species, as well as the effect of insertion elements on the expression of the *vac* gene located in pHH1, the 150 kbp major plasmid species of *H. halobium*. Experimental details have been published elsewhere.

RESULTS AND DISCUSSION

Genes encoding gas vacuole proteins

Two genes encoding a gas vacuole protein have been isolated from *Halobacterium* halobium. One of these (*p-vac*) is located in pHH1, whereas another one has been identified in the chromosomal DNA (*c-vac*)[9]. A plasmid borne *vac* gene similar to *p-vac* has been reported by DasSarma et al; however, the carboxy-terminal amino acid of the protein derived from this gene is different due to a missing cytosine residue in the DNA sequence [9,11]. The coding region of the *p-vac* and the *c-vac* gene differs by 35 nucleotide changes; however, most of these occur in the third position of the codon and result only in two amino acid changes and a three amino acid deletion near the carboxy-terminus of the *p-vac* protein (Figure 2). The other *Halobacterium* species contain only a *c-vac* gene in their chromosome [10]. The coding region of the *vac* gene characterized for *Haloferax mediterranei* also contains 35 nucleotide differences; the resulting amino acid sequence indicates only one amino acid change (pos. 2), and a deletion of three amino acids and a two amino acids insertion near the carboxy terminus (unpublished data). The three halobacterial gas vesicle proteins are highly similar to each other, and slightly larger than the gas vesicle proteins characterized for cyanobacterial species (Figure 2) [12].

Expression of *vac* genes during the growth cycle

H. halobium wild-type containing the *p-vac* and *c-vac* gene produces gas vacuoles throughout the growth cycle exclusively due to *p-vac* gene expression [10]. Analysis of the *p-vac* mRNA level during logarithmic and stationary phase of growth by Northern analysis indicates a maximum amount of *p-vac* message in samples taken during log. growth (Figure 3a, lanes 5-9). No *c-vac* mRNA can be detected using a *c-vac* gene specific probe to challenge the same Northern blot (Figure 3b, lanes 5-9). In contrast, strains containing exclusively a *c-vac* gene (i.e. *H. spec.* GN101 or *H. halobium p-vac* deletion mutants) synthesize *c-vac* mRNA (and gas vacuoles) in stationary phase of growth (see Figure 3b, lanes 1-4). Thus, gas vesicles are either synthesized by the expression of the *p-vac* gene or - in strains lacking the *p-vac* gene

```
      1         10        20        30        40        50        60        70        80

Hhc  MA   QPDSS SLAEVLDRVL DKGVVVDVWA RISLVGIEIL TVEARVVAAS VDTFLHYAEE IAKIEQAELT AGAEAPEPAP   EA  79 aa

Hmc  .V   ..... .......... .......... .......... .......... .......... .......... .....      .. TP..  78 aa

Hhpl ..   ..... G......... .......... .V........ .......... .......... .......... .....      ..    ..  76 aa

Cal  ..VEKTNS.. .....I..I. ...I...A.V .V......L. AI...I.I.. .E.Y.K...A VGLTQS.AVP .                71 aa

Ana  ..VEKTNS.. .....I..I. ...I.I.A.V .V.....QL. AI...I.I.. .E.Y.K...A VGLTQS AVP .A               71 aa

Mic  ..VEKTNS.. .....I..I. ...I.I.A.. .V......L. AI.....I.. .E.Y.K...A VGLTQX.XXA ?                70 aa
```

Figure 2. Comparison of gas vacuole proteins from *H. halobium*, *H. mediterranei* and various cyanobacteria. Hhc= *c-vac* protein, Hhpl= *p-vac* protein of *H. halobium*, Hmc= *H. mediterranei* protein, Cal= Calothrix, Ana= Anabaena, Mic= Microcystis gas vesicle proteins.

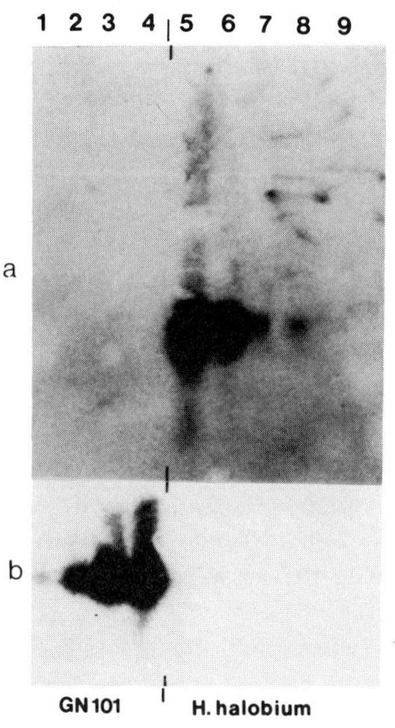

Figure 3. Northern analysis of RNA produced during the growth cycle of *H. halobium* (lanes 5-9) and *Halobacterium* species GN101 (lanes 1-4). Samples were taken during growth and the total RNA was hybridized with the *p-vac* gene (a) or the *c-vac* gene (b). Lanes 1,2,5,6 contain RNA of samples taken during log. growth, whereas lanes 3,4 and 7-9 are samples taken in stationary phase of growth.

region - by the expression of the *c-vac* gene [10]. The exclusive expression of the *p-vac* gene in *H. halobium* wild type suggests a direct or indirect suppression of the *c-vac* gene by products derived from the *p-vac* gene region. Integration of an insertion element in the *p-vac* gene itselt does not lead to *c-vac* gene expression (unpublished data). Similarly, the *vac* gene of *Haloferax mediterranei* is only expressed during the stationary phase of growth when the strain is cultured in medium containing 25% salt (unpublished data). No gas vacuole production is observed when *H. mediterranei* is grown in medium containing 15% salt [13].

Mutations in the *p-vac* gene region

The plasmid borne *p-vac* gene of *H. halobium* is affected by mutations that occur at a frequency of 1%, whereas the *c-vac* gene incurs mutations with much lower frequency [14]. In certain mutant strains, the *p-vac* gene can be affected by such high mutation frequencies that the colony contains sectors of vacuolated and non-vacuolated cells (Figure 4). Most sectors seem to originate in the centre of the colony indicating a high instability of the gas vacuole phenotype. Two types of mutations were observed that affect the expression of the *p-vac* gene. These mutations involved either the integration of an ISH-element up or downstream of the coding region ([15] and our unpublished results), or are due to deletions encompassing the entire *p-vac* gene region [16]. These mutants are easily visible as transparent colonies on agar plates, since the *c-vac* gene is not expressed during logarithmic growth.

p-vac mutants usully incur ISH-elements leading to alterations in the pHH1 restriction pattern [1] (Figure 5). The inserts in the *p-vac* gene fragment are found within the promoter region and up to 3 kbp upstream or downstream of the *p-vac* gene indicating that additional genes are necessary for gas vacuole synthesis ([15] and our unpublished data). Products of these putative genes act either as regulatory elements or are involved in the formation of the gas vesicle. At least one additional protein (*gvpC*) is known for the cyanobacterial gas vesicle that might stabilize the structure [17]. In halobacteria, each *vac* gene is transcribed as single unit, and additional proteins have not been identified so far. A more detailed analysis of the proteins found in gas vesicle preparations is in progress to identify further structural components.

p-vac deletion mutants: analysis of deletion formation

Three pHH1 deletion variants (pHH23, pHH3 and pHH4) could be isolated from *vac* mutants of *H. halobium* - all of them lost various parts of the parental plasmid (Figure 6a). All these strains start to synthesize gas vacuoles in the late stage of growth due to *c-vac* gene expression. To analyze the deletion formation and to localize the region containing functions for the pHH1 plasmid replication, further deletion derivates were obtained starting with the 36 kbp plasmid pHH4 as parental plasmid. Smaller deletion derivatives with sizes ranging from 5.7 kbp can be identified on agarose gels [16]. Four deletion derivatives were investigated in more detail (Figure 6b).

The region containing the fusion site in each of the plasmids was compared to the appropriate regions in the parental plasmid [16]. The high frequency of deletion

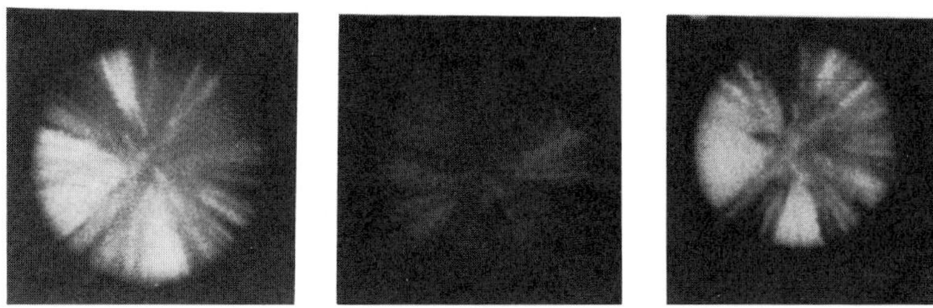

Figure 4. Sectored colonies of a mutant exhibiting a high frequency of variation in the gas vacuole phenotype. White sectors contain vacuolated cells, whereas the dark sectors are formed by gas vacuole negative cells.

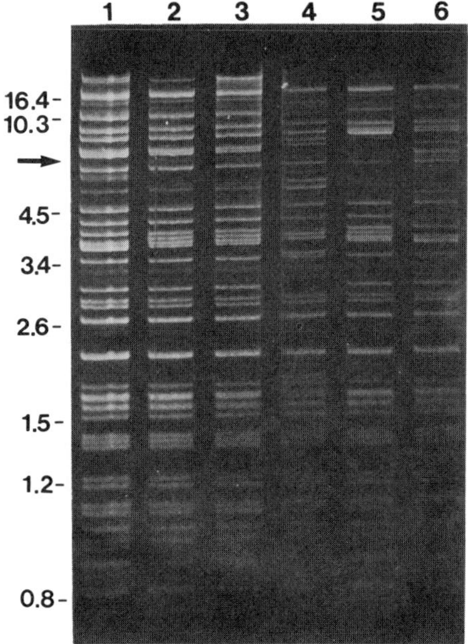

Figure 5. *Eco*RI fragment pattern of plasmids isolated from various mutants. The sizes of the fragments are given in kilobase pairs. The arrow depicts the 6.8 kbp fragment containing the *p-vac* gene.

formation suggests the involvement of ISH-elements. As shown in Figure 6b, the 17 kbp and the 16 kbp plasmids pHH6 and pHH7 contain ISH2, while pHH8 and pHH9 contain and ISH27-type element (ISH27-1) next to the fusion site. Presumably the deletion results from transposition of these ISH-elements within the same plasmid molecule. Such an intramolecular transposition leads - if the element integrates in the same orientation - to two DNA molecules, each of these containing a single copy of the ISH-element. The DNA molecule containing the functions necessary for autonomous replication and maintenance survives, whereas the non-replicative molecule will be lost. The identification of a typical ISH27 target sequence at the

290

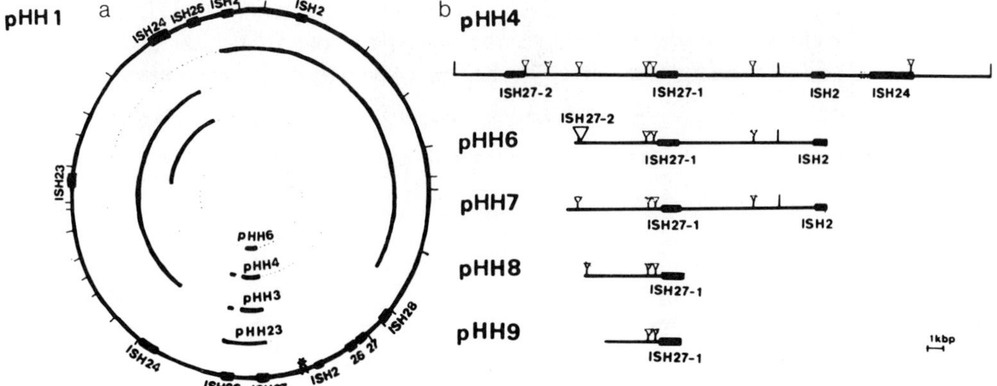

Figure 6. a) Deletion variants of pHH1. The outer circle is a *Pst*I restriction map of pHH1. The smaller variants were isolated from *vac* mutants. Black bars represent parts of pHH1 that are still present, whereas the dotted lines indicate deletions.
b) Deletion variants of pHH4. The restriction sites are *Pst*I and *Cla*I. The locations of ISH-elements within the plasmids are indicated.

fusion site of both pHH8 and pHH9 supports this conclusion [7]. Such IS-element induced deletions have been characterized in E. coli as well [18].

Thus, both types of *p-vac* mutations, insertions and deletions, are caused by the action of insertion elements. Since the transposition frequency of ISH-elements in *H. halobium* is high - especially within pHH1 - mutations in the *p-vac* gene region occur at a higher frequency than in the *c-vac* gene.

Burst of ISH27 elements

During the course of the enrichment for smaller pHH4 derivatives a burst of ISH27 elements was observed: 6 out of 23 single colonies analyzed contained a pHH4 plasmid that was altered by the integration of an additional ISH27 element [16]. Among the six elements investigated two types of ISH27 elements (designated ISH27-2 and ISH27-3) were identified. Four copies belong to the ISH27-2 class, and the other two were of the ISH27-3 type. These element types are 95% similar throughout the first 1200 bp, and only 68% similar in the last 200 bp (data not shown, see Table 2). A third ISH27 element-type (ISH27-1) is present in pHH4. The DNA sequence of ISH27-1 indicates only 82% similarity to both ISH27-2 and ISH27-3, but 90% similarity to the *Haloferax volcanii* element family ISH51 [19]. The 200 bp region of higher sequence diversity is 80% similar between ISH27-1 and ISH51 (Table 2). Thus, the ISH27 insertion elements of *H. halobium* constitute a family, and one type is more closely related to the ISH51 element family of *H. volcanii* than to other family members of the same strain. In contrast to the low activity of ISH51 in *H. volcanii*, the ISH27 elements are highly active in *H. halobium*.

Table 2. DNA sequence similarity between ISH27 elements and ISH51. The upper right half of the table gives the values for the first 1200 bp of ISH27, whereas the numbers given in the lower left half of the table are for the last 200 bp.

	ISH27-1	ISH27-2	ISH27-3	ISH51
ISH27-1	-	83	82	90
ISH27-2	64	-	95	84
ISH27-3	74	68	-	82
ISH51	80	68	68	-

Stress induces transpositions in *H. halobium*

The original sample of the *H. halobium* strain containing pHH4 derived from a liquid culture of the strain collection at the University of Würzburg kept at 4°C for several years. This culture contained only a few viable cells. The stress conditions during storage of this strain presumably induced transpositions at a higher frequency than is usually observed. The same analysis done with a pHH4 containing single clone which was repeatedly transferred to fresh medium and plated at various time points, did not indicate a high rate of transposition events (unpublished data).

These experiments suggest that there is a very high transposition rate in *H. halobium* under stressful conditions, whereas under "normal" growth conditions transpositions occur less frequently. Despite the high mutation rate *H. halobium* still survives in its habitat. One reason might be that almost all of the transposition events are confined to certain regions in the genome consisting of DNA of significantly lower guanosine plus cytosine content than the main part of the genome [20, 21]. This FII-DNA fraction contains almost all copies of the insertion elements characterized thus far, and pHH1 is part of this fraction (Table 1). All known "house keeping" genes of *H. halobium* (such as the DNA-dependent RNA-polymerase, bacterio-opsin, bacterio-opsin related protein, ferredoxin, rRNA and tRNA) are located in the more G+C rich DNA fraction that is less frequently affected by the action of transposable elements.

ACKNOWLEDGEMENTS

We wish to thank W. Zillig for discussions and support, and P. Ghahraman for valuable technical assistance. U. Santarius and W. Baumeister are thanked for help in electron microscopy. This work was financially supported by the Deutsche Forschungsgemeinschaft (SFB 145/B6) and by the Max-Planck- Gesellschaft.

REFERENCES

[1]   F. Pfeifer, G. Weidinger and W. Goebel, *J. Bacteriol.* 145:375 (1981)

[2]   M. Simsek, S. DasSarma, U. RajBhandary and H. Khorana, *Proc. Natl. Acad. Sci. USA* 79:7268 (1982)

[3]   F. Pfeifer, J. Friedman, H. W. Boyer and M. Betlach, *Nucl. Acids. Res.* 12:2489 (1984)

[4]   S. DasSarma, U. Rajbhandary and H. Khorana, *Proc. Natl. Acad. Sci. USA* 80:2201 (1983)

[5]   K. Ebert, C. Hanke, H. Delius, W. Goebel and F. Pfeifer, *Mol. Gen. Genet.* 206:81 (1987)

[6]   F. Pfeifer, H. W. Boyer and M. Betlach, *J. Bacteriol.* 164:414 (1985)

[7]   F. Pfeifer and U. Blaseio, *J. Bacteriol.* 171:5135 (1989)

[8]   W. Stoeckenius and W. Kunau, *J. Cell Biol.* 38:337 (1968)

[9]   M. Horne, C. Englert and F. Pfeifer, *Mol. Gen. Genet.* 213:459 (1988)

[10]   M. Horne and F. Pfeifer, *Mol. Gen. Genet.* 218:437 (1989)

[11]   S. DasSarma., T. Damerval, J. Jones and N. Tandeau de Marsac, *Mol. Microbiol.* 1:365 (1987)

[12]   J. Walker, P. Hayes and A. Walsby, *J. Gen. Microbiol.* 130:2709 (1984)

[13]   F. Rodríguez-Valera, G. Juez and D. Kushner, *Syst. Appl. Microbiol.* 4:369 (1983)

[14]   F. Pfeifer, U. Blaseio and M. Horne, *Can. J. Microbiol.* 35:96 (1989)

[15]   S. DasSarma, J. Halladay, J. Jones, J. Donovan, P. Giannasca and N. Tandeau de Marsac, *Proc. Natl. Acad. Sci. USA* 85:6861 (1988)

[16]   F. Pfeifer, U. Blaseio and P. Ghahraman, *J. Bacteriol.* 170:3718 (1988)

[17]   A. Walsby and P. Hayes, *J. Gen. Microbiol.* 134:2647 (1988)

[18]   P. Nisen, D. Kopecko, J. Chou and S. Cohen, *J. Mol. Biol.* 117:975 (1977)

[19]   J. Hofman, L. Schalkwyk and W. F. Doolittle, *Nucl. Acid Res.* 14:6983 (1986)

[20]   K. Ebert and W. Goebel, *Mol. Gen. Genet.* 200:96 (1985)

[21]   F. Pfeifer and M. Betlach, *Mol. Gen. Genet.* 198:449 (1985)

GENOMIC ORGANIZATION STUDIES IN HALOBACTERIA

USING PULSE FIELD GEL ELECTROPHORESIS

J. P. Abad[1], C. Smith[2] and R. Amils[1]

[1]Centro de Biología Molecular
UAM–CSIC
Universidad Autónoma de Madrid
Canto Blanco, Madrid 28049, Spain

[2]Human Genome Center, Lawrence Berkeley Laboratory
University of California, Berkeley, C.A.94720
United States of America

ABSTRACT

The interest of geneticists has been focused on *Halobacteria* due to their archaebacterial nature and genetic instability. Using pulsed field gel electrophoresis we have obtained information about the genome size, number of chromosomes, their topology, the comparative number of rRNA genes, the comparison of restriction maps and the type of extrachromosomal genetic information of halobacterial species belonging to the main taxonomic groups.

INTRODUCTION

In recent years the discovery of pulsed field gel electrophoresis (PFGE) [1] which resolves nucleic acids at the megabase size, and its application to the study of microbial genetics [2], have broadened our knowledge of the size, topology and organization of prokaryotic genomes, helping to unravel the evolutive relationships between different cell lines. Moreover, this technique can generate genetic information on new microorganisms for which there is none available and for those such as strict chemolithotrophs for which classical genetic approaches cannot be applied because of the intrinsic difficulty in isolating mutants. In this paper we report comparative results on size, topology and organization of the genome of different extremely halophilic archaebacteria generated with PFGE techniques.

MATERIALS AND METHODS

- **Strains**. *Haloarcula californiae* ATCC 33799, *"Halobacterium marismortui"*, *Haloferax gibbonsii* ATCC 33959, *Halococcus morrhuae*, *Halobacterium halobium*

NCMB 777, *Halobacterium salinarium* CCM 2148, *Haloferax mediterranei* ATCC 33580, *Natronobacterium pharaonis* NCMP 2191 (sp-1), *Halococcus sinaiiensis* ATCC 33800, were kindly provided by F. Rodriguez-Valera, A. Ventosa and H. Ross and grown in the conditions described in [3].

- **DNA preparation**. Cells were grown until mid log phase, collected by centrifugation at low speed and resuspended in the same buffer as described for *E. coli* [4] but with 2.5 M NaCl to avoid disrupting the halophilic cells prior to their inmobilization in agarose. The digestion and electrophoretic conditions are those described in references [4] and [5].

- **Restriction and hybridization**. DNA obtained from each preparation was prerun at 100 sec pulse time in order to remove plasmids and degraded DNA. Macrorestriction fragments were resolved by PFG electrophoresis in the conditions described in the legend of the figures. The PFG electrophoretic gels were blotted and hybridized with different ^{32}P labeled restriction fragments and *Hf. mediterranei* 16S and 23S rRNA which were also ^{32}P labelled [6]. The size of the chromosome according to Lee and Smith [7] and the physical map after Smith et al. [2].

RESULTS AND DISCUSSION

Since the discovery that some microorganisms formerly assigned to the prokaryotic group, based on their morphological features, formed a coherent genotypic cluster as different from the rest of the prokaryotes as from the eukaryotes [8], and their subsequent adquisition of kingdom status as archaebacteria, an incredible amount of genetic information has been generated which answers some basic evolutionary questions about the relationship between different cell lines.

The extreme halophiles as well as belonging to the archaebacterial kingdom have the additional interest of permitting the molecular bases of biochemical performance in extreme ionic conditions to be studied. There is a great deal of information concerning the genetic instability of these microorganisms, which has been correlated to the presence of insertion elements and plasmids responsible for their extremely efficient recombination [9].

A considerable amount of information also exists on the physiology, ecology and biochemistry of these microorganisms due to the intrinsic interest of their performance in extreme halophilic conditions. The taxonomy of the group is rather controversial. It has been extensively reviewed by Rodriguez-Valera et al [10] who claim the existence of six genus: *Halobacterium, Haloarcula, Haloferax, Halococcus, Natronobacterium* and *Natronococcus*, based on morphological, physiological and chemotaxonomic properties.

The use of 16S rRNA sequence comparison for three different species of *Halococcus, Halobacterium* and *Haloferax* [11] reveal a high degree of homology suggesting a low extent of diversification in this group of archaebacteria which contrasts with that of the methanogenic or sulfur dependent groups. Recently the molecular biology of this group has gained importance due to the obtention of ribosomal subunit cristals which, in theory can allow highly resolved difraction patterns to be obtained [12]. The correlation of structural and functional data on the translational apparatus and the resulting increase in information on the regulation of transcription and genomic organization of these microorganisms suggests that the halophilic system will be among the first to be understood at the molecular level. Following this line we decided to use PFGE techniques to gain insight into the genomic characteristics of this important group of archaebacteria. This report describes the current status of the research in progress in our lab.

Extrachromosomal genetic information

Since the genetics of halophilic archaebacteria was first studied, the high degree of genetic variability in relation to bacteriorhodopsin and gas vacuole gene expression has been related to the presence of insertion elements and plasmids [9]. Great advances have been made in the characterization of halophilic plasmids and the study of their high level of recombination [13]. We have applied the properties of PFGE electrophoresis to a comparative study of the extrachromosomal elements in halobacteria. When intact halophilic DNA is obtained after disruption and selective digestion of the RNA and proteins in conditions which avoid the damage by physical shearing, and then run in PFG at high pulse times (100 seconds), a series of discrete bands appear with different relative mobilities which are very reproducible. The number of these bands and their relative mobilities are peculiar to each species (Figure 1A).

The susceptibility of this material to DNAse and restriction enzymes suggests that they correspond to DNAs smaller than the main chromosome(s) which remain at the origin of the gel in these electrophoretic conditions. In order to examine the topology of these extrachromosomal elements, electrophoresis at different pulse times were performed to evaluate their relative electrophoretic mobility in relation to lineal DNA markers like *S. cerevisiae* chromosomes and/or lambda phage concatemers. The results obtained show that they are not linear and that their topology must be complex because their relative mobility changes with the pulse time resulting in inversions in some cases which make their study difficult (Figure 1B). Further analysis is needed in order to determine the size and topology which produces their abnormal behaviour in the electrophoretic field.

Genome size

Until the appearance of PFGE this important evolutionary query could only be answered by indirect methods based on renaturation kinetics, with a marked bias produced by the GC content of the DNA. The obvious application of PFG techniques to obtain absolute values of genome size generates extremely valuable data on the differential amount of genomic information present in microorganisms (Table I). This table shows that the prokaryotes exhibit an interesting variability in the genome size, from one to ten megabases, which might be related to the amount of information required to perform under different ecological conditions: smaller genome size for those microorganisms living in symbiotic relationships with higher eukaryotic systems and one order of magnitude higher size for those microorganisms living in variable habitats. The halophilic archaebacteria exhibit a rather intermediate size between *E. coli* and *Mycoplasma*. For *Hf. mediterranei* the average size calculated from *Bam*H I restriction maps is around 3.5 Mb (Figure 2). *Nb. pharaonis* shows a smaller chromosome of around 2.8 Mb. The rest of the halobacterial species cannot be measured accurately due to the obtention of overlapping fragments of small molecularweight (smaller than 40 Kb) with the restriction enzymes used so far.

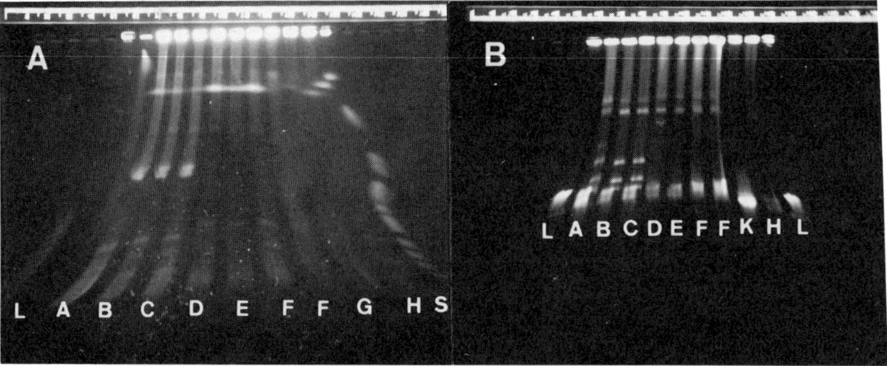

Figure 1.   Pulsed field gel electrophoresis of intact DNA from different halobacteria. Run A, 40 hr., 330 volts, 100 sec. pulse time. Run B, 40 hr., 330 volts, 1 sec. pulse time. The samples in the different wells are the following: A) *Hc. californiae*, B) *Hb. marismortui*, C) *Hb. salinarium*, D) *Hf. gibbonsii*, E) *Hc. sinaiinsis*, F) *H. mediterranei*, G) *Nb. pharaonis*, H) *Hf. volcanii*, L) concatemers of lambda phage and S) *S. cerevisiae* chromosomes.

Topology of the main chromosome

PFGE allows the topology of genomes smaller than ten megabases [18] to be ascertained. Long runs at large pulse times which resolve the linear chromosomes of *Schizosaccharomyces pombe* (3.5, 4.6 and 5.7 megabases respectively) do not allow the chromosomes of different halobacterial species to enter into the gel (Figure 1), suggesting a circular structure. Partial linearization of *H. mediterranei* chromosome shows that the linear form runs near the 3.5 megabase marker (Figure 3), which agrees with the apparent size calculated by restriction mapping, while the main part of the intact chromosomes remains at the origin as expected for circular DNA structures of this size.

298

Table 1. Genome size of different microorganisms determined by PFG.

Microorganism	Size	Topology	Reference
Escherichia coli	4.700 Kb	circular	(2)
Haemophilus influenzae	1.980 Kb	?	(14)
Mycoplasma	900–1.280 Kb	circular	(15)
Chlamydia	1.450 Kb	?	(16)
Rickettsia	1.720–2.100	linear	(16)
Boorrelia burgdorfei	950 Kb	linear	(17)

Figure 2.   *Bam*H I macrorestriction fragments of different halobacterial chromosomes separated by PFG. DNA was prepared according to the materials and methods section. The run was done at 500 volts for 3 days at 25 sec. pulse time. The species in the different wells are: A) *Hc. californiae*, B) *Hb. marismortui*, C) *Hb. salinarium* D) *Hf. gibbonsii*, E) *H. mediterranei*, F) *Hc. sinaiiensis*, G) *Nb. pharaonis*, L) concatemers of lambda phage, S) *S. cerevisiae* chromosomes.

299

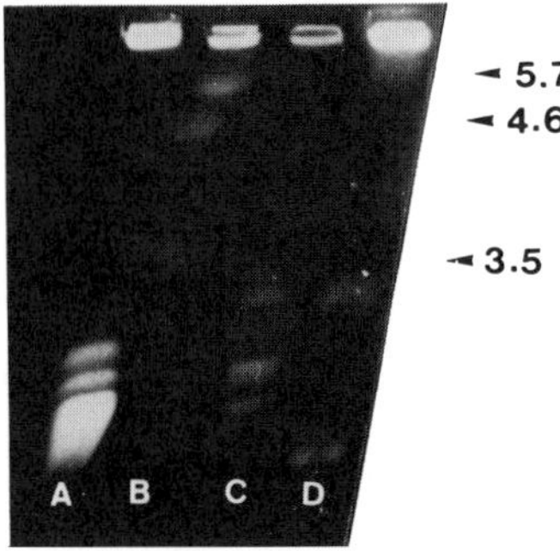

Figure 3.   Partial linearized chromosomes of *H. mediterranei* separated by PFG. The electrophoretic conditions were the following: 3 volt/cm., 7 days run, 4.500 sec. pulse time. DNA samples: A) *S. cerevisiae*, B) *S. pombe*, C) *Picchia sinaiensis*, D) *H. mediterranei*. The numbers on the right mark the size of the standards in megabases.

Restriction patterns comparison

During the search for appropriate restriction enzymes capable of sizing different halophilic chromosomes into discrete pieces of adequate size, the striking observation was made that most of the halobacteria exhibited different restriction patterns suggesting a rather divergent genomic primary structure in contrast to the results obtained with 16S rRNA sequence comparison [11]. Figure 2 shows the restriction patterns of several halobacterial chromosomes digested with *Bam*H I. Of course the observed differences suggesting dissimilar numbers of recognition sites in each chromosome and different relative positions could be generated by the differential modification of the sequence in different cell lines. Preliminary experiments performed to evaluate the amount of sequence homology existing between halophiles using hybridization with specific restriction fragments, suggest a rather low level of homology between the species studied (data not shown).

Physical maps and genome organization

The restriction patterns obtained for *H. mediterranei* and *Nb. pharaonis* with *Bam*H I enable us to construct physical maps of those halophiles. The "top down" approach consisting of the orientation of the macrorestriction fragments generated with a rare cutter by hybridization with fragments obtained by another restriction enzyme; combined with the "bottom up" approach based on the linkage of clones by the identification of overlaps and the correspondent generation of large scale maps from the arrangements of individual clone maps, will allow the physical maps of taxonomically interesting halophilic archaebacteria to be obtained. Using the "top down" approach 80% of the *H. mediterranei* and 70% of the *Nb. pharaonis* physical maps have been obtained in our lab so far.

300

Table 2. rRNA genes detected in halobacterial chromosomes using PFGE.

Microorganism	Minimum gene number
Haloarcula californiae	4
Haloferax gibbonsii	4
Halobacterium halobium NCMB 777	3
Haloferax mediterranei	4
"Halobacterium marismortui"	3
Halococcus morrhuae	2
Halobacterium salinarium	1

As mentioned above the obtention of restriction patterns for different enzymes using taxonomically different halophilic archaebacteria together with the corresponding hybridization experiments designed to answer basic questions on the genome organization of halobacteria, such as the importance of the insertion elements in generating genomic instability, or the identification of homologous genes and the correspondent sequence comparison, creates a controversial picture that deserves further clarification. On one hand the differential restriction patterns suggest a highly variable degree of genomic organization which contrasts with the homology exhibited by the 16S rRNA sequence. Obviously, we cannot rule out the possibility that differential restriction patterns are due to variations in the level of regulation (methylation) and/or to the recombinatory effect produced by insertion elements and plasmids, which abound in halobacteria. We addressed the question of variable genome organization in halophiles by studying the distribution of rRNA genes in halophilic species. Several probes were used to generate data on the organization of the three species of rRNA present in halophilic ribosomes and to facilitate the interpretation of the hybridization experiments [19].

Table II shows the minimum numbers of rRNA gene clusters obtained with halobacteria. The heterogeneity observed in the number of rRNA genes contrasts with the high degree of conservation reported for their sequence. Earlier studies showed that the sulfur dependent branch of archaebacteria is uniform in having only one copy of its rRNA genes, although some differences in their arrangements were reported [20]. The situation changes in the methanogenic–halobacterial branch which is highly variable. This differential characteristic in gene organization in archaebacteria may be of phylogenetic relevance. Current studies in rRNA gene organization in eubacteria are also creating controversy. The report that the extreme chemolithotroph *Thiobacillus ferroxidans* has only two rRNA genes (I. Marín, personal communication) and the versatile phototroph *Rhodobacter sphaeroides* has two rRNA genes in the main chromosome and one in a putative plasmid (S. C. Dryden, personal communication), opens up an interesting discussion on the meaning of the high copy number of rRNA genes observed in *E. coli* (seven) and *B. subtilis* (nine-ten) [21]. The differences observed in halophilic species could be the result of gene amplification promoted by insertion elements. Preliminary experiments performed with *H. mediterranei* show that the restriction pattern of the main chromosome seems stable after three hundred generations (G. Juez et al., unpublished results). Work in progress will ascertain the genomic stability of this biotechnologically important halophilic archaebacteria in ecological conditions.

Unfortunately there is as yet not much information available on the comparative genome organization for other halophilic genes, but the obtention of physical maps using PFGE will rapidly create a body of data in this important aspect of the study of the evolutionary genetics of microorganisms.

ACKNOWLEDGEMENTS

We wish to thank the CICYT for grant support for these studies (grant No. PB87-0092-C02-00).

REFERENCES

[1] D. C. Schwartz and C. R. Cantor, *Cell*, 37: 67 (1984).

[2] C. L. Smith, J. Econome, A. Schutt, S. Klco and C. R. Cantor, *Science* 236: 1448 (1987).

[3] J. L. Sanz, I. Marín, M. A. Balboa, D. Urea and R. Amils, *Biochemistry*, 27: 8194 (1988).

[4] C. L. Smith and C. R. Cantor, *in* R. Wu, ed., Academic Press, New York (1987)

[5] C. L. Smith, S. K. Lawrence, G. A. Gillespie, C. G. Cantor, S. M. Weessman and F. S. Collins, *in* "Methods in Enzymology 151", Mo. Gottesman, ed. Academic Press, New York, (1987).

[6] E. J. Southern, *J. Mol. Biol.*, 98: 503 (1975).

[7] J. J. Lee and H. O. Smith, *J. Bacteriol.*, 170, 4402 (1988).

[8] C. R. Woese, *Microbiol. Rev.* 51: 221 (1987).

[9] W. F. Doolittle in "The Bacteria", Archaebacteria. Vol. III, C. R. Woese and R. S. Wolfe, eds. Academic Press Inc., New York, (1985).

[10] M. Torreblanca, F. Rodriguez-Valera, G. Juez, A. Ventosa, M. Kamekura and M. Kates, *System. Appl. Microbiol.* 8: 89 (1986).

[11] E. Huysmans and R. De Wachter, *Nucl. Acids. Res.*, 14, supplement r73-r118 (1986).

[12] A. Yonath, W. Bennett, H. Hansen, U. Evers, Z. Berkovitch- Yellin and H. G. in "The structure, Function and Evolution of Ribosome", W. Hill, ed. ASM, in press.

[13] F. Pfeifer, U. Blaseio and P. Ghahraman, *J. Bacteriol.* 170: 3718 (1988).

[14] L. Kauc, M. Mitchell, S. H. Goodgal, *J. Bacteriol.* 171: 2474 (1989).

[15] L. E. Pyle, L. N. Corcoran, B. G. Cock, A. D. Bergenann, J. C. Whitley and L. R. Finch, *Nucl. Acids Res.* 16: 6015 (1988).

[16] R. Frutos, M. Pages, M. Bellis, G. Roizes and M. Bergoin, *J. Bacteriol.* 171: 4511 (1989).

[17] M. S. Ferdows and A. G. Barbour, *Proc. Natl. Acad. Sci.* 86: 5969 (1989).

[18] C. L. Smith, O. Matsumoto, J. B. Niwa, M. Fan, Yanagida and C. R. Cantor, *Nucl. Acids Res.* 15: 4481 (1987).

[19] J. L. Sanz, I. Marín, L. Ramirez, J. P. Abad, C. L. Smith and R. Amils. *Nucl. Acids Res.* 16: 7827 (1988).

[20] A. Bock, H. Hummel, M. Jarsch and G. Wich, *in* "Biology of Anaerobic Bacteria", H. C. Dubourquier et al. eds., Elsevier Science Publ., Amsterdam. pp. 206-225 (1986).

[21] G. La Fauci, R. L. Widom, R. L. Eisner, E. D. Jarvis and R. Rudner, *J. Bacteriol.* 165: 20477214 (1985).

PHYSICAL AND GENETIC MAPPING OF THE UNSTABLE GAS VESICLE PLASMID IN *HALOBACTERIUM HALOBIUM* NRC-1

Wai-lap Ng and Shiladitya DasSarma

Department of Microbiology
University of Massachusetts
Amherst, MA 01003
USA

ABSTRACT

The extremely halophilic archaebacterium, *Halobacterium halobium* strain NRC-1, contains a heterogeneous collection of large covalently closed circular (CCC) DNAs, the most abundant of which is the unstable gas vesicle plasmid, pNRC100. We have used a modified alkaline procedure to purify pNRC100 in quantity for physical and genetic mapping. Restriction mapping of pNRC100 using three restriction enzymes, *Dra*I, *Hin*dIII and *Sfi*I and the CHEF pulsed-field gel system shows it to be 190 kb in size. Three minor circular DNAs have also been purified and shown to be derivatives of pNRC100. The genes encoding the buoyant gas vesicles of *H. halobium* are clustered in a unique region of pNRC100, flanked by a pair of IS*H8* (iso-IS*H26*) elements. At least three other families of IS elements are also present in pNRC100.

INTRODUCTION

Extremely halophilic archaebacteria contain a large variety of extra-chromosomal circular DNAs, from less than 2 kb to more than 100 kb in size. Several small plasmids, e.g., the 1736 bp plasmid pHSB1 in *Halobacterium* strain SB3 and the 6354 bp plasmid pHV2 from *Halobacterium volcanii*, have been completely sequenced and used for DNA-mediated transformation [1-4]. Restriction maps for a large 150 kb *H. halobium* plasmid, pHH1, and its deletion derivatives have also been reported and several insertion sequences (IS elements) have been mapped to these plasmids [5 - 7].

General and Applied Aspects of Halophilic Microorganisms
Edited by F. Rodriguez-Valera, Plenum Press, New York, 1991

We have initiated a detailed study into the structure of another large plasmid, pNRC100, which contains a cluster of genes specifying the buoyant gas vesicles of *H. halobium* strain NRC-1 [8]. Our interest stems from the observed high mutation frequency to gas vesicle deficiency, about 1%, exhibited by some *H. halobium* strains [9]. Our results thus far indicate that rearrangements in pNRC100, including insertions and deletions, and imcompatibility between pNRC100 and mutant derivatives are responsible for the hypermutable property of *H. halobium* apparent in the gas vesicle phenotype [9, 10]. For precise molecular analysis of some complex rearrangements observed in the mutants, a detailed physical and genetic map of the parent plasmid is required. In this paper, we present a progress report on our efforts to map pNRC100.

MATERIALS AND METHODS

Preparation of pNRC100

Plasmid was extracted from late logarithmic cultures of *H. halobium* NRC-1 by a modified alkaline-SDS procedure [11, 12]. One litre of culture was chilled on ice and cells were pelletted by centrifugation at 8000 x g for 15 minutes at 4°C. Cells were gently resuspended in 20 ml basal salt solution (4.3 M NaCl, 81 mM $MgSO_4$) and added dropwise using a buret into 400 ml lysis buffer {50 mM Tris-HCl, pH 12.45, 20 mM EDTA, 1% (w/v) SDS} stirring at 50 rpm in a 1 l beaker. The lysate was adjusted to pH 12.45 with 1 M NaOH and stirred for 10 additional minutes. The pH was adjusted to 8.5 using 2 M Tris-HCl (pH 7.0) with a concomitant drop in viscosity. Precipitated material was removed by filtration through cheese cloth. The lysate was thoroughly but gently mixed with 200 ml of phenol which had been saturated with 3% NaCl (w/v) and the phases separated by centrifugation at 8000 x g for 10 min. About 80% of the aqueous layer was collected (with care to avoid the interface) using a wide bore pipette and extracted with 200 ml chloroform. The aqueous layer was collected and mixed with 0.015 volumes of 1 M $MgSO_4$, 0.025 volumes of a sodium phosphate buffer containing 0.1 M Na_2HPO_4 and 0.1 M NaH_2PO_4 and 0.7 volumes of cold ethanol and the mixture chilled at -20°C for 2 hours. The precipitate was collected by centrifugation at 8000 x g for 15 minutes, air dried, dissolved in a solution of 10 mM Tris-HCl (pH 8.0) and 1 mM EDTA. The plasmid DNA was further purified by CsCl-ethidium bromide gradient centrifugation.

CHEF gel electrophoresis

Plasmid DNA digested by restriction enzymes using conditions recommended by the manufacturer was analyzed by CHEF gel electrophoresis [13]. DNA was resolved on 0.7% agarose gels in 45 mM Tris base, 45 mM boric acid, and 1 mM EDTA running buffer with buffer circulation at 16°C for 10-12 hours. Electrohporesis was at 120 volts, with 0.05 sec initial switching interval and 0.05 sec switching increment with 30-60 steps.

Southern hybridization analysis

Following electrophoresis, CHEF gels were treated first with 0.25 N HCl for

20 minutes, followed by 0.5 N NaOH/1.5 M NaCl for 45 minutes, and finally by 1 M Tris-HCl (pH 8.0)/1.5 M NaCl for 45 minutes before transferring DNA to nitro-cellulose by blotting [14]. DNA was immobilized by heating at 80°C for 2 hours under vacuum and the membranes were prehybridized overnight at 68°C in a solution containing 0.45 M NaCl, 0.045 M sodium citrate, 0.2% (w/v) each Ficoll, polyvinyl-pyrrolidone, and bovine serum albumin, and 100 μg/ml denatured and sheared salmon sperm DNA. Hybridization was carried out in the same solution containing ^{32}P-labelled probe. Filters were washed for 1 hour with 0.45 M NaCl and 0.045 M sodium citrate

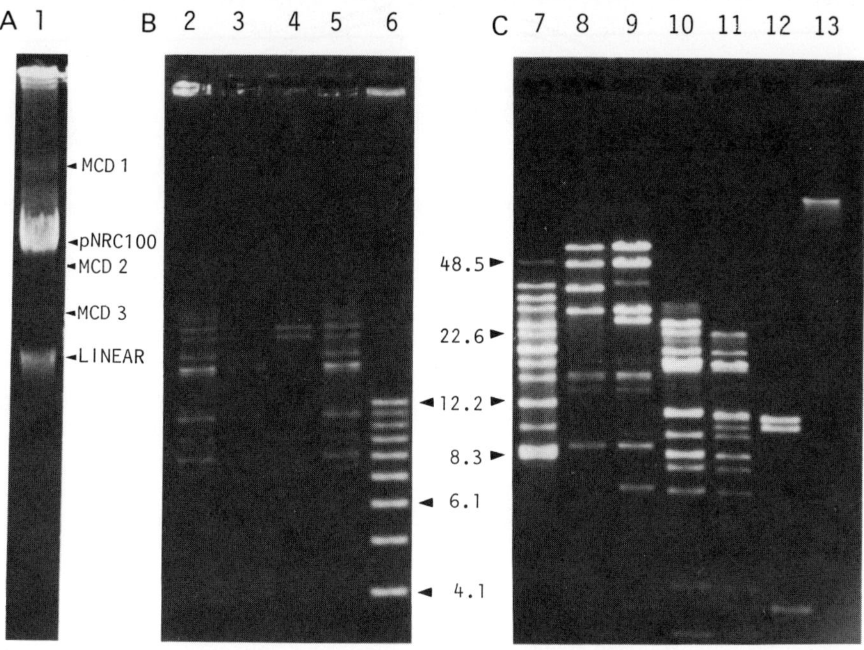

Figure 1. Plasmids in *H. halobium* strain NRC-1. (A) Standard 0.4% agarose gel electrophoresis of CCC DNA from strain NRC-1 (lane 1). (B & C) Restriction mapping analysis of pNRC100 and minor circular DNAs using CHEF gel electrophoresis. Comparison of *Hind*III digests of MCD1 (lane 2), MCD2 (lane 3), MCD3 (lane 4) and pNRC100 (lane 5). Comparison of pNRC100 digested with *Dra*I (lane 8), *Dra*I + *Sfi*I (lane 9), *Hind*III (lane 10), *Hind*III + *Sfi*I (lane 11), and *Sfi*I (lane 13). Lane 12 contains the largest pNRC100 *Hind*III fragment A cloned in pTZ19 digested with *Hind*III + *Sfi*I and lanes 6 and 7 contain molecular weight markers (in kb).

with three changes in wash solution. The hybridization and washing temperature was 68°C for nick translated probes and 25°C for the oligodeoxynucleotide (see below).

Probes

The following IS element probes were used for analysis: for IS*H1*, a 1 kb *Tth*111I fragment internal to IS*H1* was gel purified from a bacterio-opsin gene clone, pSD17 [15]. The IS*H2* probe was a 170 bp internal *Bgl*II fragment cloned in pBR322 [16]. The IS*H3* (iso-IS*H51*) probe was a 1726 bp *Hin*dIII-StuI fragment cloned in pKS(+) which contains an entire IS*H3* element and most of the *gvpA* gene from strain R1 [9]. The IS*H4* (iso-IS*H50*) probe was a 15 nucleotide long synthetic oligodeoxynucleotide (5'-ATGTCTTGATTGGCA-3') homologous to a region near the left end of the element [17]. The IS*H8* (iso-IS*H26*) probe was an ≈ 1470 bp *Bgl*II fragment containing 128 bp of flanking *gvpE* sequence cloned in M13mp19. The IS*H1*, IS*H2*,

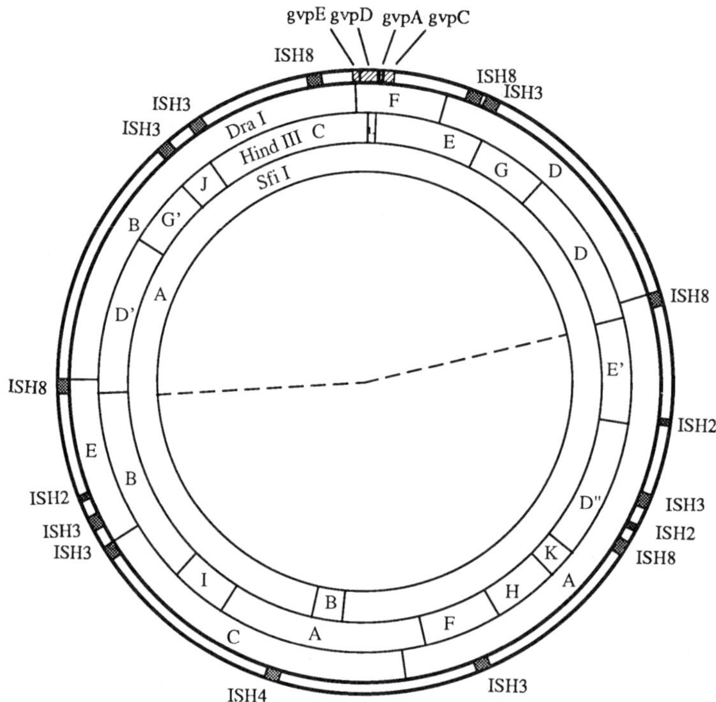

Figure 2. Restriction and genetic map of pNRC100. Gas vesicle genes and IS elements are indicated by hatched and shaded boxes on the outer circle and the *Dra*I, *Hin*dIII and *Sfi*I restriction maps are drawn in the internal circles. The restriction fragments are identified alphabetically by decreasing size. The dashed line indicates an ambiguity in the map (see text) and the order of the IS elements in fragment *Hin*dIII-D" has not been established.

IS*H3*, and IS*H8* probes were ^{32}P-labelled by nick translation [18] while the IS*H4* oligodeoxynucleotide was ^{32}P-labelled using T4 polynucleotide kinase [19].

RESULTS AND DISCUSSION

Isolation of CCC DNA and physical mapping of pNRC100

Using a modified alkaline-SDS procedure for large plasmid preparation (see Materials and Methods), we were able to reproducibly purify 100-200 μg of supercoiled CCC DNA per litre culture of *H. halobium* strain NRC-1. The CCC DNA electrophoresed on standard agarose gels as one major band (named pNRC100) but several faster and slower migrating minor species were also detectable (Figure 1A, lane 1). When pNRC100 and three major circular DNA species (MCDs) were purified from a low melting point agarose gel, digested with *Hin*dIII, and compared on a pulsed-field gel, the restriction fragments from the MCDs were found to be the same size as those from pNRC100 (Figure 1B, *cf.* lanes 2-4). The largest minor species, MCD1, had an identical restriction pattern to pNRC100, indicating that it is a multimer form of pNRC100, while MCD2 and MCD3 contained exclusive subsets of the pNRC100 pattern, and thus appeared to be two smaller circles resulting from intramolecular recombination in pNRC100.

A physical map of pNRC100 was constructed using three restriction enzymes, *Dra*I, *Hin*dIII, and *Sfi*I, utilizing pulsed-field gel electrophoresis for separation of large fragments. *Dra*I generated 6 restriction fragments, *Hin*dIII generated 16, and *Sfi*I generated 2 (Figure 1C, lanes 8, 10 and 13). Restriction mapping analysis by double digestion of pNRC100 and also of several gel purified or cloned restriction fragments of pNRC100 was used to construct a tentative map (Figure 1C, lanes 9, 11, 12 and data not shown). Fragment order was verified by Southern hybridization analysis using

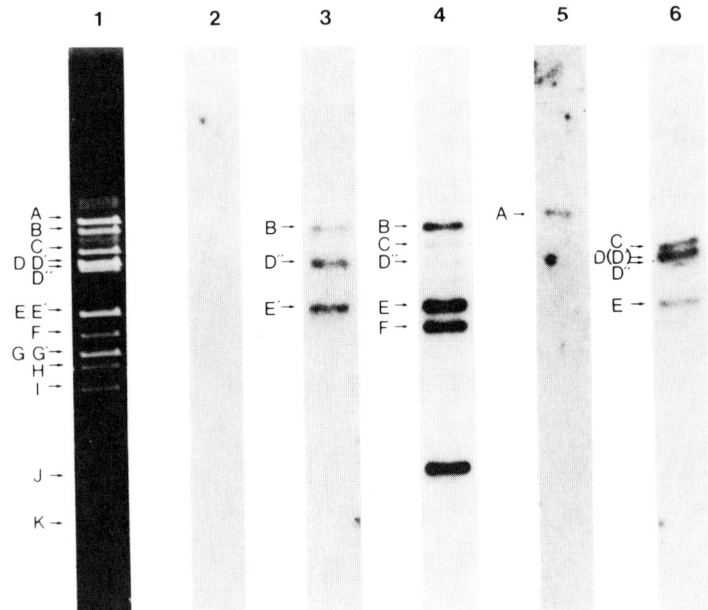

Figure 3. Southern hybridization analysis using ISH element probes. pNRC100 was digested with *Hin*dIII, fractionated on a CHEF gel (ethidium stained lane, lane 1), and hybridized with ISH1 (lane 2), ISH2 (lane 3), ISH3 (iso-ISH51) (lane 4), ISH4 (iso-ISH50) (lane 5), and IS*H8* (iso-IS*H26*) (lane 6). The identity of each hybridizing band is indicated.

*Hind*III end-labelled probes which were hybridized to *Eco*RI, *Pst*I and *Sst*I digests of pNRC100 (data not shown). The map inferred from this data is shown in Figure 2. The only ambiguity remaining in the map is the linkage between the *Hind*III D and D' to either B or E'. The two possible alternative structures are related by inversion of about half of pNRC100 (at the dashed lines in Figure 2).

Genetic mapping of pNRC100

To begin construction of a genetic map for pNRC100, we carried out Southern hybridization analysis using halobacterial genes and IS elements (Figure 3). The major gas vesicle protein (*gvp*) gene, *gvpA*, was shown to map to a unique segment of pNRC100 in the *Hind*III E fragment, while three other genes, *gvpC, gvpD* and *gvpE*, which have been identified by extensive sequencing and restriction mapping analysis of class II gas vesicle mutants, map near *gvpA* (refs. 8, 20, Figure 2). Interestingly, the *gvp* gene cluster is flanked by a pair of IS*H8* (iso-IS*H26*) elements, suggesting transposition as a mechanism for mobilization of the genes between the plasmid and chromosome and perhaps between different bacteria [20, 21]. Two or three additional copies of IS*H8* are also present in pNRC100. pNRC100 contains several other IS elements, including three copies of IS*H2*, six to eight copies of IS*H3* (iso-IS*H51*), and a single copy of IS*H4* (iso-IS*H50*) [22-25]. One well-characterized element, IS*H1*, implicated in many bacterio-opsin gene mutations and also occurring on the pNRC100-like plasmid in the purple membrane overproducer strain S9 [10], is absent from pNRC100.

CONCLUSIONS

The large gas vesicle plasmid from *H. halobium* NRC-1, pNRC100, has been isolated and mapped using three restriction enzymes and several IS elements and gas vesicle gene probes. Three minor circular DNAs co-purifying with pNRC100 have been examined and shown to be derivatives of pNRC100. pNRC100 exhibits extreme instability, resulting in frequent gas vesicle mutations, but sufficient stability such that a discernable 190 kb structure can be mapped.

ACKNOWLEDGEMENTS

We wish to thank the NSF (DMB-8703486) and the NIH (GM41980) for grants supporting these studies and Dr. Neil R. Hackett for stimulating discussions.

REFERENCES

[1]   K. Ebert, W. Goebel and F. Pfeifer, *Mol. Gen. Genet.* 194:91 (1984).

[2]   N. R. Hackett and S. DasSarma, *Can. J. Microbiol.* 35:86 (1989).

[3]   R. L. Charlebois, W. L. Lam, S. W. Cline and W. F. Doolittle, *Proc. Natl. Acad. Sci. USA* 84:8530 (1987).

[4]   W. L. Lam and W. F. Doolittle, *Proc. Natl. Acad. Sci. USA* 86:5478 (1989).

[5]   G. Weidinger, G. Klotz and W. Goebel, *Plasmid* 2:377 (1979).

[6]   F. Pfeifer, G. Weidinger and W. Goebel, *J. Bacteriol.* 145:375 (1981).

[7]   F. Pfeifer, U. Blaseio and P. Ghahraman, *J. Bacteriol.* 170:3718 (1988).

[8]   S. DasSarma, T. Damerval, J. G. Jones and N. Tandeau de Marsac, *Mol. Microbiol.* 1:365 (1987).

[9]   S. DasSarma, J. T. Halladay, J. G. Jones, J. W. Donovan, P. J. Giannasca and N. Tandeau de Marsac, *Proc. Natl. Acad. Sci. USA* 85:6861 (1988).

[10]   S. DasSarma, *Can. J. Microbiol.* 35:65 (1989).

[11]   P. L. Bergquist, Incompatibility, *in*: "Plasmids: a practical approach", K. G. Hardy, ed., IRL Press, Oxford (1987).

[12]   F. Casse, C. Boucher, J. S. Julliot, M. Michel and J. Dénarié, *J. Gen. Microbiol.* 113:229 (1979).

[13]   G. Chu, D. Vollrath and R. W. Davis, *Science*, 234:1582 (1986).

[14]   E. M. Southern, *J. Mol. Biol.* 98:503 (1975).

[15]   M. Simsek, S. DasSarma, U.L. RajBhandary and H. G. Khorana, *Proc. Natl. Acad. Sci. USA* 79:7268 (1982).

[16]   S. DasSarma, U. L. RajBhandary and H. G. Khorana, *Proc. Natl. Acad. Sci. USA* 80:2201 (1983).

[17]   W.-L. Xu and W. F. Doolittle, *Nucleic Acids Res.* 11:4195 (1983).

[18]   P. W. J. Rigby, M. Dieckmann, C. Rhodes and P. Berg, *J. Mol. Biol.* 113:237 (1977).

[19]   A. M. Maxam and W. Gilbert, *Methods Enzymol.* 65:499 (1980).

[20]   J. G. Jones, N. R. Hackett, J. T. Halladay, D. J. Scothorn,, C.-F. Yang, W.-L. Ng and S. DasSarma, *Nucleic Acids Res.* 17: 7785 (1989).

[21]   M. Horne, C. Englert and F. Pfeifer, *Mol. Gen. Genet.* 213:459 (1988).

[22]   F. Pfeifer, H. Boyer and M. Betlach, *J. Bacteriol.* 164:414 (1985).

[23]   J. D. Hofman, L. C. Schalkwyk and W. F. Doolittle, *Nucleic Acids Res.* 14:6983 (1986).

[24]   F. Pfeifer, M. Betlach, R. Martienssen, J. Friedman and H. W. Boyer, *Mol. Gen. Genet.* 191:182 (1983).

[25]   K. Ebert, C. Hanke, H. Delius, W. Goebel and F. Pfeifer, *Mol. Gen. Genet.* 206:81 (1987).

ENZYMOLOGY AND GENETICS OF AN ALPHA-LIKE DNA POLYMERASE FROM *HALOBACTERIUM HALOBIUM*

Irène Sorokine[1], Kamel Ben-Mahrez[1], Masashi Nakayama[2] and Masamichi Kohiyama[1]

[1]Institut Jacques Monod
Université Paris VII
2 Place Jussieu
75251 Paris Cedex 05
France

[2]University of Osaka Kyoiku
Department of Biology
Minami-Kawahori-Cho
Tennoji-Ku
Osaka 543
Japan

ABSTRACT

The DNA polymerase alpha of *Halobacterium halobium* is constituted of two pairs of Mr 70,000 and 60,000 subunits [1]. We have previously isolated a 70 kD protein immunologically related to v-*myc* oncogene product [2]. This v-*myc* like protein, purified with a totally different method from that used for the polymerase alpha, is identical with the Mr 70,000 subunit of the polymerase. In order to study the degree of homology of this polymerase with eucaryotic DNA polymerase alpha or with v-*myc* protein, we have started the cloning of the corresponding gene using the yeast DNA *pol*I gene as probe.

MATERIALS AND METHODS

Halobacterium halobium

The strain CCM2090 was grown in Sehgal and Gibbons' complex medium [3].

Antibodies

Polyclonal anti-v-*myc* antibodies raised against a synthetic peptide were possesses an associated primase activity. It is constituted of two pairs of Mr 70,000 and 60,000 subunits [1]. We have also isolated, with a totally different method from that used for DNA polymerase alpha, a 70 kD protein which is immunologically related to the MC29 v-*myc* oncogene product [2]. This purified v-*myc*-like protein stimulates *in vitro* DNA synthesis carried out by the DNA polymerase alpha of *H. halobium* [6].

In this report, we present evidences that the 70 kD protein immunologically related to v-*myc* oncogene product and the Mr 70,000 subunit of DNA polymerase alpha are identical. In order to study the degree of homology of this polymerase with eucaryotic alpha polymerase or with v-*myc* protein, we have started the cloning of the corresponding gene using the yeast DNA *pol*I as a probe.

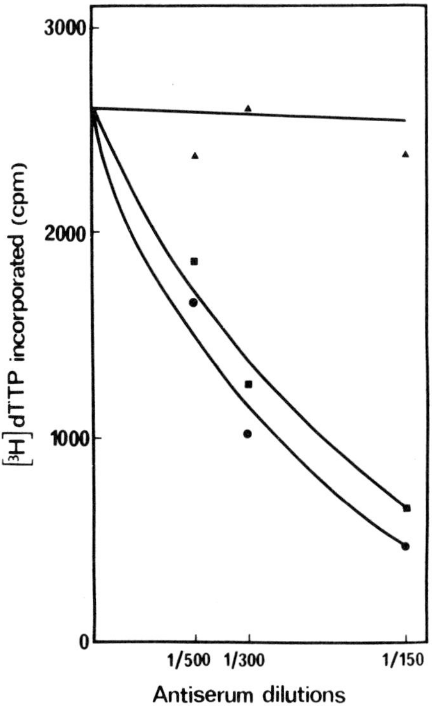

Figure 1. Effect of anti-(v-*myc*-like protein) antiserum on DNA polymerase alpha activity: 0.2 μg DNA polymerase was preincubated for 60 min on ice with antiserum at various concentrations before being assayed in standard conditions for 30 min with activated DNA and [³H]dTTP in the presence of anti-(DNA polymerase alpha) antiserum (●) or anti-(v-*myc*-like-protein) antiserum (■) or control antiserum (▲).

314

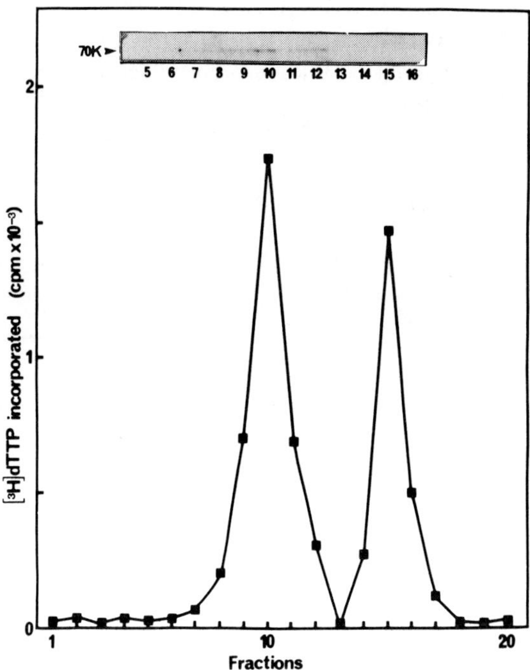

Figure 2.   Western blot analysis using anti-(v-*myc*-like protein) antiserum: glycerol gradient fractions were assayed for DNA polymerase activities in standard conditions for 60 min with activated DNA and [³H]dTTP. 15 μl-aliquots of fractions 5 to 16 were tested by Western blot analysis using anti-(v-*myc*-like protein) antiserum (1/200 dilution).

EFFECT OF ANTI-(V-*MYC*-LIKE PROTEIN) ANTISERUM ON DNA POLYMERASE ACTIVITIES

We have previously shown that the v-*myc* like protein of *H. halobium* stimulated the *in vitro* DNA synthesis carried out by the DNA polymerase alpha [6]. To define the relationship between the 70 kD protein and the DNA polymerase alpha we have examined the effect of anti-(vm-*myc*-like protein) antibodies on the activity of the DNA polymerase alpha by preincubating the DNA polymerase with various concentrations of the antiserum. As shown in Figure 1, the polymerase activity of the DNA polymerase alpha was clearly inhibited by the anti-(v-*myc*-like protein) antibodies. Similarly, when the DNA polymerase alpha was preincubated with the anti-(v-*myc*-like protein) antibodies, its primase activity is inhibited (not shown). Thus, the two activities (primase and polymerase) of the *H. halobium* DNA polymerase alpha are inhibited by the antiserum prepared against the v-*myc*-like protein. The kinetics of DNA polymerase alpha inhibition by the anti-(v-*myc*-like protein) is comparable to the inhibition by the anti-(DNA polymerase alpha) antibodies.

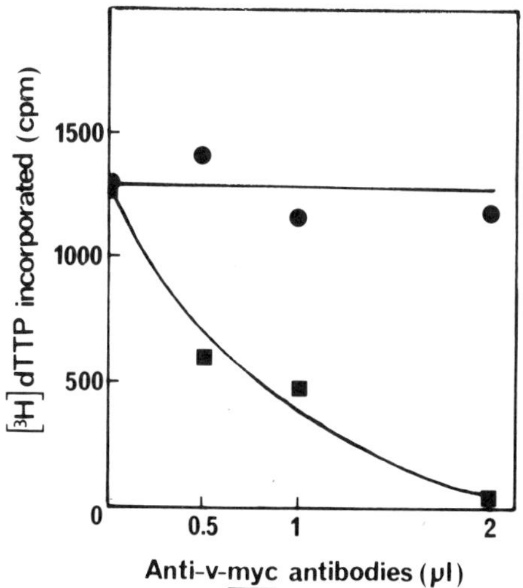

Figure 3. Effect of anti-v-*myc*-antibodies on DNA polymerase alpha (■) and beta (•) activities: DNA polymerases alpha and beta were preincubated in presennce of various concentrations of "Oncor" anti-v-*myc*-antibodies and then assayed in standard conditions with activated DNA and [^{3}H]dTTP.

In the last step of purification, the DNA polymerase alpha was separated from DNA polymerase beta of *H. halobium* by sedimentation on glycerol gradient [1]. Fractions of the gradient were assayed for DNA polymerase activities and cross-reaction with anti-(v-*myc*-like protein) antibodies. Figure 2 shows that only the Mr 70,000 subunit present in the alpha peak reacted with the anti-(v-*myc*-like protein) antibodies.

Thus, the anti-(v-*myc*-like protein) antibodies inhibited the DNA polymerase alpha activities by reacting with its Mr 70,000 subunit.

EFFECT OF "ONCOR" ANTI-V-*MYC* ANTIBODIES ON DNA POLYMERASE ALPHA.

DNA polymerases alpha and beta were preincubated with anti-v-*myc* antibodies and then assayed for DNA synthesis. As shown in Figure 3, contrary to DNA polymerase beta, the DNA polymerase alpha was inhibited by the anti-v-*myc* antibodies. In order to define the target of the anti-v-*myc* antibodies on the DNA polymerase alpha, we carried out a Western blot analysis. Figure 4 shows that the Mr 70,000 subunit of the polymerase alpha was recognized by the anti-v-*myc* antibodies in almost the same way as the v-*myc*-like protein.

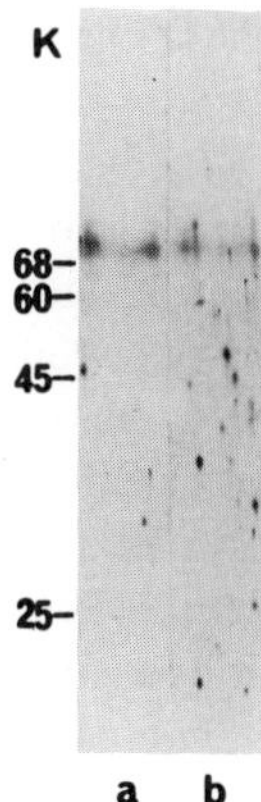

Figure 4. Western blot analysis using anti-v-*myc* antiserum: 2 μg DNA polymerase alpha (a) and 1 μg v-*myc*-like protein (b) were subjected to Western blot analysis using anti-v-*myc* antiserum (1/20 dilution). Positions of molecular-mass standards (values in kDa) are indicated on the left.

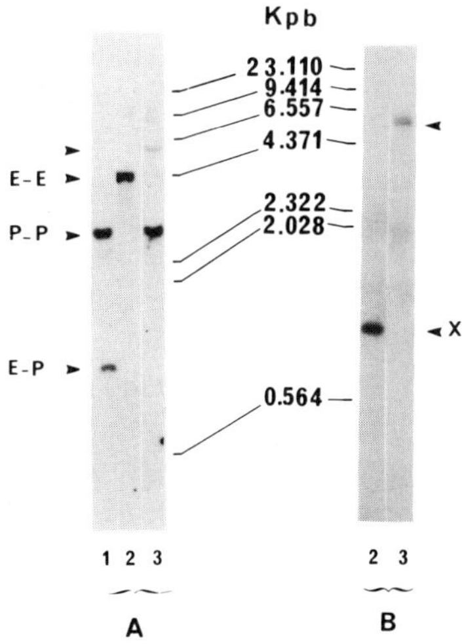

Figure 5. Hybridization of *H. halobium* DNA with the DNA polymerase I probe of yeast. Samples (10 μg) of purified DNA from exponentially growing *H. halobium* were digested with 50 units of *Eco*RI (2) and *Pst*I (3) restriction endonucleases and electrophoresed in 0.8% agarose gels. The E-E fragment was electroeluted from gel and digested with 15 units of *Pst*I restriction endonuclease before electrophoresis (1). The resulting fragments were blotted onto Biodyne membrane sheets and hybridized with 32P-labelled probes (106 cpm/ml) from yeast DNA polymerase I gene: (A) *Sal*I/*Bam*HI; (B) *Bgl*II/*Eco*RI.

SEARCH FOR DNA SEQUENCES RELATED TO YEAST DNA POLYMERASE I GENE

Purified DNA from *H. halobium* was digested with *Eco*RI or *Pst*I restriction endonucleases and the resulting fragments were electrophoresed, blotted onto Biodyne membrane and hybridized with ^{32}P-labelled probes. As shown in Figure 5-A, hybridization with the 3' part of DNA polymerase I gene of yeast (*Sal*I/*Bam*HI fragment) led to the detection of one fragment in *Eco*RI digestion (4.2 kb) and two fragments in *Pst*I digestion (5.5. kb and 2.7 kb). The 4.2 kb band was eluted from gel and digested by *Pst*I restriction endonuclease before electrophoresis, blotting onto Biodyne membrane and hybridization with the same probe (Figure 5-A-1). The hybridization shows that the *Eco*RI/*Eco*RI (E-E) fragment is constituted by at least 2 fragments, one *Pst*I/*Pst*I (P-P) of 2.7 kb and one probably *Pst*I/*Eco*RI (P-E) of approximately 0.9 kb. Hybridization with the 5' part of the DNA polymerase gene (*Bgl*II/*Eco*RI fragment) led to the detection of another fragment with *Eco*RI digestion (fragment X, 0.8 kb) and one fragment with *Pst*I digestion (5.5 kb). The different fragments were eluted from gels and cloned in lamda gt11 or in pUC 8 DNAs for immunological screening and/or screening with yeast DNA probes.

DISCUSSION

We have previously shown that the *H. halobium* DNA synthesis in inhibited *in vivo* by aphidicolin [7]. The DNA polymerase target of this inhibitor is the DNA replicase which we have previously purified and identified as an eucaryotic alpha-like DNA polymerase [1].

The results described in the present work show that the Mr 70,000 subunit of *H. halobium* DNA polymerase alpha is immunologically related to v-*myc* oncogene product and contains probably both polymerase and primase activities. This conclusion derived from the following facts: (i) Polymerase and primase activities of the DNA polymerase alpha cosedimented during centrifugation on a glycerol gradient [1]. (ii) Both activities were in the same way inhibited by aphidicolin and resistant to dideoxyribonucleotides [1]. (iii) Anti-(v-*myc*-like protein) antiserum inhibited the two activities with the same kinetic by reacting with only the Mr 70,000 subunit of the polymerase. In fact, the Mr 70,000 subunit has the same electrophoretic motility as the v-*myc*-like protein, and the two polypeptides are recognized in Western blot by polyclonal anti-(v-*myc* oncogene product) antibodies. Thus, these two polypeptides obtained by two different purification protocols are identical. Accordingly the anti-v-*myc* antibodies inhibited the *in vitro* DNA synthesis carried out by the DNA polymerase alpha and had no effect on the DNA polymerase beta of *H. halobium*. Three groups have proposed that c-*myc* protein is involved directly in mammalian DNA replication. Addition of antibodies against the human c-*myc* protein to nuclei isolated from human cells inhibited DNA synthesis and DNA polymerase activity of these nuclei [8]. Classon et al. [9] have shown that elevated c-*myc* expression facilitates the replication of SV40 DNA in human lymphoma cells. Moreover, it was suggested that the c-*myc* protein may promote cellular DNA replication by binding to replication origins [10]. We propose that the role of *myc* protein in DNA replication may be mediated in *H. halobium* by DNA polymerase alpha regions shared with the *myc* protein. In order to define these regions, we have started the cloning of the DNA polymerase gene using yeast DNA *pol*I gene as probe. Hybridization experiments showed the presence of DNA fragments in *H. halobium* genome which hybridized with DNA probes from yeast DNA *pol*I gene. The DNA *pol*I of yeast is the equivalent to mammalian DNA polymerase alpha [4]. Sequences of *H. halobium* DNA fragments and their relationship to DNA polymerase alpha of the archaebacterium remain to be determined.

REFERENCES

[1] M. Nakayama, K. Ben-Mahrez and M. Kohiyama, DNA primase activity found in an alpha-like DNA polymerase obtained from *Halobacterium halobium. Eur. J. Biochem.* 175: 265 (1988)

[2] K. Ben-Mahrez, B. Perbal, C. Kryceve-Martinerie, D. Thierry and M. Kohiyama, A protein of *Halobacterium halobium* immunologically related to v-*myc* gene product, FEBS Letters 227: 56 (1988)

[3] S. N. Sehgal and N. E. Gibbons, Effect of some metal ions on the growth of *Halobacterium cutirubrum, Can. J. Microbiol.* 6: 165 (1960)

[4] L. M. Johnson, M. Snyder, L. M. S. Chang, R. W. Davis and J. Cambell, Isolation of the gene encoding yeast DNA polymerase I, *Cell* 43: 369 (1985)

[5] W. Zillig, R. Schnabel and K. O. Stetter, Archaebacteria and the origin of the eukaryotic cytoplasm, *Curr. Top. Microbiol. Immunol.* 114: 1 (1985)

[6] K. Ben-Mahrez, W. Sougakoff, M. Nakayama and M. Kohiyama, Stimulation of an alpha-like DNA polymerase by v-*myc* related protein of *Halobacterium halobium, Arch. Microbiol.* 149: 175 (1988)

[7] P. Forterre, C. Elie and M. Kohiyama, Aphidicolin inhibits growth and DNA synthesis in halophilic archaebacteria, *J. Bacteriol.* 159: 800 (1984)

[8] G. P. Studzinski, Z. S. Brelvi, S. C. Feldman and R. A. Watt, Participation of c-*myc* protein in DNA synthesis of human cells, *Science* 234: 467 (1985)

[9] M. Classon, M. Henriksson, J. Sünegi, G. Klein and M. L. Hammaskjöld, Elevated c-*myc* expression facilitates the replication of SV40 DNA in human lymphoma cells, *Nature* 330:272 (1981)

[10] S. M. M. Igushi-Ariga, T. Itami, Y. Kiji and H. Ariga, Possible function of the c-*myc* product: promotion of cellular replication, *EMBO J.* 6:2365 (1987)

PHYLOGENY OF DNA-DEPENDENT RNA POLYMERASES:

TESTIMONY FOR THE ORIGIN OF EUKARYOTES

Wolfram Zillig, Peter Palm, Hans-Peter Klenk, Gabriela Pühler, Felix Gropp and Christa Schleper

Max-Planck-Institut für Biochemie, D-8033 Martinsried, Fed. Rep. of Germany

ABSTRACT

The organization of the genes for the large components of DNA-dependent RNA polymerase in archaebacteria resembles that in eubacteria.

A phylogenetic dendrogram derived by various algorithms from the comparison of aligned sequences of the genes of the A plus C components of archaebacterial, the A components of eukaryotic and the β' or β' plus β'' components of eubacterial DNA-dependent RNA polymerases shows the archaebacteria beside eukaryotic pol2 and/or 3. The latter two are separated from pol1 which shares a bifurcation with the eubacteria. This topology is invariant to several corrections and appears significant when checked by bootstrapping.

This branching order most probably implies that eukaryotes are bi- or oligophyletic chimeras which arose by some sort of genome fusion from archaebacterial and eubacterial ancestors.

The data confirm the unity of the archaebacteria and clearly exclude both the eocyte and the photocyte hypothesis.

INTRODUCTION

Extensive comparative analysis of the nucleotide sequences of rRNA genes has lead Carl Woese and collaborators to divide the living world into three mono-phyletic holophyletic urkingdoms including two divisions of prokaryotes, eubacteria and archaebacteria, beside the eukaryotes [1-2]. Phylogenetic dendrograms derived from the comparison of sequences of several other single genes showed trifurcation points in accord with this view [3-10] except for some peculiarities, e.g. the position of *E. coli* in the GAPDH (glyceraldehyde phosphate dehydrogenase) tree and the positions of *Thermus* in both the MDH (malate dehydrogenase) and the PGK (phospho glycerate kinase) trees as lowest branch of the eukaryotes rather than among the other eubacteria (9-10; R. Hensel and coworkers, unpublished). With different alignments and a novel algorithm for the construction of rate invariant trees James A. Lake proposed a different tree topology from the rRNA data essentially viewing the archaebacteria as a paraphyletic group giving rise to the eukaryotes our of their extremely thermophilic ("eocyte") branch and to the eubacteria from their extremely halophilic branch (combined with the eubacteria as "photocytes") [11-15].

The apparent similarity between DNA-dependent RNA polymerases of archaebacteria, especially extreme thermophiles, and eukaryotes [16-17] and the existence of three specialized polymerases in eukaryotes has prompted us to include this ubiquitous highly conserved molecule into phylogenetic analysis.

The sequences of genes for the large components A and C or the RNA polymerases of *H. halobium* [18], *S. acidocaldarious* [19], *M. vannielii* (I. Arnold and P. Palm, unpublished) and *T. celer* (H.-P. Klenk, unpublished) have been determined and compared with the corresponding sequences from *M. thermoautotrophicum* [20], with A component genes of RNA polymerase II (pol2) of *S. cerevisiae* [21], mouse [22], and *T. brucei* [23] or pol3 of *S. cerevisiae* [21] and *T. brucei* [24], and of pol1 of *S. cerevisiae* [25] *Schizosaccharomyces pombe* [26] and *T. brucei* [27-30]. Phylogenetic dendrograms obtained by various algorithms show the archaebacteria as a coherent group including the *Halobacteriales* thus confirming the concept of Carl Woese and disproving the proposals of J. A. Lake. But in disagreement with the assumption of a monophyletic origin of the eukaryotes the lineages of the different eukaryotic RNA polymerases do not spring from a common stem. A thorough examination of the significance of the tree topology was therefore desirable.

RESULTS

Gene Organization

The genes encoding the large components of the DNA-dependent RNA polymerases of archaebacteria, B and B" plub B' respectively, A and C are arranged, in this order, in reading units resembing the rpoBC operon of *E. coli* (Figure 1). The B genes are homologous and colinear to the genes for the second largest components of eukaryotic polymerases, which we therefore term B, and to the *E. coli* β gene rpoB [19]. They do not however contain the terminal heptapeptide repeats of eukaryotic B components [31-32]. The A genes correspond to about the first two thirds and the C genes to about the last third of the genes for the largest components, A, of eukaryotic polymerases and to the *E. coli* β' gene, rpoC. Thus archaebacteria exhibit a characteristic AC gene split where both eukaryotes and eubacteria possess contiguous A (β') genes. Extremely thermophilic sulfur archaebacteria of the orders *Sulfolobales*,

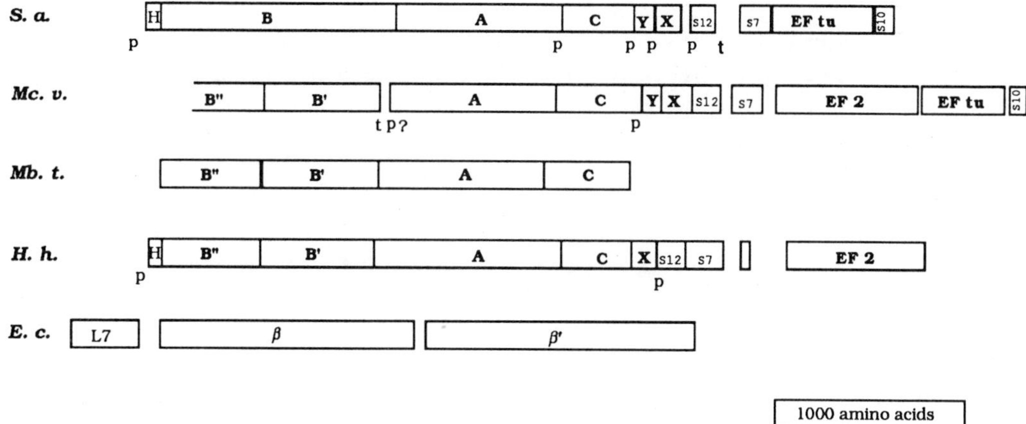

Figure 1. Organization of large component genes of DNA-dependent RNA polymerases of archaebacteria and *E. coli*. Abbreviations as in Table 1 except for *Mc.v.* = *Methanococcus vannielii* and *Mb.T.* = *Methanobacterium thermoautotrophicum*. Y = ORF homologous to yeast ribosomal protein L30 and X = unassigned homologous ORFs occurring in *S.A., Mc.v.* and *H.h.* The order of the genes to the right of the polymerase operons was taken from [8].

Thermoproteales, Thermococcales and *Thermoplasmales* contain contiguous B genes, whereas the *Methanococcales, Methanobacteriales* and *Halobacteriales* share a distinct B"B' split.

Whereas the order of the genes in the operons corresponds to that in *E. coli*, the neighbourhood is different: the gene immediately upstream of rpoB (*S. acidocaldarious*) and rpoB" (*H. halobium*) respectively encodes one of the around 8 smaller polymerase components, H, in the case of *S. acidocaldarious* containing 88 aminoacid residues. The open reading frames following downstream comprise a gene for a ribosomal protein corresponding to yeast L30, which has no counterpart in *E. coli*, an unidentified open reading frame (ORF X) genes for the ribosomal proteins S12 and S7 and the translation factors EFG and EFtu which in *E. coli* are combined in the strep operon situated far upstream and in opposite orientation to the rpoBC operon. One or the other gene is absent in one or the other case, e.g. the correspondent to yeast L30 in *H. halobium*, the gene for EFG in *S. acidocaldarius* and the gene for EFtu in *H. halobium*.

Transcription Units

In the case of *S. acidocaldarius*, a promoter is situated upstream of the gene for component H, and an additional promoter in front of the C gene (Figure 1). Further, stronger promoters precede the 3 ORFs immediately downstream. The common terminator for all these promoters is found downstream of the S12 gene. In *M. vannielii*, an additional terminator followed by a promoter is found between the B' and the A gene and the only promoter found downstream is in front of ARF1 encoding

Table 1. Similarities of A+C (archaebacteria) A (eukaryotes) and ß' (eubacteria) genes of DNA- dependent RNA polymerases. Upper right triangle identities, lower left triangle distance values calculated from iudentities according to [38]. Abbreviations: *M.t., Methanobacterium thrmoautotrophicum; Tc., Thermococcus celer; S.a.. Sulfolobus acidocaldarius; H.h., Halobacterium halobium; M.m., Mus musculus; S.c., Saccharomyces cerevisiae; T.b., Trypanosoma brucei; S.p., Schizosaccharomyces pombe; T.m., Thermotoga maritima; S.C., Spinach Chloroplast; M.p.C., Marchantia polymorpha Chloroplast; N.c., Nostoc commune; P.p., Pseudomonas putida; E.c., Escherichia coli.*

	M.t.	T.c.	S.a.	H.h.	M.m.II	S.c.II	T.b.II	T.b.III	S.c.III	S.c.I	T.b.I	S.p.I	T.m.	S.C.	M.p.C.	N.c.	P.p.	E.c.
M.t.	X	75.2	70.7	72.9	57.9	58.1	49.2	46.7	48.5	38.9	36.1	34.2	31.6	33.0	33.7	36.7	37.1	37.1
T.c.	28.6	X	76.7	73.7	56.4	58.3	52.8	48.1	56.2	42.6	37.1	38.3	44.9	38.0	39.5	41.0	39.8	40.6
S.a.	34.8	26.6	X	69.2	58.6	59.4	53.6	50.8	57.0	41.4	37.9	37.1	54.5	39.8	39.8	41.0	39.1	41.0
H.h.	31.6	30.6	36.9	X	52.6	55.6	45.7	45.4	49.4	39.5	35.1	35.9	40.8	36.5	38.0	39.1	39.1	40.6
M.m.II	56.5	57.4	53.5	64.4	X	68.5	57.5	43.3	48.9	41.6	38.2	38.1	33.8	33.0	33.0	34.5	33.0	34.8
S.c.II	54.5	54.2	52.2	58.8	37.9	X	56.8	43.7	48.5	41.6	38.6	36.6	35.7	36.0	36.7	34.1	34.5	33.6
T.b.II	71.0	64.0	62.6	78.6	55.5	56.8	X	44.4	43.8	36.6	34.7	34.4	37.4	33.1	34.2	33.8	33.8	34.6
T.b.III	76.3	73.5	68.0	79.3	84.0	83.1	81.4	X	51.9	39.2	37.0	37.8	32.7	33.0	33.7	32.2	34.5	35.2
S.c.III	72.6	57.7	56.4	70.7	71.8	72.6	82.9	65.7	X	41.0	33.9	36.3	34.2	33.1	33.1	33.1	33.1	34.6
S.c.I	94.7	85.7	88.5	93.3	87.9	87.9	100.9	93.9	89.4	X	46.5	71.0	27.7	27.2	30.4	28.8	26.8	28.4
T.b.I	102.1	99.5	97.4	105.2	96.7	95.7	106.3	99.7	108.7	76.9	X	46.4	30..2	29.3	31.3	31.7	29.7	29.7
S.p.I	107.6	96.4	99.5	102.7	96.8	100.9	107.2	97.6	101.6	34.3	77.0	X	27.3	29.2	29.2	28.0	27.2	27.6
T.m.	98.2	80.3	81.2	90.1	108.8	103.3	98.8	112.2	107.7	128.8	120.1	130.2	X	56.0	59.4	69.5	62.0	66.2
S.C.	111.4	97.2	92.3	101.2	111.4	102.7	111.0	111.4	111.0	130.6	123.2	123.7	58.1	X	81.6	65.2	59.2	62.2
M.p.C.	109.1	93.3	92.3	97.2	111.4	100.6	107.7	109.1	111.0	119.7	116.6	123.7	52.2	20.3	X	68.9	61.4	64.4
N.c.	100.6	89.5	89.5	94.2	106.9	108.0	108.8	113.8	111.0	125.0	115.3	127.8	36.4	42.9	37.3	X	70.4	75.7
P.p.	99.6	92.3	94.2	94.2	111.4	106.9	108.8	106.9	111.0	132.1	121.9	130.6	47.9	52.6	48.9	35.2	X	90.6
E.c.	99.6	90.4	89.5	90.4	105.8	103.7	106.6	104.7	106.6	126.4	121.9	129.2	41.4	47.6	44.1	28.0	9.9	X

the correspondent of yeast L30 [8]. In *H. halobium*, the one downstream promoter is in front of the S12 gene. The transcription punctuation is thus somewhat variable between different archaebacteria.

Sequence Similarity

The alignment of the derived aminoacid sequences of corresponding components presents no major difficulties in some 10 conserved regions but remains uncertain, especially between the eubacterial and all other sequences, in the connecting stretches. The phylogenetic analysis has therefore been restricted to positions without gaps in any of the aligned sequences. Because representatives of all three eukaryotic polymerases were available, an extensive comparison has so far only been performed for the A plus C (archaebacteria), A (eukaryotes) and β' (eubacteria) components [28-30].

Whereas the gene organization resembles that in *E. coli*, the archaebacterial genes show much higher similarities to their eukaryotic correspondents, especially pol2 and pol3, than to the corresponding eubacterial genes (Table 1). The similarity between the archaebacterial components and those of pol2 and pol3 is even higher than between the latter two eukaryotic components themselves.

Phylogenetic Dendrograms

Unrooted phylogenetic trees have been constructed by various algorithms [29-30] including the distance matrix method of Fitch and Margoliash [33], DNA parsimony analysis employing a computer program of Felsenstein [34], Felsenstein's maximum likelihood method [35-36] and the evolutionary parsimony algorithm of J. A. Lake [13-14] which has been designed as rate invariant method. Except for the latter tree in which the archaebacteria *H. halobium* and *S. acidocaldarius* have changed positions, the topologies of all these dendrograms are the same (Figure 2). As expected the archaebacteria form a coherent group in the immediate neighbourhood of the eukaryotic RNA polymerases 2 and 3 each of which also form coherent branches. The next ramification leads to a coherent pol1 branch that shares a bifurcation with the branch containing the eubacteria. The unexpected feature of this topology is the separation of the three eukaryotic polymerase lineages of which pol2 appears beside the archaebacteria and pol1 beside the eubacteria. The far reaching implications of this branching order call for examination of its significance.

As reported previously [29-30] the same tree topology has been obtained from an identity and from a similarity matrix [37]. Neither correction for multiple exchanges in the long branches as proposed by Feng et al. [38] nor restriction of the considered positions to increasingly conserved regions changed the tree. An attempt to evaluate the significance of various possible trees by comparing the numbers of assumed nucleotide exchanges required for their construction by the DNA parsimony algorithm (employing a program of Felsenstein [34]) found the same topology preferred [28]. However alternative topologies in which the pol2 and 3 lineages had changed positions or shared a common stem were hardly less probable. In contrast, the bifurcation of pol1

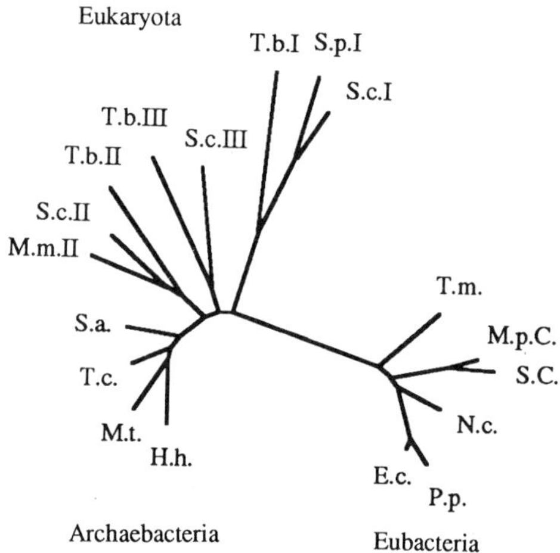

Figure 2.   Phylogenetic dendrogram of RNA polymerase component genes calculated from the distance matrix in Figure 1 according to [33].

with the eubacteria and the separation of poll from the other eukaryotic polymerase lineages appeared highly significant and the eocyte photocyte tree proposed by J. A. Lake seemed least probable.

In order to increase the significance of the tree we have since included several supposedly sort and deep branching organisms, *Thermococcus celer* for the archaebacteria and *Thermotoga maritima* for the eubacteria (I. Arnold, H.-P. Klenk, P. Palm, C. Schleper, V. Schwass and W. Zillig, unpublished), and further sequences determined by other laboratories (see introduction) in the analysis. The unexpected features of the topology remained invariant though all these lineages joined the tree in the expected positions (Figure 2).

A bootstrapping analysis [39-40] of 1660 ungapped positions of an alignment of the genes from 11 species yielded the same topology in 183 of 200 tree replicas and the topology in which the three eukaryotic polymerase lineages share a common stem in only two replicas (Table 2).

Phylogenetic Implications: The Fusion Hypothesis

Since the topology thus apears significant we must consider two alternative explanations:

1)   Assuming the three specialized eukaryotic polymerases had arisen within the ancestral eukaryote, the urkaryote, by two duplication-diversification events, one or the other or both prokaryotic kingdoms should have originated from within the eukaryotes by reduction events. This would impy that reduction was not restricted to conditions which make a fraction of the genome

326

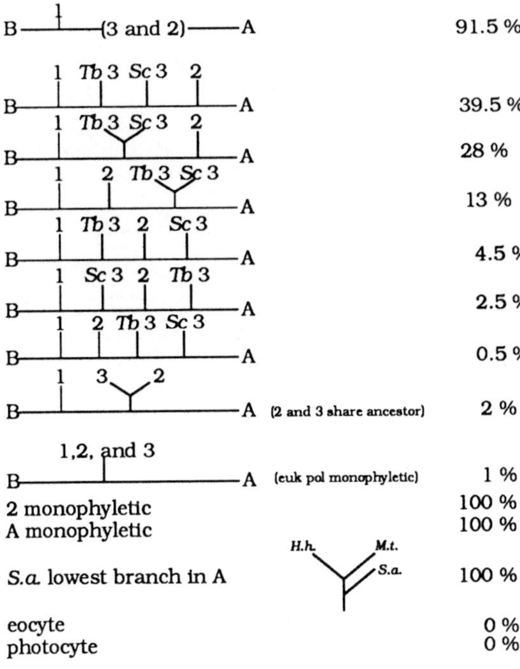

Table 2. Significance analysis of phylogenetic dendrogram considering 1660 ungapped positions of alignment of RNAP ß', A, and A+C genes, resp.; DNA Boot (40), 200 replicas.
B, eubacteria; A, archaebacteria; *T.b.*, *Trypanosoma brucei; S.c., Saccharomyces cerevisiae; H.h. Halobacterium halobium; M.t., Methanobacterium thermoautotrophicum; S.a., Sulfolobus acidocaldarius*, 1, 2, and 3, eukaryotic RNA polymerases 1, 2 and 3.

superfluous. It contradicts the current assumptions regarding a rather late arisal of eukaryotes in Earth's history. It would require the remaining of the previously specialized enzymes to regain omnipotence. Thus, this possibility appears highly improbable.

2) The genes for the three different eukaryotic enzymes or at least for the ancestor of pol2 and 3 on the one hand and pol1 on the other entered the urkaryote from different ancestral lineages, either upon its genesis by a fusion event or by the early acquisition of one set of genes by a recipient already harbouring the other e.g. via an early endosymbiont (different from the ancestors of mitochondria) or some other sort of horizontal gene transfer (Figure 3). This is in line with old ideas on a chimeric origin of the eukaryote based on the existence of a nuclear in addition to the cytoplasmic membrane [41]. According to this interpretation the eukaryote is a biphyletic or oligophyletic chimera.

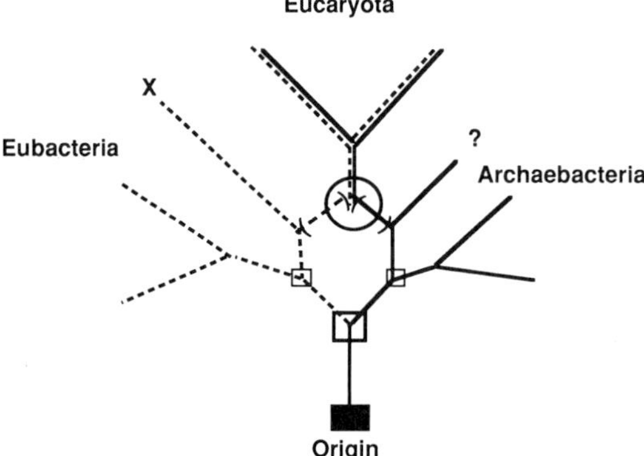

Figure 3. Schematic presentation of the fusion hypothesis. The black box symbolizes the progenote stage, the transparent box the bifurcation between eubacteria and archaebacteria, the circle is around the point of fusion. The right of the small boxes appears as the trifurcation point between the kingdoms in a dendrogram of a gene entering the eukaryote via its archaebacterial ancestor, the left of the small boxes as the trifurcation point for a gene entering the eukaryote via its eubacterial ancestor. The left of the lines in brackets (entering the "fusion circle") would not be visible when looking at a eukaryotic gene of archaebacterial origin. The right of these lines would not be visible when looking at a eukaryotic gene of eubacterial origin. X designates the lineage of the eubacterial fusion partner (e.g. *Thermus* in the MDH and PGK trees). ? designates the lineage from which the archaebacterial fusion partner arose.

The tree topology is certainly in accord with this hypothesis. Yet the poll sequences behaved ambivalently in resembling the eubacterial sequences in some regions but those of pol2 and pol3 in others. The subunit complexity of poll corresponds to that of pol3 and pol2 rather than the "streamlined" composition of eubacterial polymerases. This could however have resulted from a mutual adjustment in the course of the coexistence of pol1, 2 and 3 within the eukaryotes only restricted by the differences in function.

Further evidence?

If the genes for different polymerases had been acquired by some sort of fusion rather than single gene transfer one should find further examples of genes of ancestral archaebacterial but also of ancestral eubacterial nature within the eukaryotic genome. Amino acid sequences of ribosomal proteins and translation factors but also of ATPase α and β components or archaebacteria have much more similarity to those of eukaryotes (in the case of ATPases the vacuolar type) than the corresponding sequences of eubacteria. In two cases of gene duplication observed in all three kingdoms, namely the ATPase α and β components, and EFtu and EFG, where the trees of the paralogous duplicands could be rooted by determining their intersection points, the eubacteria were found to branch off prior to the ramification between the eukaryotes and the archaebacteria [42-43]. Several eukaryotic ribosomal proteins have homologs in archaebacteria but not in eubacteria [8; 18-19; 44]. In contrast, in the case of the genes of three enzymes of intermediary metabolism, glyceraldehyde phosphate dehydrogenase (GAPDH), malate dehydrogenase (MDH) and phosphoglycerate kinase (PGK), the similarity is much greater between eukaryotes and eubacteria than between eukaryotes and archaebacteria [9-10; R. Hensel and coworkers, unpublished]. Not only does the latter appear at a large distance from the other closely related two, but one eubacterium branches off at the basis of the eukaryotes rather than within the other eubacteria (in the case of GAPDH *E. coli* and in the case of MDH and PGK *Thermus*). This is in perfect accord with a presumed eubacterial origin of these genes in eukaryotes, even more specific with their origin from two distinct eubacterial ancestors. Several consecutive fusion or acquisition events might therefore have been involved in the generation of the eukaryotic chimera. With the exception of ATPase, all "archaebacterial" genes in eukaryotes known so far are parts of the genetic machinery whereas the three "eubacterial" genes code for important metabolic enzymes, suggesting that packages of genes of similar or related organisms were selected from one or the other ancestral genome.

Roots of the Universal Tree

These results confirm the unity of the archaebacteria which was deduced by Woese and coworkers from the rRNA sequence data[45]. Assuming the roots were situated outside one of the kingdoms, they are further proof of their monophletic nature. They clearly show *H. halobium* to be an archaebacterium even though in the tree derived by evolutionary parsimony in contrast to all other trees they appear as the lowest branch of the kingdom. We consider this to be a consequence of an insufficient number of positions in the alignment and/or of the high GC content of the *H. halobium* sequence [46]. Several details of our results furnish additional evidence: *H. halobium* shares the characteristic AC split in the polymerase operon, the situation of "Strep operon genes" immediately downstream of the polymerase operon and a small subunit (H) gene at its 5' end with other archaebacteria but not with *E. coli*. The split of the rpoC (β') gene in cyanobacteria and chloroplasts is, however, about 600 nucleotides upstream. *H. halobium* shares the B"B' split with *M. thermoautotrophicum* and *M. vannielii* in accord with its position in the rRNA tree. *Sulfolobus* RNA polymerase recognizes *H. halobium* but not *E. coli* promoters (U. Hüdepohl and W.-D. Reiter, unpublished). In accord, *H. halobium* promoters show the archaebacterial consensus [47].

If our interpretation of the data were correct, both archaebacteria and eubacteria would form paraphyletic rather than holophyletic kingdoms because both participated in generating the urkaryote. Because of their generation by synthesis, the eukaryotes would form a bi- or oligophyletic rather than monophyletic kingdom. The only meaningful branching event between kingdoms would be the bifurcation between archaebacteria and eubacteria such that proof of the hypothesis would make the determination of the universal root unnecessary. A branching order in which the eubacterial lineage arose prior to the bifurcation between archaebacteria and eukaryotes has been determined from two different gene duplications which lead to the formation of pairs of paralogous proteins [42-43]. We consider this a tree topology for these genes rather than for organisms and kingdoms. If our hypothesis were correct, one should also find the opposite case: duplications of genes which were later contributed to the urkaryote from eubacterial ancestors should yield a topology in which the archaebacteria branch off before the bifurcation between eubacteria and eukaryotes. Such a case could possibly be found among enzymes of intermediary metabolism.

Several observations tetify to the possibility of the postulated fusion event. In the course of their history, eukaryotic cells were at least twice subject to the immigration of endosymbionts leading to the evolution of mitochondria and chloroplasts. Members of certain groups of bacteria show a high potential for specific interaction with eukaryotes. The primitive eukaryote *Pelomyxa palustris* contains a eubacterial and an archaebacterial endosymbiont [48]. Sexuality of eukaryotes involves the fusion of germ cells. Both eubacteria and archaebacteria [49] have developed mechanisms for genetic exchange.

The Paradox of Genome Organization

Archaebacteria and eubacteria represent two modes of prokaryotic life distinguished in many details but sharing "economized" ("streamlined") small genomes with little size variation, the use of operons for the coordination of gene expression and the response to changing environments by adaptation via mutation. In contrast eukaryotes possess genomes which are at least one order of magnitude larger. They appear to keep growing in the course of evolution which seems to proceed by invention rather than streamlined adaptation. Genes are usually single and not in recognized order. They are dissolved in exons separated by introns. It appears attractive to speculate that this fundamental difference arose as a consequence of the proposed genome fusion event(s) creating the eukaryotic chimera. The original chimera would have possessed a completely duplicated set of essential genes furnishing a first playground for functional diversification-invention. Such a mode of evolution would become practically irreversible after only a few steps. Addition of further playgrounds, e.g. by sexuality involving diploidia, or via recombination rather than duplication and possibly also the exon intron organization of eukaryotic genomes, could have increased the invention potential.

REFERENCES

[1]   G.E. Fox, E. Stackebrandt, R.B. Hespell, J. Gibson, J. Maniloff, T.A. Dyer, R. S. Wolfe, W.E. Balch, R. S. Tanner, L. J. Magrum, L. B. Zablen, R.

Blakemore, R. Gupta, L. Bonen, B. J. Lewis, D. A. Stahl, K. R. Luehrsen, K. N. Chen and C. R. Woese, *Science* 209:457 (1980)

[2]   C. R. Woese, *Microbiol. Reviews* 51:221 (1987)

[3]   H. Leffers, J. Kjems, L. Ostergaard, N. Larsen and R. A. Garrett, *J. Mol. Biol.* 195:43 (1987)

[4]   H. Hori, T. Itoh and S. Osawa, *Zentralbl. Bakteriol. Hyg. I. Abtr. Orig.* C3:18 (1982)

[5]   G. E. Fox, K. R. Luehrsen and C. R. Woese, *Zentralbl. Bakteriol. Hyg. I. Abt. Orig.* C3:330 (1982)

[6]   P. Willekens, K. O. Stetter, A. Vandenberghe, E. Huysmans and R. De Wachter, *FEBS Lett.* 204:273 (1986)

[7]   R. Gupta, *System. Apl. Microbiol.* 7:102 (1986)

[8]   J. Auer, K. Lechner and A. Böck, *Can. J. Microbiol.* 35:200 (1989)

[9]   S. Fabry, J. Lang, T. Niermann, M. Vingron and R. Hensel, *Eur. J. Biochem.* 179:405 (1989)

[10]   R. Hensel, P. Zwickl, S. Fabry, J. Lang and P. Palm, *Can. J. Microbiol.* 35:81 (1989)

[11]   J. A. Lake, E. Henderson, M. Oakes and M. W. Clark, *Proc. Natl. Acad. Sci. USA,* 81:3786 (1984)

[12]   J. A. Lake, *Nature,* 319:626 (1986)

[13]   J. A. Lake, *J. Mol. Evol.* 26:59 (1987)

[14]   J. A. Lake, *Mol. Biol. Evol.* 4:167 (1987)

[15]   J. A. Lake, *Nature,* 331:184 (1988)

[16]   J. Huet, R. Schnabel, A. Sentenac and W. Zillig, *EMBO J.* 2:1291 (1983)

[17]   R. Schnabel, M. Thomm, R. Gerardy-Schahn, W. Zillig, K. O. Stetter, and J. Huet, *EMBO J.* 2:751 (1983)

[18]   H. Leffers, F. Gropp, F. Lottspeich, W. Zillig and R. A. Garrett, *J. Mol. Biol.* 206:1 (1989)

[19]   G. Pühler, F. Lottspeich and W. Zillig, *Nucl. Acids Res.* 17:4517 (1989)

[20]   B. Berghöfer, L. Kröckel, C. Körtner, M. Truss, J. Schallenberg and A. Klein, *Nucl. Acids Res.* 16:8113 (1988)

[21]   L. A. Allison, M. Moyle, M. Shales and C. J. Ingles, *Cell,* 42:599 (1985)

[22]   J. L. Corden, D. L. Cadena, J. M. Ahearn and M. E. Dahmus, *Proc. Natl. Acad. Sci. USA,* 82:7934 (1985)

[23]   R. Evers, A. Hammer, J. Köck, W. Jess, P. Borst, S. Memet and A.W.C.A. Cornelissen, *Cell,* 56:585 (1981)

[24]   J. Köck, R. Evers and A.W.C.A. Cornelissen, *Nucl. Acids Res.* 16:8753 (1988)

[25]   S. Memet, M. Gouy, C. Marck, A. Sentenac and J.-M. Buhler, *J. Biol. Chem.* 263:2830 (1988)

[26]   M. Yamagishi and M. Nomura, *Gene,* 74:503 (1988)

[27]   W. Jess, A. Hammer and A.W.C.A. Cornelissen, *FEBS Lett.* 249:123 (1989)

[28]   G.Pühler, H. Leffers, F. Gropp, P. Palm, J.-P. Klenk, F. Lottspeich, R.A.Garrett and W. Zillig, *Proc. Natl. Acad. Sci. USA,* 86:4569 (1989)

[29]   W. Zillig, J.-P. Klenk, P. Palm, G. Pühler and F. Gropp, *Can. J. Microbiol.* 35:73 (1989)

[30]   W. Zillig, J.-P. Klenk, P. Palm, H. Leffers, G. Pühler, F. Gropp and R. A. Garrett, *Endocyt. C. Res.* 6:1 (1989)

[31] M. Nonet, D. Sweetser and R.A. Young, *Cell*, 50:909 (1987)

[32] L. A. Allison, J.K.-C. Wong, V.D.Fitzpatrick, M. Moyle and C.J.Ingles, *Mol. Cell Biol.* 8:321 (1988)

[33] W.M. Fitch and E. Margoliash, *Science*, 15:279 (1967)

[34] J. Felsenstein, The statistical approach to inferring evolutionary trees and what it tells us about parsimony and compatibility, *in*: "Cladistics: perspectives in the reconstruction of evolutionary history", T. Ducan and T. F. Stuessy, eds. Colombia University Press, New York, (1984) pp. 169-191

[35] J. Felsenstein, *Syst. Zool.* 22:240 (1973)

[36] J. Felsenstein, *J. Mol. Evol.* 17:368 (1981)

[37] M. O. Dayhoff, R. M. Schwartz and B.C.Orcutt, A model of evolutionary change in proteins, *in*: "Atlas of Protein Sequence and Structure", Vol. 5, Suppl. 3. National Biomedical Research Foundation, Washington, D.C. pp. 345-358 (1978)

[38] D. F. Feng, M. S. Johnson and R. F. Doolitle, *J. Mol. Evol.* 21:112 (1985)

[39] J. Felsenstein, *Evolution* 39:783 (1985)

[40] J. Felsenstein, PHYLIP version 3.21 (c) Copyright 1986, University of Washington and J. Felsenstein (1989)

[41] C. Mereschkowsky, *Biolog. Zentralbl.* 30:278-303, 321-347, 353-367 (1910)

[42] J. P. Gogarten, H. Kibak, P. Dittrich, L. Taiz, E. J. Bowman, B. J. Bowman, M. F. Manolson, R. J. Poole, T. Date, T. Oshima, J. Konishi, K. Denda and M. Yoshida, *Proc. Natl. Acad. Sci. USA* 86:6661 (1989)

[43] N. Iwabe, K. Kuma, M. Hasegawa and S. Osawa, *Proc. Natl. Acad. Sci. USA* (in press)

[44] P. P. Denis (1990) this volume.

[45] C. R. Woese and G. J. Olsen, *Syst. Appl. Microbiol.* 7:161 (1986)

[46] M. Gouy and W.-H. Li, *Nature* 339:145 (1989)

[47] W.-D. Reiter, P. Palm and W. Zillig, *Nucl. Acids Res.* 16:1 (1988)

[48] C. K. Stumm, J. J. A. van Bruggen, K. B. Zwart and G.D. Vogels, Methanogenic bacteria as endosymbionts of anaerobic protozoa, *in*: "Archaebacteria '85", O. Kandler and W. Zillig, eds. Gustav Fischer Verlag, Stuttgart, New York, (1986)

[49] M. Mevarech and R. Werczberger, *J. Bacteriol.* 162:461 (1985)

DNA TOPOLOGY IN HALOBACTERIA

Patrick Forterre, Daniele Gadelle, Franck Charbonnier
and Mouldy Sioud

Institut de Génétique et Microbiologie
Université Paris-Sud
91 405 Orsay Cedex
France

ABSTRACT

Inhibitors of eubacterial and eucaryotic DNA topoisomerases II induce topological changes and/or DNA cleavage in the plasmids of halobacteria. As in eubacteria, novobiocin halts DNA replication and induces positive supercoiling of plasmids in halobacteria. This positive supercoiling is prevented by actinomycin D, indicating that it may be generated by transcription as in eubacteria.

INTRODUCTION

Supercoiling influences DNA functions by several mechanisms: i) it helps the assembly of nucleoprotein complexes by increasing DNA bendability, ii) it determines the exact pitch of the double helix and therefore modulates the binding of proteins to specific DNA sequences, iii) it stabilizes (positive supercoiling) or destabilizes (negative supercoiling) the double helix.

Supercoiling is generated during transcription of the DNA molecule because untwisting of the double-helix generates waves of positive and negative superturns ahead of and behind the transcription fork, respectively [1]. In addition, two DNA topoisomerases can invert the sign of supercoiling at the expense of ATP: the eubacterial DNA gyrase (from positive to negative) [2], and the reverse gyrase of extremely thermophilic archaebacteria (from negative to positive) [3, 4]. Other topoisomerases relax these superturns (Figure 1). Therefore, the level of DNA supercoiling in vivo is probably determined by the extent of superturns formation through transcription and of their processing by topoisomerases.

General and Applied Aspects of Halophilic Microorganisms
Edited by F. Rodriguez-Valera, Plenum Press, New York, 1991

333

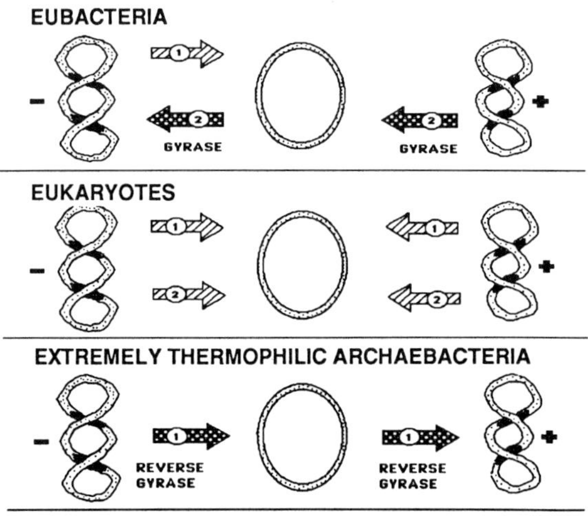

Figure 1.   Variety of topoisomerases in different urkingdoms,
1: topo I, 2: topo II.

DNA topoisomerase activities are different in eubacteria, eucaryotes and archaebacteria (Figure 1). All cells contain a type I enzyme (topo I), which catalyses the passage of a DNA strand via a transient topo I-linked DNA nick, and type IIenzyme (topo II), which catalyses the passage of a DNA duplex via a transient topo II-linked double-stranded break. However, whereas in eubacteria topo I and II have antagonistic activities (topo I relaxes only negative superturns whereas topo II, DNA gyrase, produces negative superturns), in eucaryotes, both topo I and II relax either negative or positive superturns. Still another situation exists in the extremely thermophilic archaebacteria: these organisms contain the unusual topo I, reverse gyrase [4-6]. This activity has not been found in methanogens, halobacteria or Thermoplasma acidophilum [6, 7]; thus, different mechanisms to control DNA supercoiling may exist among archaebacteria.

EFFECT OF DNA TOPOISOMERASE INHIBITORS ON THE GROWTH OF HALOBACTERIA: IDENTIFICATION OF THEIR TARGET

Table I shows that halobacteria are resistant to the inhibitor of eucaryotic topo I, campthotecin, but are sensitive to both inhibitors of eubacterial and eucaryotic topo II [8, 9]. Most of these drugs are useful clinical agents: the fluoroquinolones pefloxacin and norfloxacin are very potent antibiotics against a wide range of pathogenic eubacteria whereas the epipodophyllotoxines, etoposide and teniposide, are antitumoral drugs in humans.

Several lines of evidence indicate that the target of these inhibitors in halobacteria is a topo II: i) novobiocin inverts the sign of supercoiling of halobacterial plasmids: whereas they are normally negatively supercoiled [9], they become positively supercoiled after novobiocin treatment [10], indicating that novobiocin disrupts the

Table 1. Effects of DNA topoisomerases inhibitors on halobacteria.

drugs	target	inhibition of growth	effect on DNA	ref.
camptothecin	I(ek)	-	-	(unpublished)
novobiocin	II(eb)	+	Positive supercoiling	[10]
ciprofoxiacin[a]	"	+	cleavage	[12]
etoposide	II(ek)	+	cleavage	[8. 11]
teniposide	"	+	"	"

I(ek): Inhibitor of eukaryotic DNA topoisomerase I, II(eb): Inhibitors of eubacterial DNA topoisomerase II (DNA gyrase), II(ek): Inhibitors of eukaryotic DNA topoisomerase II.

[a]Only active on Natronobacteria, DNA cleavage with ciprofloxacin is observed in neutrophilic halobacteria when the drug is added in a magnesium-depleted medium.

balance between antagonistic supercoiling activities in halobacteria, probably by inhibiting their topo II. As in eubacteria, novobiocin also inhibits DNA replication in halobacteria rapidly and specifically [8]. ii) ciprofloxacin, etoposide and teniposide produce single- and double-stranded breaks in halobacterial DNA [9, 11, 12]. DNAcleavage induced by these drugs is prevented by novobiocin. In the case of etoposide we have found that a protein is covalently linked to the 5' ends of the DNA breaks [11]. These drugs should stabilize a cleavable complex between DNA and topo II in halobacteria as in eubacteria and eucaryotes.

DNA cleavage by ciprofloxacin and etoposide did not occur randomly: a definite set of DNA fragments was obtained after a secondary cleavage by a restriction enzyme with a single cutting site on the plasmid pGRB-1 of *Halobacterium* GRB. The same fragments were produced with both drugs [12]. In that case, with the use of an archaebacterial system, it was demonstrated that fluoroquinolone antibiotics and antitumoral drugs of the epipodophyllotoxin family trap the same DNA topo II complexes in vivo.

ORIGIN OF POSITIVE SUPERCOILING IN HALOBACTERIA

Positive supercoiling of halobacterial plasmids induced by the inhibition of topo II with novobiocin can be due either to an halobacterial reverse gyrase or to residual transcription in the presence of an eubacterial-like topo I [7]. In eubacteria, the transcriptional hypothesis was demonstrated by the inhibition of

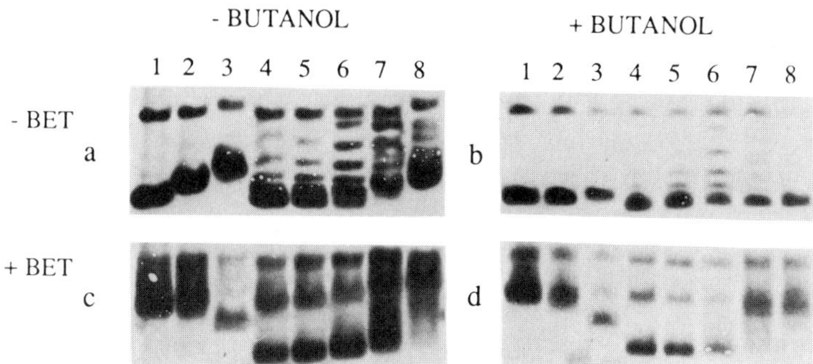

Figure 2. Effect of combining novobiocin and actinomycin D on plasmid topology in halobacteria: various actinomycin D concentrations were added to cells of *Halobacterium* GRB at OD_{600} = 0,3 (0, 1, 12, 0, 0.4, 1, 4, 8 µg/ml from lanes 1 to 8, respectively). In lanes 4 to 8, 2 µg/ml of novobiocin were added 2 hours after the time of actinomycin D addition. Cells were lysed after 48 hours; preparation of extracts and gel run were as previously described [9,10]. The plasmids were detected after Southern transfer using a sulfonated pGRB-1 probe.

novobiocin-induced positive supercoiling with rifampycin [13]. Unfortunately, the same experiment cannot be performed in halobacteria since their RNA polymerase is not sensitive to this drug. In the hope of bypassing this problem, we have checked the effect of actinomycin D on novobiocin- induced positive supercoiling in halobacteria. The doses of actinomycin D which inhibit transcription in eubacteria correspond to the doses which inhibit their growth and are ten times lower than doses inhibiting DNA replication [14]. However, actinomycin D is not as specific as rifampycin for transcription since it binds to the DNA and may interfere with other DNA-directed processes.

We found that actinomycin D inhibits the growth of *Halobacterium* GRB with an ID50 of 1.5 ug/ml (not shown). This dose is in the range of those inhibiting transcription by 50% in eubacteria. We have tested the effect of combining actinomycin D and novobiocin on pGRB-1 topology. After drug treatment, the plasmids were run on two gels, with and without ethidium bromide (0.04 ug/ml), to monitor their sign of supercoiling. Figure 2 (a, c) shows that the migration of pGRB1 on agarose gel was disturbed at doses of actinomycin D which inhibit cell growth. To avoid this problem, actinomycin D was extracted with butanol prior to the electrophoresis (Figure 2, b and d). The plasmids isolated from cells treated by novobiocin were composed by a majority of positively supercoiled DNA (not relaxed by ethidium bromide) and by a minority of negatively supercoiled DNA (relaxed by ethidium bromide) (Figure 2-c, lanes 4-6). In contrast all plasmids isolated from cells treated by both novobiocin and doses of actinomycin D which completely inhibited cell growth (4 and 8 ug/ml) were relaxed by ethidium bromide (lanes 7-8), indicating that they were completely negatively supercoiled. Therefore, actinomycin D prevents novobiocin-induced positive supercoiling of pGRB-1. It should be noticed that actinomycin D alone changes the migration pattern of some pGRB-1 plasmids at 12 ug/ml (the band with a higher electrophoretic mobility in lane d3). We have found that these plasmids are not linearized (not shown) and the determination of their sign of supercoiling is underway.

The inhibition of novobiocin-induced positive supercoiling by actinomycin D in halobacteria suggests that this supercoiling is generated by transcription as in eubacteria. This hypothesis also fits well with one of our previously unexplained observations: topo II inhibition by etoposide or ciprofloxacin did not induce positive supercoiling in the population of plasmids which have escaped DNA cleavage [11, 12]. Since these two drugs have no effect on topo I, including reverse gyrase, they probably prevent positive supercoiling by forming topo II poisons which inhibit residual transcription. The transcriptional hypothesis for positive supercoiling in halobacteria also concurs with the absence of detectable reverse gyrase activity in these organisms and the related archaebacteria, methanogens and thermoplasma.

DISCUSSION

Our results indicate that DNA supercoiling in archaebacteria is probably produced both by transcription and topoisomerase activities as in the other two urkingdoms. Studies on DNA topology in extremely thermophilic archaebacteria have revealed an unique archaebacterial feature, the existence of reverse gyrase. In contrast, the data we have obtained so far on DNA topology in halobacteria are very reminiscent of those obtained in eubacteria:

i)   as in eubacteria, the halobacterial topo II is absolutely required for DNA replication, probably to remove the positive superturns which otherwise would accumulate during DNA chain elongation. In contrast, either topo I or topo II can permit DNA replication in eucaryotes [15, 16]. This is probably due to the fact that, unlike its eubacterial counterpart, the eucaryotic topo I can relax positive superturns.

ii)   topo II inhibition by novobiocin produces positive supercoiling. This also indicates that halobacteria have no eucaryotic-like topo I activity relaxing positive superturns. Positive supercoiling of plasmids has been observed in yeast cells only when they contain the exogeneous gene encoding eubacterial topo I and when both their own topo I and II are inactivated by mutations [17]. The absence of a eucaryotic topo I in halobacteria is also in line with their resistance to camptothecin (Table I).

Recently, another striking similarity emerges between eubacteria and archaebacteria: the common gene organization of RNA polymerase and ribosomal protein genes into similar operons [18, 19]. This suggests to us that eubacteria and archaebacteria shared a common procaryotic ancestor, distinct from their common ancestor with eucaryotes. The existence of eucaryotic features in archaebacteria, such as the sensitivity of their topo II to antitumoral drugs, may indicate that these microorganisms resemble the common primitive procaryotic ancestor more than eubacteria; in that sense the term archaebacteria has probably been a good choice.

ACKNOWLEDGEMENTS

This work was supported by a grant from the "Association de la Recherche contre le Cancer". M Sioud is recipient of a fellowship of the "Association de la Recherche contre le Cancer".

REFERENCES

[1]   L. F. Liu and J. C. Wang, *PNAS* 84:7024 (1987).

[2]   M. Gellert, M. H. O'Dea, T. Itoh and J. Tomizawa, *PNAS* 73:4474 (1976).

[3]   A. Kikuchi and K. Asai, *Nature*, 309:677 (1984).

[4]   P. Forterre, G. Mirambeau, C. Jaxel, M. Nadal and M. Duguet, *EMBO J.* 4:2123 (1985).

[5]   A. I. Slesarev, *Eur. J. Biochem.* 173:395 (1988).

[6]   R. G. Collin, H. W. Morgan, D. R. Musgrave and R. M. Daniel, *FEMS Lett.* 55:235 (1988).

[7]   P. Forterre, C. Elie, M. Sioud and A. Hamal, *Can. J. Microbiol.* 35:228 (1989).

[8]   M. Sioud, G. Baldacci, P. Forterre and A. M. De Recondo, *Eur. J. Biochem.* 169:231 (1987).

[9]   M. Sioud, O. Possot, C Elie, L. Sibold and P. Forterre, *J. Bacteriol.* 170:946 (1988).

[10]   M. Sioud, G. Baldacci, A. M. De Recondo and P. Forterre, *Nucl. Acids Res.* 16:1379 (1988).

[11]   M. Sioud, G. Baldacci, P. Forterre and A. M. De Recondo, *Nucl. Acids Res.* 15:8217 (1987).

[12]   M. Sioud and P. Forterre, *Biochemistry*, 28:3638 (1989).

[13]   H. Y. Wu, S. Shyy, J. C. Wang and L. F. Liu, *Cell*, 53:433 (1988).

[14]   J. Hurwitz, J. J. Furth, M. Malamy and M. Alexander, *PNAS*, 48:1222 (1962).

[15]   L. Yang, M. S. Wold, J. J. Li, T. J. Kelly and L. F. Liu, *PNAS*, 844:950 (1987).

[16]   T. Uemura and M. Yanagida, *EMBO J.* 3:1734 (1984).

[17]   G. N. Giaever and J. C. Wang, *Cell*, 55:849 (1988).

[18]   L. C. Shimmin and P. P. Dennis, *EMBO J.* 8:1225 (1989).

[19]   G. Puhler, H. Leffers, F. Gropp, P. Palm, H. P. Klenk, F. Lottspeich, R. A. Garret and W. Zillig, *PNAS*, 86:4569 (1989).

PART V

APPLIED ASPECTS OF HALOPHILIC ORGANISMS

AN APPLICATION OF A BIOREACTOR WITH FLOCCULATED CELLS OF HALOPHILIC *MICROCOCCUS VARIANS* SUBSP. *HALOPHILUS* WHICH PREFERENTIALLY ADSORBED HALOPHILIC NUCLEASE H TO 5'-NUCLEOTIDE PRODUCTION

Hiroshi Onishi[1], Haruhiko Yokoi[2], and Masahiro Kamekura[3]

[1]Department of Agricultural Chemistry
Kagoshima University
Kagoshima-shi
Japan

[2]Hiratsuka Research Laboratory
Sumitomo Heavy Industries, Ltd.
Hiratsuka-shi
Kanagawa-ken
Japan

[3]Noda Institute for Scientific Research, Noda-shi
Chiba-ken
Japan

ABSTRACT

A bioreactor system was designed with the flocculated cells of halophilic *Micrococcus varians* subsp. *halophilus* which immobilized halophilic nuclease H on the surface, for the production of flavouring agent, 5'-guanylic acid (5'-GMP). As the bacterium produced an extracellular 5'-nucleotidase as well as the nuclease H, selective inactivation of the 5'-nucleotidase contaminated in the column was necessary to obtain high yields of 5'-GMP from RNA. A fact that the exogenous nuclease H was preferentially adsorbed on the flocculated cells over the 5'-nucleotidase was advantageous to the production. Desalting treatment of the flocculated cells in the presence of 80 mM $MgSO_4$ preferentially inactivated 5'-nucleotidase, and the RNA degradation in the presence of 0.25 mM $ZnSO_4$ greatly improved the yield. Furthermore, Ca-enzyme-complex which was formed by adding $CaCl_2$ to the supernatant of the culture at 25°C was applied to the bioreactor.

General and Applied Aspects of Halophilic Microorganisms
Edited by F. Rodriguez-Valera, Plenum Press, New York, 1991

INTRODUCTION

Kamekura and Onishi [1, 2] reported that a moderate halophile *Micrococcus varians* subsp. *halophilus*, produced an extracellular halophilic nuclease (nuclease H) which showed the highest activity at 2 to 3 M NaCl or KCl, and that the enzyme degraded RNA and DNA to produce 5'-mononucleotides exonucleolytically. This fact suggested the use of the nuclease H to the production of 5'-GMP and 5'-inosinic acid (5'-IMP) as flavouring agents, which are produced commercially by enzymatic RNA degradation or direct fermentation. Commercial application of halophiles and halophilic enzymes has rarely been reported. In one application of nuclease H, Kamekura et al. [3] reported the production of 5'-GMP from yeast RNA in a batch system, using a supernatant of the culture of this halophile. On the other hand, it was also demonstrated [4, 5] that flocculation of the cells occurred during the growth in medium containing 3 M NaCl and a concentration of $MgSO_4$ and KH_2PO_4 greater than 40 and 14 mM, respectively, and that the extracellular enzymes produced were fully adsorbed on the surface of the flocculated cells.

This paper describes the design of a bioreactor with a column of flocculated cells of the halophile which preferentially adsorbed nuclease H and some trials of the application of the bioreactor to 5'-nucleotide production.

MATERIALS AND METHODS

Bacterial strains, cultural conditions and enzyme assay. The moderate halophile *Micrococcus varians* subsp. *halophilus* ATCC 21971 was used. Basal medium CM [5] consisted of 1% Casamino acids (Difco), 1% yeast extract (Difco), and 3 M NaCl, pH 7.0. The bacterium was cultivated aerobically for 4 days at 30°C, and the supernatant obtained by centrifuging was used as the enzyme solution. Nuclease and 5'-nucleotidase activities were assayed as described previously [5, 6]. To obtain flocculated cells, the halophile was cultivated in 3 M NaCl-CM supplemented with 80 mM $MgSO_4$ and 9 mM KH_2PO_4 (Mg-CM). The flocculated cells were harvested by centrifuging and suspended in 25 mM Tris-HCl buffer containing 2 M NaCl and 80 mM $MgSO_4$, pH 8.0, to give a Klett-Summerson unit of 520 ± 5. The enzyme activities adsorbed on cells were assayed with the supernatants of deflocculated cell suspension after repeated suspension in 3 M NaCl-10 mM Tris-HCl buffer, pH 8.0 [7].

RNA degradation with a column of flocculated cells. The flocculated cells were enriched with nuclease H by suspension in the enzyme solution with 80 mM $MgSO_4$ added. A 12.5 ml portion of the suspension was mixed with 1 g (wet weight) of Celite 545 which had been washed with distilled water. The slurry was poured into a column fitted with a water jacket, giving a column of 11 by 140 mm. After the column was washed with 10 mM Tris-HCl buffer containing 3 M NaCl and 80 mM $MgSO_4$, pH 8.0, 10 ml of 1% RNA in the buffer was charged and eluted with the same buffer. The flow rate was kept constant at 10 ml/h and the effluents (2 g per tube) were collected. The degradation was also carried out in the presence of 0.25 and 0.5 mM $ZnSO_4$.

Desalting treatment. The enzyme solution was concentrated at 0°C to one tenth of its initial volume with an ultrafiltration module. The concentrated enzyme solution was diluted to the initial volume with 25 mM Tris-HCl buffer, pH 8.0, containing 0 to 0.5 M NaCl with or without 80 mM $MgSO_4$ and concentrated again with the module. After this procedure was repeated three times, the enzyme solutions were incubated at 30°C and the activities were measured periodically. The flocculated cells with increased activity by adsorption were suspended in 25 mM Tris-HCl buffer, containing 0 to 0.5 M NaCl and 80 mM $MgSO_4$. After a few minutes, the cells were collected by centrifuging and suspended in the same solutions. This procedure was repeated three times, and then the suspension was incubated at 30°C.

Preparation of Ca-enzyme complex. 136 mM $CaCl_2$ was added to the enzyme solution from 3 M NaCl CM culture at 25°C for 4 days, and then incubated at 30°C for 1 h to form a complex of enzyme with calcium phosphate. The resulting precipitate was collected by centrifuging and resuspended in 25 mM Tris-HCl buffer, pH 8.0 containing 3 M NaCl, 80 mM $MgSO_4$ and 2 mM $CaCl_2$.

Analytical methods. RNA content was determined by Schneider's method [8]. The degree of RNA degradation was determined by measuring the increase in 75% ethanol-soluble nucleotides [3]. RNA degradation products were analysed by high-performance liquid chromatography [9] after removal of salt from the samples with a small column of active charcoal [3].

RESULTS AND DISCUSSION

Adsorption of nuclease H on flocculated cells. When *Micrococcus varians* subsp. *halophilus* was cultivated in CM medium with added magnesium and phosphate for

Table 1. Adsorption of nuclease H on flocculated cells [7]

Vol. of enzyme soln. (ml) added to flocullated cells[a]	Enzyme activity (U) adsorbed on cells	
	Nuclease H	5'-nucleotidase
0	154	2,616
5	343	2,396
6.3	343	2,210
8.3	363	1,986
12.5	392	2,040
22	405	1,806
50	401	1,436

[a]Flocculated cells in 10 ml of culture were collected by centrifuging and mixed with various volumes of enzyme solution to which was added 80 mM $MgSO_4$ to prevent deflocculation. The activities of the enzyme solution were 20.4 U of nuclease H and 34.1 U of 5'-nucleotidase per ml.

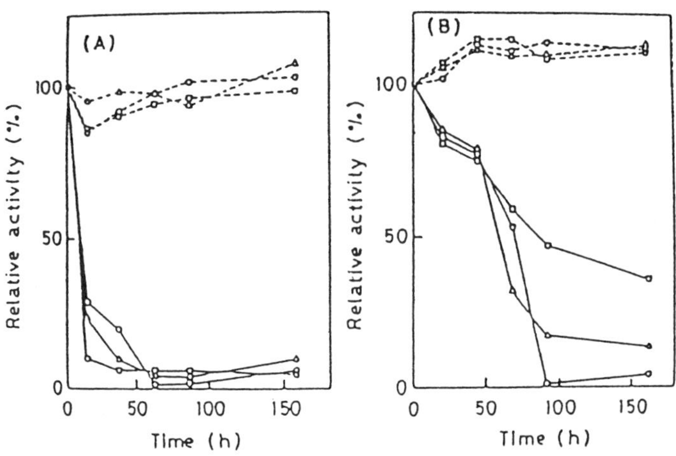

Figure 1.   Loss of 5'-nucleotidase activity during desalting treatment of enzyme solution (A) and flocculated cell suspension (B) [7]. After incubation at 30°C in 25 mM Tris-HCl buffer, pH 8.0, containing 80 mM $MgSO_4$ and 0 to 0.5 M NaCl, residual activities were measured periodically. Initial activities of the enzyme solution were 24.9 U of nuclease H and 77.0 U of 5'-nucleotidase per ml. Those of the flocculated cells were 23.1 U of nuclease H and 70.0 U of 5'-nucleotidase/cells per ml. Symbols: ...., nuclease; ________, 5'-nucleotidase; o, 0 M NaCl; Δ, 0.1 M NaCl; □, 0.5 M NaCl.

flocculation, the enzyme was adsorbed on the flocculated cells (15 U of nuclease H/cells per ml and 262 U of 5'-nucleotidase/cells per ml). Then we measured the further adsorption of exogenous enzymes added to the flocculated cells. The increase of the exogenous enzyme solution added to the flocculated cells resulted in an increase of the total activity of the nuclease H of the cells, whereas that of the 5'-nucleotidase decreased considerably (Table 1). This fact favoured the application of the flocculated cells to the production of 5'-nucleotides from RNA.

RNA degradation by using a bioractor. Some 70 and 31% of RNA was degraded at 40 and 60°C respectively, using the floc column bioreactor. Degradation products at 60°C were 5'-adenylic acid (5'-AMP), 5'-uridylic acid (5'-UMP) and 5'-cytidylic acid (5'-CMP) but 5'-GMP was not detected. Degradation at 40°C resulted in the absence of 5'-AMP as well as 5'-GMP. Comparison of the sum of each nucleotide and its corresponding nucleoside suggested a further degradation of nucleosides to bases and ribose.

Selective inactivation of 5'-nucleotidase by desalting. The effect of desalting treatment in the presence of 80 mM $MgSO_4$ on activities of the enzyme solution is shown in Figure 1A. The nuclease was stable in the absence of NaCl for as long as 160 h, whereas 5'-nucleotidase was inactivated rapidly under the same conditions. In the

344

absence of $MgSO_4$, however, the nuclease activity was also lost rapidly at 0 to 0.5 M NaCl. Desalting treatment of the enzyme adsorbed on the flocculated cells was also carried out in the presence of 80 mM $MgSO_4$. The nuclease was stable in the absence of NaCl, while the 5'-nucleotidase activity was gradually lost with remaining activity of 4, 14 and 38% at 0, 0.1 and 0.5 M NaCl respectively, after 164 h (Figure 1B). Thus, it was demonstrated that a desalting treatment of the flocculated cells with 25 mM Tris-HCl buffer, pH 8, containing 80 mM $MgSO_4$ was quite effective for selective inactivation of 5'-nucleotidase without loss of nuclease activity.

RNA degradation by a column bioreactor after desalting treatment. The degree of RNA degradation at 50°C by a treated and untreated bioreactor was 43.9 and 37.0% respectively. Quantitative analysis of the products showed that the amount of 5'-nucleotides produced by the treated bioreactor was larger than that produced by the untreated one, except 5'-UMP, in spite of the low RNA degradation rate (Table 2). In particular, a flavouring agent, 5'-GMP, was produced by the treated bioreactor but not by the untreated one. The amount of 5'-AMP, the precursor of the flavouring agent 5'-IMP [10], in the product produced by the treated bioreactor was about 10-fold larger than that produced by the untreated one. Thus, it was demonstrated that, after desalting treatment, the bioreactor was effective for production of 5'-nucleotides, especially 5'-GMP and 5'-AMP.

Table 2. Analysis of RNA degradation products at 50°C with a bioreactor after desalting treatment[a] [7]

Product	Degradation (mg/litre) with:	
	Treated bioreactor[b]	Untreated bioreactor[c]
5'-AMP	305.7	31.7
5'-GMP	120.9	O
5'-CMP	124.0	48.7
5'-UMP	228.1	447.4
Adenosine	19.7	55.5
Guanosine	0	290.7
Cytidine	10.8	16.2
Uridine	0	310.3

[a]Effluents of the highest A_{260} were subjected to analysis.
[b]Activity of the column after desalting treatment: 201 U of nuclease H and 60 U of 5'-nucleotidase per ml of column.
[c]Activity of the untreated column: 208 U of nuclease H and 546 U of 5'-nucleotidase per ml of column.

Effects of metal ions on the activities of nuclease and 5'-nucleotidase. For the selection of a preferential inhibitor of 5'-nucleotidase, the effects of the following

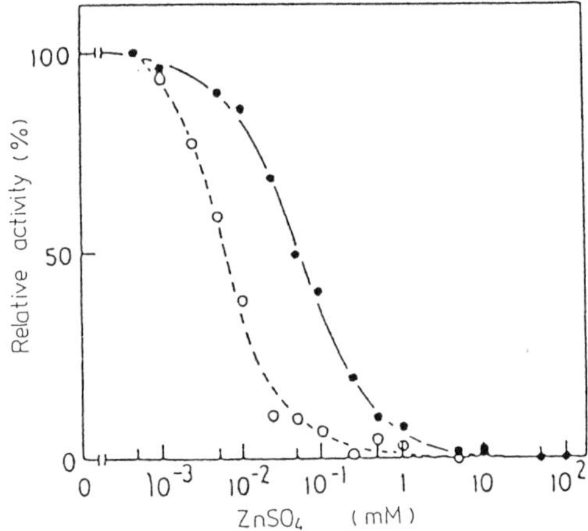

Figure 2. Effect of ZnSO$_4$ on the activities of nuclease H ( ) and 5'-nucleotidase (o) adsorbed on flocculated cells [11]. The assay was carried out with the flocculated cells in the presence of various concentrations of ZnSO$_4$. The activities of nuclease H and 5'-nucleotidase in the absence of ZnSO$_4$ were taken as 100%.

metal ions on the activities of nuclease and 5'-nucleotidase in the crude enzyme solution were examined; NiCl$_2$, MnCl$_2$, CuCl$_2$, MgSO$_4$, CoCl$_2$ and ZnSO$_4$. Among them, Cu^{2+} and Zn^{2+} showed the selective inhibition for the enzymes. With 5 mM Cu^{2+}, the activity of 5'-nucleotidase was almost completely inhibited, while 30% of the nuclease activity remained. In the case of Zn^{2+}, the activity of 5'-nucleotidase was completely inhibited at 0.05 mM, whereas more than 50% of the nuclease activity still remained. It was suggested that Zn^{2+} was more suitable for practical use because it was effective at as low a concentration as 0.05 mM and it caused less inhibition of the nuclease. The effect of Zn^{2+} on the activities of both enzymes adsorbed on the flocculated cells is shown in Figure 2. Although 5'-nucleotidase showed only 5% activity in the presence of 0.1 mM Zn^{2+}, more than 40% of the nuclease activity still remained [11].

RNA degradation with a column bioreactor in the presence of ZnSO$_4$. The degree of RNA degradation in the presence of 0, 0.25 and 0.5 mM ZnSO$_4$ were 94.3%, 82.1% and 78.9% respectively [11]. As shown in Table 3, the amount of 5'-nucleotides produced in the presence of ZnSO$_4$ was greater than that in its absence, with the exception of 5'-UMP. In particular, 5'-GMP was produced in the presence of ZnSO$_4$. Also, the amount of 5'-AMP was about 5-fold greater in the presence of ZnSO$_4$.

RNA degradation with a column bioreactor of enriched Ca-enzyme complex. A separate experiment has shown that the cultivation at low temperature such as 25°C repressed the production of 5'-nucleotidase without decrease in the nuclease H production. However, the cells grown under such a condition could not be applicable to a bioreactor because of poor flocculation of the cells in Mg-CM. When the CM

Table 3. Analysis of RNA degradation products with a bioreactor in the presence of $ZnSO_4$[a] [11]

Product	Degradation (mg/litre) in the presence of $ZnSO_4$		
	0 mM	0.25 mM	0.5 mM
5'-AMP	33.9	163.8	136.8
5'-GMP	0	117.9	110.3
5'-CMP	52.0	106.3	110.1
5'-UMP	478.6	270.1	277.5
Adenosine	59.3	38.2	27.7
Guanosine	310.9	81.9	119.4
Cytidine	17.3	17.9	13.8
Uridine	331.9	39.1	64.7

[a]Effluents of the highest A_{260} were subjected to analysis.
Activity of the column: 208 U nuclease and 546 U 5'-nucleotidase per ml of column.

Table 4. Analysis of RNA degradation products with the Ca-enzyme complex bioreactor[a]

Product	Degradation (mg/litre) at:			
	1st run	2nd run	3rd run	4th run
5'-AMP	225.1	208.2	187.5	185.9
5'-GMP	92.2	96.3	58.7	79.5
5'-CMP	148.3	97.1	83.5	78.4
5'-UMP	224.6	175.6	210.8	169.8
Adenosine	18.2	17.3	24.6	16.6
Guanosine	29.7	41.5	63.1	63.7
Cytidine	6.4	8.0	14.8	10.5
Uridine	2.1	3.0	4.6	3.9

[a]Effluents of the highest A_{260} were subjected to analysis.

supernatant of 4 day-culture at 25°C was added with $CaCl_2$ at the final concentration of 136 mM and incubated at 30°C for 1 h, the nuclease H was found to be separated as a white precipitate. Then, the complex was suspended further in the extra CM supernatant added with 25 mM $CaCl_2$ and incubated at 30°C for 1 h. After adsorption of the exogenous enzymes, the total activity of the nuclease H of the complex increased from 5.3 U/ml to 11.8 U/ml, while that of the 5'-nucleotidase increased only slightly from 2.2 U/ml to 2.7 U/ml, showing preferential adsorption of the nuclease H. A 300-ml portion of the suspension of the enriched Ca-enzyme complex was mixed with 7 g (wet weight) of Celite 545 and a column bioreactor was prepared in the same manner as described above. After the column was washed with 25 mM Tris-HCl buffer, pH 8.0 containing 3 M NaCl, 80 mM $MgSO_4$ and 2 mM $CaCl_2$, 10 ml of 1% RNA in the buffer was charged and eluted with the same buffer. The flow rate was kept constant at 10 ml/h. The continuous degradation experiments, 6 h each, were run four times at 40°C. Intervals between each run were 13 to 14 h. Enzyme activity of the column was 266 U nuclease H and 61 U 5'-nucleotidase per ml of the column. The degree of RNA degradation in the 1st, 2nd, 3rd and 4th runs was 81.7%, 63.8%, 44.1% and 46.7% respectively. Quantitative analysis of the products showed that the higher yield of 5'-nucleotides was obtained compared with the untreated flocculated cell bioreactor (Table 4). It seems quite reasonable because the ratio of nuclease H activity to 5'-nucleotidase of the Ca-enzyme complex was much higher than that of the untreated flocculated cell bioreactor. It is noted that an appreciable amount of 5'-GMP was produced though the ratio of GMP to GMP + guanosine gradually decreased as the run was repeated. Strain improvement would be useful for further increase in the yield of 5'-nucleotides. Selection of 5'-nucleotidase-poor or -less mutants by UV-irradiation, penicillin enrichment and replica plating on a phosphate-free synthetic medium containing 0.5% 5'-CMP are now in progress.

REFERENCES

[1]   M. Kamekura and H. Onishi, Halophilic nuclease from a moderately halophilic *Micrococcus varians* subsp. *halophilus*. *J. Bacteriol.* 119: 339 (1974)

[2]   M. Kamekura and H. Onishi, Properties of the halophilic nuclease of a moderate halophile, *Micrococcus varians* subsp. *halophilus*. *J. Bacteriol.* 133:59 (1978)

[3]   M. Kamekura, T. Hamakawa and H. Onishi, Application of halophilic nuclease H of *Micrococcus varians* subsp. *halophilus* to commercial production of flavouring agent 5'-GMP. *Appl. Environ. Microbiol.* 44: 994 (1982)

[4]   M. Kamekura and H. Onishi, Effect of magnesium and some nutrients on the growth and nuclease formation of a moderate halophile, *Micrococcus varians* subsp. *halophilus*. *Can. J. Microbiol.* 22: 1567 (1976)

[5]   M. Kamekura and H. Onishi, Flocculation and adsorption of enzymes during growth of a moderate halophile, *Micrococcus varians* subsp. *halophilus*. *Can. J. Microbiol.* 24: 703 (1978)

[6]   H. Onishi, T. Kobayashi and M. Kamekura, Purification and some properties of an extracellular halophilic 5'-nucleotidase from a moderate halophile, *Micrococcus varians* subsp. *halophilus*. *FEMS Microbiol. Lett.* 24:303 (1984)

[7]   H. Onishi, M. Kamekura, H. Yokoi and T. Kobayashi, Production of 5'
nucleotide by using halophilic nuclease H preferentially adsorbed on
flocculated cells of the halophile *Micrococcus varians* subsp. *halophilus*. *Appl.
Environ. Microbiol.* 54: 2632 (1988)

[8]   W. C. Schneider, Phosphorous compounds in animal tissues. I. Extraction and
estimation of deoxypentose nucleic acid and pentose nucleic acid. *J. Biol.
Chem.* 161: 293 (1945)

[9]   H. Yokoi, T. Watanabe and H. Onishi, Simultaneous analysis of 5'-
mononucleotides and nucleosides in saline solution by high performance liquid
chromatography. *Agric. Biol. Chem.* 51: 3147 (1987)

[10]   A. Kuninaka, M. Kibi and K. Sakagushi, History and development of flavor
nucleotides. *Food Technol.* 19: 287 (1964)

[11]   H. Yokoi and H. Onishi, Effects of several metal ions, especially zinc ions,
on RNA degradation by halophilic nuclease H in solution or adsorbed on
flocculated cells of halophilic *Micrococcus. Agric. Biol. Chem.* 53: 1817 (1989)

NOVEL COMPATIBLE SOLUTES AND THEIR POTENTIAL

APPLICATION AS STABILIZERS IN ENZYME TECHNOLOGY

Erwin A. Galinski and Karin Lippert

Institut für Mikrobiologie and Biotechnologie
Rheinische Friedrich-Wilhelms-Universität
Meckenheimer Allee 168
D-5300 Bonn 1
Federal Republic of Germany

ABSTRACT

A novel class of compatible solutes, the "ectoines", are evidently widespread in nature and seem to serve both osmotic and enzyme protective functions. The ectoines are characterized by the following structural features: - hydrogenated pyrimidine carboxylic acid in the conformation of a half chair, - delocalized π-bonding (and positive charge) between nitrogen 1 and 3, - permanent zwitterionic properties at physiological pH. Using an unidentified actinomycete (strain A5-1) as a producer strain and a chemical extraction method followed by chromatographic purification procedures we are presently able to gain approximately 10 g of ectoines from 100 g of bacterial dry mass. A preliminary investigation has shown that ectoines exert a remarkable protective effect on a number of labile enzymes, which in many cases surpasses that of betaine.

INTRODUCTION

Besides their function as an osmoticum compatible solutes are also required to maintain the enzymes' hydration shell in a low water environment and to counterbalance the effect of elevated ionic strength within the cytoplasm. In view of this second physiological function compatible solutes may well find a technological application as stabilizers and protective agents in enzyme technology.

Our investigation into halophilic heterotrophic eubacteria has shown that betaine, formerly thought to be the dominant compatible solute [1], is only accumu-

General and Applied Aspects of Halophilic Microorganisms
Edited by F. Rodriguez-Valera, Plenum Press, New York, 1991

lated when the organisms are grown on complex media constituents [2]. Screening of the compatible solute spectrum of these eubacteria, when grown on <u>synthetic</u> media, has revealed that the vast majority of aerobic chemoheterotrophic eubacteria do not synthesize betaine but employ other compatible solutes as osmotica.

The most widespread among these novel compatible solutes proved to be the "ectoines", which form an unusual class of cyclic amino compounds originally detected in the genus *Ectothiorhodospira* [3]. These findings were confirmed by a thorough investigation of members of the family *Halomonadaceae* [4] and other halophilic and halotolerant strains available from type culture collections (Severin et al., in preparation).

We therefore aimed at a biotechnological production of "ectoines" in order to test their enzyme protective effect against a number of stress factors like freeze--thawing and heat denaturation.

MOLECULAR CHARACTERISTICS OF ECTOINE

The novel cyclic amino acid ectoine was originally isolated and identified from extremely halophilic species of the bacterial genus *Ectothiorhodospira* [3]. Ectoine (1,4,5,6 -Tetrahydro-2-methyl-4-pyrimidine carboxylic acid) is structurally related to a pyrimidine derivative where all except one double bond have been hydrogenated. Its zwitterionic structure in aqueous solution probably resembles that of the crystal structure resolved by X-ray analysis [5] (Figure 1).

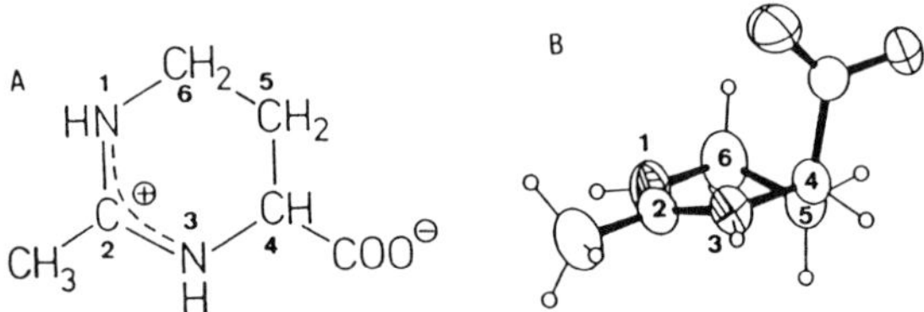

Figure 1. Structural formula (A) and ORTEP-plot (B) of 1,4,5,6-Tetrahydro-2-methyl-4-pyrimidine carboxylic acid. This novel cyclic amino acid was given the vernacular name ectoine due to its discovery in the bacterial genus *Ectothiorhodospira*.

The molecule adopts the conformation of a half chair where all ring atoms, except atom 5, lie in a plane to which the carboxyl group takes an axial position. The most remarkable structural feature is a delocalized π-bonding in the N–C–N group which results in a permanent zwitterionic structure due to mesomeric stabilization of the N–H protons. This is reflected in the titration curve of ectoine which displays only one pK value in aqueous solution (Figure 2).

352

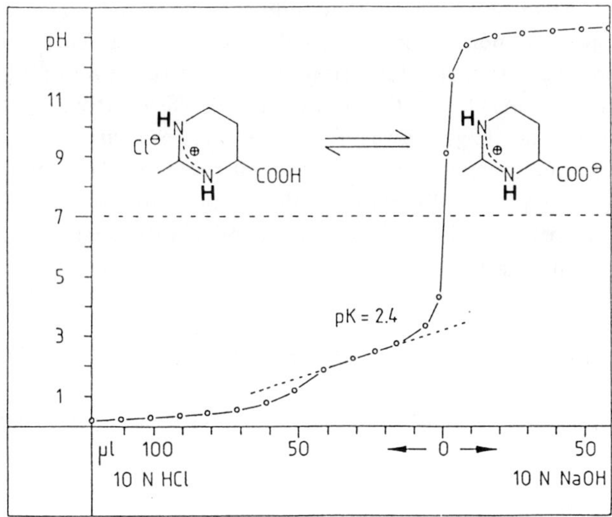

Figure 2. Titration curve of ectoine in aqueous solution presenting an apparent pK of 2.4 (carboxyl group). N-H protons do not dissoziate under the conditions employed.

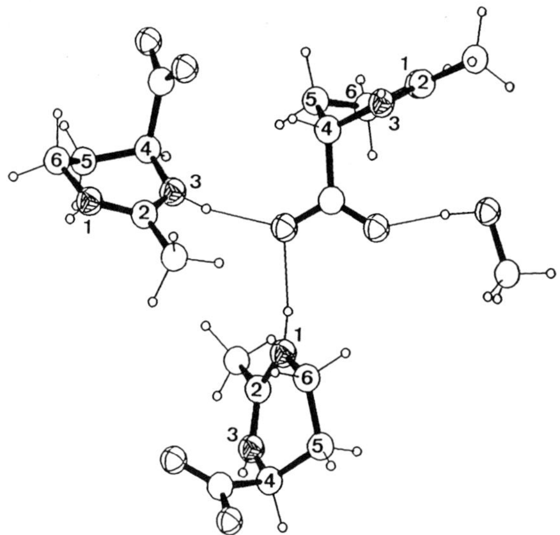

Figure 3. Crystal structure of ectoine displaying hydrogen bondings between N-H groups, carboxyl oxygen and crystallized solute.

As can be seem in Figure 3, ectoine carries polar N- and O-groups which are able to form 4 or more hydrogen bonds per molecule. This structural feature probably accounts for its high solubility in water (6 mol/kg at 4°C) and good solute properties. It is also noteworthy that the centre of the positive charge lies close to carbon atom 2 (3 bond lengths away from the carboxyl carbon) and carries a low charge density due to mesomerization. The first characteristic compares to that of ß-amino acids, which generally show a higher solubility than α-amino acids (no disturbance of hydration shells), whereas the second quality seems to resemble a typical structural feature of betaines.

BIOTECHNOLOGICAL PRODUCTION OF SOLUTES

An unidentified halophilic actinomycete strain A5-1 isolated from mud samples of the salinas in Alicante (Spain) served as a model organism for the biotechnological production of novel compatible solutes. As can be perceived from the NMR spectrum presented in Figure 4 this organism produces ectoine and another novel compound Y as its major compatible solutes. The new substance is also classed with the ectoine family. This finding supports our conception that ectoines represent an important group of solutes employed for osmotic and enzyme protective purposes.

Compatible solutes were extracted from bacterial cell mass by a Soxhlet extraction procedure (Figure 5) and subsequently isolated and purified using a combination of chromatographic techniques. We are presently able to gain

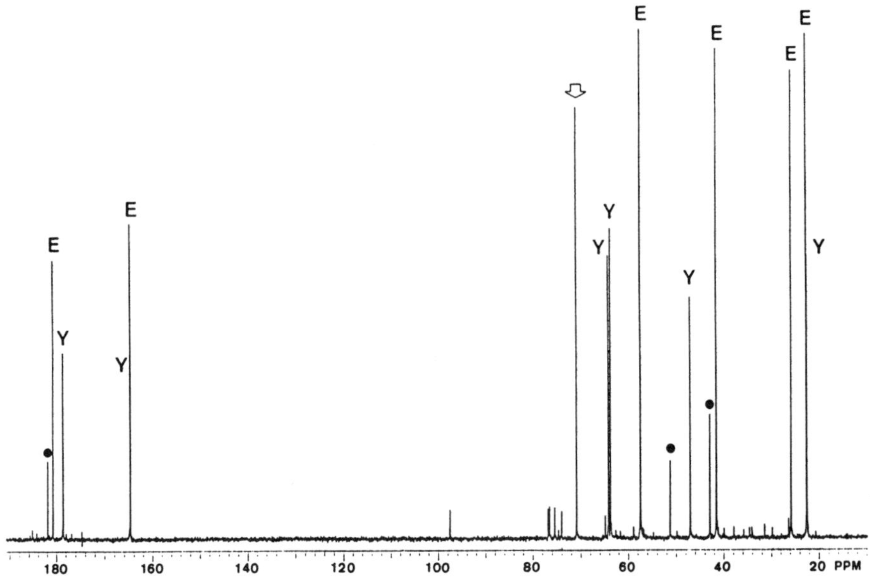

Figure 4. ^{13}C NMR spectrum of a crude extract from actinomycete A5-1
E = ectoine, Y = novel compound, • = unidentified compound.

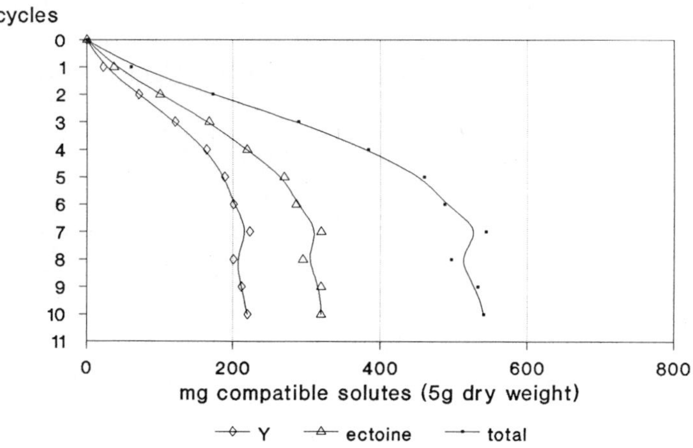

Figure 5. Recovery of compatible solutes from bacterial cell mass by means of a Soxhlet extraction method.

approximately 10 g of ectoines from 100 g bacterial dry mass. As the relative proportion of different types of compatible solutes varies considerably with changing physiological parameters [6] we have obtained a means of specifically directing biosynthesis towards the production of certain compounds.

ENZYME PROTECTIVE EFFECT

Until now work on the stabilizing effect of compatible solutes has mainly centered on the relief of salt inhibition. On the other hand there is an abundance of experimental data concerning heat- and cryoprotection by sugars, polyols and certain stabilizing salts [7]. A preliminary investigation has shown that the most common solutes (betaine, ectoine, Y) exert remarkable stabilizing effects on heat- and cryolabile enzymes. Unexpectedly however, it is evidently impossible to draw any general conclusions as such, since the different solutes displayed a very diverse pattern of action. As depicted in Figure 6 A, for example, the novel compound Y proved an excellent stabilizer against heat inactivation of lactate dehydrogenase, whereas ectoine destabilized the enzyme. Protection against freeze-thaw stress on the other hand is achieved with all three mentioned solutes but to varying degrees (Figure 6 B). As the order of effectiveness as a protective agent may change completely with different enzymes it is at present not possible to present a uniform concept of the stabilizing action of compatible solutes.

MODELS OF COMPATIBLE SOLUTE ACTION

The mechanism of action of compatible solutes is at present poorly understood. Proposed conceptions like the water replacement hypothesis [8], the hydrophobic interaction theory [9] and the preferential exclusion model [10] all seem

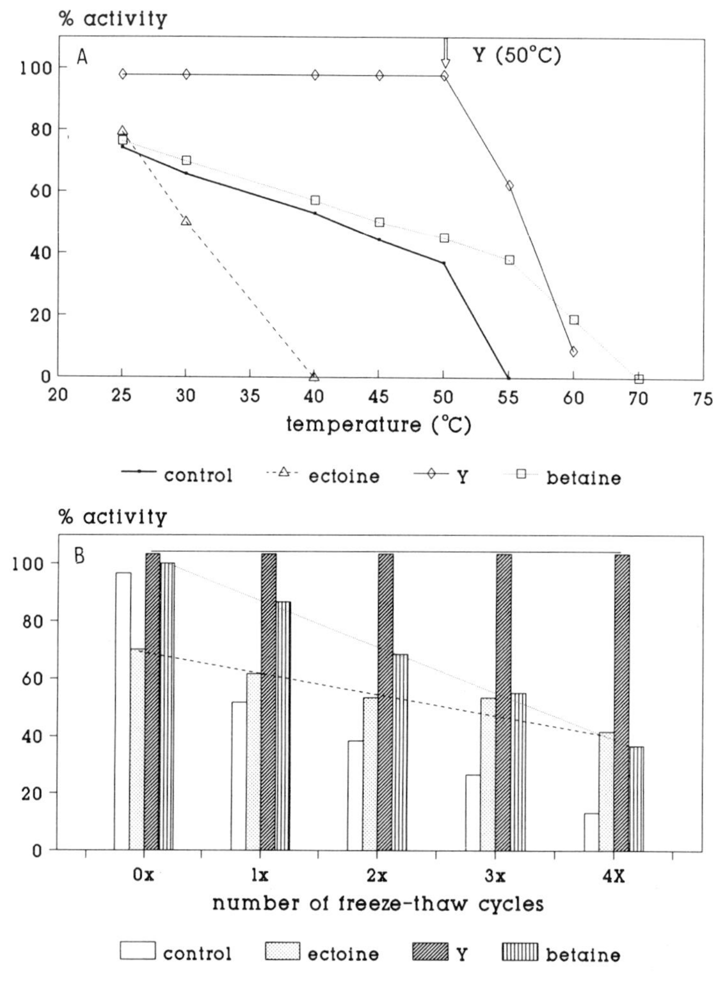

Figure 6. Heatprotection (A) and cryoprotection (slow freeze-thaw stress) (B) of lactate dehydrogenase.

to fit for certain compatible solutes only or even lead to conflicting results depending on the system under investigation. The preferential exclusion model has been most extensively studied and at first seemed to offer a universal concept for the understanding of compatible solute action (as well as Hoffmeister series effects) on native enzymes [11]. This model attributes the stabilizing effect to changes of the surrounding water structure rather than direct protein-solute interactions. The maintenance of the native state of a protein is an entropy driven process which results in the exclusion of hydrophobic moieties from contact with water.

$$\Delta G_{total} (unfolding) = \Delta G_c + \Delta G_s$$

Whereas ΔG_c (always negative) represents the free energy change due to conformational changes of the protein, the term ΔG_s describes the free energy change due to solvent protein interactions. ΔG_s may become negative or positive (depending on the solute) and is therefore responsible for the stabilization of the native structure. The decrease of the entropy term upon unfolding of the protein is caused by the formation of a highly organized water structure close to the surface of hydrophobic moieties (the so called Franks-Evans icebergs). Chaotropes (like certain salts, urea, etc.) break the structure and decrease the degree of solvent organization enforced by hydrophobic groups, whereas kosmotropes (like SO_4^{2-}, PO_4^{3-}, compatible solutes) enforce a water organization on the interface to exposed hydrophobic groups. The former will therefore destabilize and the latter stabilize the native structure of a protein.

Although there is now increasing evidence that protection against a variety of different stress factors like freeze-thaw treatment, drying, heat and salt denaturation may be explained by similar molecular mechanisms, the picture is far from clear and it is probably not justified to conclude that compatible solutes comply with any single one of the above mentioned models. As the solutes under investigation (all structure makers) exert very dissimilar protecting (or even destabilizing) effects depending on the test enzyme and the stress factor applied, there seems to be no simple explanation at hand. Possibly a combination of the different mechanisms mentioned above will provide the clue for the interpretation of compatible solute action on enzymes.

REFERENCES

[1]   J. F. Imhoff and F. Rodriguez-Valera, *J. Bacteriol.* 160:478 (1984).

[2]   A. Wohlfarth, J. Severin and E. A. Galinski, *in* "Microbiology of Extreme Environments and its Potential for Biotechnology", M. S. Da Costa, J. C. Duarte and R. A. D. Williams (eds), p. 421, Troia, Portugal, Sept. 18-23 1988, Elsevier Applied Science, London (1989).

[3]   E. A. Galinski, H. P. Pfeiffer and H. G. Trüper, *Eur. J. Biochem.* 149:135 (1985).

[4]   A. Wohlfarth, J. Severin and E. A. Galinski, *J. Gen. Microbiol..* 136: 705 (1990).

[5]   W. Schuh, H. Puff, E. A. Galinski and H. G. Trüper, *Z. Naturforsch.* 40c:780 (1985).

[6]   E. A. Galinski, *in* "Microbiology of Extreme Environments and its Potential for Biotechnology", M. S. Da Costa, J. C. Duarte and R. A. D. Williams (eds), p. 375, Troia, Portugal, Sept. 18-23, 1988, Elsevier Applied Science London (1989).

[7]   J. F. Carpenter and J. H. Crowe *Cryobiol.* 25:244 (1988).

[8]   J. S. Clegg, P. Seitz, W. Seitz and C. F. Hazlewood, *Cryobiol.* 19:306 (1982).

[9]   B. Schobert and H. Tschesche, *Biochim. Biophys. Acta* 541:270 (1978).

[10]   T. Arakawa and S. N. Timasheff, *Biophys. J.* 47:411 (1985).

[11]   K. D. Collins and M. W. Washabaugh, *Quat. Rev. Biophys.* 18:323 (1985).

LYSIS OF HALOBACTERIA WITH BILE ACIDS AND
PROTEOLYTIC ENZYMES OF HALOPHILIC ARCHAEOBACTERIA

Masahiro Kamekura and Yukio Seno

Noda Institute for Scientific Research
399 Noda, Noda-shi, Chiba-ken
Japan 278

ABSTRACT

Several authentic bile acids and related compounds were tested for their ability to lyse the cells of a halophilic archaeobacterium *Halobacterium cutirubrum*. They showed different lytic activities depending on the structure (number and configuration of OH group). For example, chenodeoxycholic acid caused complete lysis of the cells in a few minutes, while ursodeoxycholic acid showed very little activity. Bile acids which showed no lytic activity did not cause any decrease of viable cell numbers of *H. cutirubrum* by incubation for 10 minutes.

A proteolytic enzyme F-II of an unidentified halophilic archaeobacterium 172 P1 caused lysis of all tested strains of halobacteria at a concentration of 1 - 1.5 mg/ml in the presence of 31% NaCl, though at varying degree from strain to strain. Subtilisin Carlsberg also showed slight lytic activity on *H. cutirubrum* under similar conditions. Another protease preparation of *Haloferax mediterranei* caused spheroplast formation of some strains of halobacteria, while some were not affected.

INTRODUCTION

Extremely halophilic rod-shaped bacteria require high concentrations of NaCl for growth and cell integrity, and lyse immediately when exposed to low salt concentration [1]. However, Kamekura et al. [2] have shown that halobacteria were lysed in a 25% NaCl solution containing 1% Bacto-Peptone from Difco and they identified the lytic substances as taurocholic acid and glycocholic acid as the main components. They also demonstrated that Bacto-Peptone contained 9 different bile acids, and individual bile acids had a varying lytic activity. In this paper the authors extended the study further on the lytic activities of various bile acids.

General and Applied Aspects of Halophilic Microorganisms
Edited by F. Rodriguez-Valera, Plenum Press, New York, 1991

Table 1. Lysis of *H. cutirubrum* with bile acids and related compounds.

Bile acid (49 μg/ml)	Configuration[a]	ΔA_{578}/min	% activity
cholanic acid	(5ßB)	0.00	0.0
lithocholic acid	(5ßB-3α-ol)	0.00	0.0
cholanic acid-3-one	(5ßB-3-one)	0.01	0.7
hyodeoxycholic acid	(5ßB-3α, 6α-ol)	0.43	25
chenodeoxycholic acid	(5ßB-3α, 7α-ol)	1.74	100
ursodeoxycholic acid	(5ßB-3α, 7ß-ol)	0.01	0.7
deoxycholic acid	(5ßB-3α, 12α-ol)	1.74	100
cholanic acid-3 α-ol-6-one	(5ßB-3α-ol, 6-one)	0.02	1.2
cholanic acid-3,6-dione	(5ßB-3,6-dione)	0.00	0.0
cholic acid	(5ßB-3α, 7α, 12α-ol)	0.02	1.3
hyocholic acid	(5ßB-3α, 6α, 7α-ol)	0.02	1.0
dehydrocholic acid	(5ßB-3, 7, 12-one)	0.00	0.0
cholestane-3ß,5α,6ß-triol		0.00	0.0
cholesterol		0.00	0.0

Note: *H. cutirubrum* was grown in Sehgal and Gibbons complex medium (SGC) containing 25% NaCl [6]. The lysis experiments were carried out with cell suspension of an initial absorbance at 578 nm of 2.34. The decrease of the absorbance was monitored for 5 minutes at 25ºC.

[a] Configuration at C5 and of hydroxy groups is indicated by Greek letters.

On the other hand, extreme halophiles are characteristic in that they lack peptidoglycan layer in their cell wall [3]. Despite the lack of a rigid peptidoglycan, halobacteria maintain a stable rod-shaped morphology when grown at optimal salt concentration. The glycoprotein is the main component of the cell envelope, accounting for 40 to 50% of the total protein and all of the non-lipid carbohydrates. The primary structure and the location of glycosylation sites of the glycoprotein were determined by cloning and sequencing of the gene of *Halobacterium halobium* [4]. Recently Kessel et al. [5] proposed a three dimensional structure model of the glycoprotein layer of *Haloferax volcanii*. During experiments on proteolytic enzymes of an extreme halophile strain 172 P1 the authors noticed the protease caused lysis of the cells of *H. volcanii*.

In order to know whether the primary structure and the three dimensional model proposed apply generally to all strains of halobacteria, the author tried to detect difference, if any, in the structures of protein moieties of glycoproteins by measuring susceptibilities of the cells to bile acids and proteolytic enzymes.

LYSIS OF HALOBACTERIA WITH VARIOUS AUTHENTIC BILE ACIDS

The lysis of *H. cutirubrum* NRC 34001 was measured with various authentic bile acids. Three mg of bile acids were dissolved in 1 ml of 99.5% ethanol, 50 μl aliquots were mixed with 3 ml of the cell suspension (final concentration of 49 μg/ml), and the lysis was followed for 5 minutes at 25ºC by measuring the decreasing turbidity at 578 nm. The results in Table 1 showed that bile acids with 2 OH groups, deoxycholic acid and chenodeoxycholic acid exhibited very strong lytic activities, whereas less than 1.5% of these activities was detected with cholic acid or hyocholic acid containing 3 OH groups, and no activity with cholanic acid with no hydroxy group and lithocholic acid containing one OH group. Hyodeoxycholic acid with an α-OH at C6 showed lower lytic activity than chenodeoxycholic acid with an α-OH at C7.

On the other hand, chenodeoxycholic acid caused immediate lysis of all of the 13 strains listed in Table 2, but almost no lysis was observed with ursodeoxycholic acid.

Viable cell numbers of *H. cutirubrum* were measured after treatment with various bile acids. The cells were taken at various growth phases, from early exponential to stationary phase, mixed with bile acid (49 μg/ml) at 4 different cell densities (absorbance of 0.19 to 0.52 at 578 nm), and incubated at 37ºC for 10 minutes.

Table 2. Proteolysis of extreme halophiles with proteases.

Strain	Protease	
	F-II of 172 P1	*H. medi-terranei*
Halobacterium cutirubrum NRC 34001 (rods)	+	s (++)
Halobacterium halobium CCM 2090 (rods)	+	s (++)
Halobacterium saccharovorum ATCC 29252 (rods)	+	s (++)
Haloarcula vallismortis ATCC 29715	+++	(++)
Haloarcula trapanicum NCMB 767 (short rods)	++	s -
Haloferax mediterranei ATCC 33500	++	-
Haloferax volcanii NCMB 2012	+++	(++)
Haloferax gibbonsii ATCC 33959	++	-
Unidentified strains:		
Y12 (rods)	+	s (++)
B 1T)rods)	++	-
L-11	++	-
172-T (rods)	+	-
172 P1 (rods)	+	-

Note: Cells grown in 25% salts-water medium were suspended in 25% NaCl, 10 mM Tris-HCl, pH 7.6, mixed with equal volume of F-II protease or *H. mediterranei* protease of a concentration of 2.5 mg/ml and 2.3 mg/ml, respectively, and incubated at 37ºC.
The cells incubated with the F-II protease lysed after incubation for 0.5 h (+++), 2 h (++) or several hours (+).
The cells which turned coccoid form by incubating with the protease of *H. mediterranei* are indicated by s. Lysis of the spheroplasts by diluting the NaCl concentration to 20% is shown by (++)..

The cell suspensions were then diluted appropriately with 25% NaCl solution containing 0.1 M $MgCl_2$, and plated out on agar plates of SGC medium containing 25% NaCl [6]. The non-lytic bile acids, lithocholic acid and dehydrocholic acid, did not affect the viability of the cells at all, whereas deoxycholic acid caused decrease of the viable cells to less than 10^{-5} in 10 minutes with concomitant complete lysis of the cells.

In the previous paper [2], Kamekura et al. showed that eight strains of *Halococcus morrhuae* were resistant to lysis with cholic acid. Reinvestigation in this study also showed that *Halococcus* species were not susceptible to cholic acid at concentrations which caused complete lysis of *H. cutirubrum*. Detailed examination showed that deoxycholic acid (49 μg/ml) caused a drastic loss of viability (10^{-4} to 10^{-5}) of the cells of three strains of *H. morrhuae* in 30 minutes, although the decreases of the cell densities were less than 10% during the incubation.

LYSIS OF HALOBACTERIA WITH HALOPHILIC PROTEOLYTIC ENZYMES

A. Preparation of proteolytic enzymes of *Haloferax mediterranei* and the strain 172 P1

H. mediterranei strain R4 ATCC 33500 obtained from Dr. G. Juez, University of Alicante, Spain, was cultivated in a 30 l jar fermentor containing 20 l of a chemically defined synthetic medium of the following composition (g/l): 20 g sodium glutamate monohydrate, 10 g sucrose, 20 g $MgSO_4 \cdot 7H_2O$, 4 g KCl, 0.3 g K_2HPO_4, 160 g NaCl, pH 6.4. After cultivation at 37ºC for 3 days at an agitation rate of 240 rpm with an aeration of 20 l per min, cells were harvested by centrifugation and the supernatant was concentrated with an ultrafiltration module with a molecular weight cut off of 3000 (Asahikasei Industries Ltd., Tokyo, Japan).

The concentrate was centrifuged to get rid of insoluble materials, and solid ammonium sulfate was added at a concentration of 15 g per 100 ml, pH 7. The enzyme solution was loaded to a column of Butyl Toyopearl 650 C (Tosoh MFG Co. Ltd., Japan), which had been equilibrated with 10 mM Tris-HCl, pH 7.6 containing 25% NaCl and 15% $(NH_4)_2SO_4$. After washing the column with the same buffer, the protease was eluted with 10 mM Tris-HCl, pH 7.6 containing 25% NaCl and 5% $(NH_4)_2SO_4$. The eluate was concentrated, dialyzed against 10 mM Tris-HCl, 25% NaCl, pH 7.6, and subjected to gel permeation chromatography using Toyopearl HW-50 Fine (Tosoh MFG Co. Ltd.). Active fractions were pooled, concentrated, and used as the protease preparation.

An unidentified halophilic archaeobacterium strain 172 P1 was cultivated in 18% salts medium containing 0.5% yeast extract [7] and the protease F-II was prepared as described elsewhere [8].

B. Proteolysis of halobacteria with the proteolytic enzymes

As described in the INTRODUCTION, halobacterial cells are surrounded by glycoprotein layer. Authors were interested in seeing what would happen when the protein moieties of the glycoproteins were degraded with proteolytic enzymes in the presence of high NaCl concentration. Kamekura had a partially purified preparation of a proteolytic enzyme from culture supernatant of *H. mediterranei*, which required at least 15% NaCl for stability of the activity. In the present study we noticed that when the cells of *H. cutirubrum* grown in Sehgal and Gibbons complex medium (25% NaCl) were incubated with the protease preparation of *H. mediterranei* in the presence of 25% NaCl, they turned into completely cocoid forms after incubation for 2 days at 37ºC. Furthermore several percents of the spheroplasts retained the ability to form colonies just by plating out on an agar plate of 25% NaCl Sehgal and Gibbons complex medium. Similar experiments on some other strains of rod-shaped halobacteria also showed a transformation of rod cells into coccoid forms.

On the other hand, treatment of the cells of *H. cutirubrum* with the proteolytic enzyme F-II of the strain 172 P1 in the presence of 25% NaCl resulted in gradual lysis of the cells. Microscopic observation showed that the long rod-shaped cells of *H. cutirubrum* turned into coccoid forms upon suspension in a solution of the protease F-II and then lysed gradually upon further incubation.

Table 2 summarises some of the lysis experiments with the preparations of the F-II protease and the *H. mediterranei* enzyme. Both enzyme preparations contained 31% NaCl to prevent inactivation and autodigestion during incubation at 37ºC.

Apparently there was marked difference in the susceptibilities of the strains to proteolysis by F-II protease; some were relatively resistant to F-II, while some strains easily lysed, e.g. *H. volcanii* lysed completely in 30 minutes with the F-II protease of the strain 172 P1.

No strain lysed even after incubation for 48 hours in the protease solution of *H. mediterranei*, which showed caseolytic activity approximately 85% of that of the F-II protease of 172 P1 in the presence of 31% NaCl. Cells of some strains, in addition to *H. cutirubrum*, turned to spheroplasts completely during the incubation at 37ºC for 48 hours. When the spheroplast suspensions were diluted with 10 mM Tris-HCl, pH 7.6, to reduce the NaCl concentration to 20% the spheroplasts of some strains lysed immediately, as indicated by(++).

Since *H. volcanii* was the most sensitive strain to proteases, viable cell numbers were measured during the proteolysis. Incubation with F-II protease for 1 hour reduced the cell density to 37% of the initial, while remaining viable cell number decreased to less than 0.1%. Incubation with 10 mM Tris-HCl containing 25% NaCl, pH 7.6 or with the *H. mediterranei* protease did not cause any decrease of the cell density and the viable cell numbers.

Strains B1T, 172 P1 and 172T were extraordinary in that they remained rod shaped in the *H. mediterranei* protease, although they lysed in 2 or 4 hours incubation with the F-II protease in the presence of 31% NaCl. It was also noticed that cells of *H. mediterranei* grown in the chemically defined synthetic medium mentioned above (16% NaCl) lysed quite easily with the protease of *H. mediterranei* itself in the presence of 25% NaCl, compared to the cells grown in 25% salts-water medium.

LYSIS OF HALOBACTERIA WITH OTHER PROTEASES

Finally, the authors measured the lysis of halobacterial cells with some other protease preparations; subtilisin Carlsberg, trypsin, α-chymotrypsin, and alkaline serine protease of *Aspergillus sojae*, a soy sauce mold, at concentrations of 2-3 mg/ml. None of them caused lysis of *H. cutirubrum* or *H. volcanii* in 2-3 h at 37ºC in the presence of 25% NaCl. Incubation for 24 h with subtilisin, however, caused slight lysis of *H. cutirubrum*, but almost no lysis was observed in *H. volcanii*.

Measurement of viable cell numbers of *H. cutirubrum* during the treatment with subtilisin showed a drastic death of the cells; viable cells decreased to less than 10^{-4} in 1 h incubation at 37ºC while the cell density remained the same.

DISCUSSION

In this paper the authors showed that all strains of halobacteria tested lysed with chenodeoxycholic acid (CDCA), while almost no lytic activity was observed with ursodeoxycholic acid (UDCA) on those 13 strains. Although the site of action or binding of bile acids remains to be determined, it seems that all of the non-coccoid form halobacteria contain relatively similar active sites.

An interesting feature of the lysis of halobacteria is that UDCA, which showed lytic activity of less than 0.7% of that of CDCA at a concentration of 49 μg/ml, differs from the latter only in the configuration of OH attached to C7 (Table 1). Although both bile acids were soluble in ethanol, CDCA turned turbid upon addition to 25% NaCl solution at a final concentration of 60 μg/ml, while UDCA remained soluble. At present we still do not have a reasonable explanation on the correlation of structure (particularly stereoconfiguration of hydroxy groups), and the lytic activity of bile acids. The facts that taurine-conjugate of chenodeoxycholic acid had stronger lytic activity than the free acid, while tauro- and glycocholic acid were less active than cholic acid [2] make the situation more complex. More extensive study is required on the mechanism and specificity of the lysis with various bile acids.

On the other hand, halobacteria showed different susceptibility to proteolysis, suggesting that individual halobacterial strains have glycoprotein with different protein moieties. It might be reasonable to deem that the proteases might attack the glycoproteins at some specific peptide bonds according to their substrate specificities. Thus strains with such peptide bonds at easily accessible positions might lose rigidity of the glycoprotein layer, thus changing their morphology.

A notable difference in the time course of the turbidity decrease in the lysis experiments was that the lysis with the F-II protease occurred after a considerable lag period depending on the protease concentration, whereas bile acids caused the lysis without detectable lag time. We still do not know why the spheroplasts formed in the protease solution of *H. mediterranei* did not lyse, while they lysed in the F-II protease solution. Characterization of the protease of *H. mediterranei*, in particular its substrate specificity, might explain the reason.

Jarrell and Sprott [9] have shown that cells of *H. cutirubrum* and *Halobacterium salinarium* turned spheroplasts upon suspension in 0.1 M MES buffer, pH 7.0, containing 0.5 M sucrose, 0.25 M NaCl, and 0.01 M MgCl$_2$. More recently Cline and Doolittle [10] described a method of transfection of *H. halobium* using spheroplasts formed in a medium containing 2 M NaCl. The spheroplasts formed by the proteolysis in the present study were stable for a considerable period in the presence of 31% NaCl, thus they might be useful in the transformation experiments of halobacteria.

ACKNOWLEDGEMENTS

The authors express sincere thanks to Dr. Donn J. Kushner, Department of Microbiology, University of Toronto, for his valuable suggestions during the work.

REFERENCES

[1] D. J. Kushner, The halobacteriaceae, *in* "The bacteria: a treatise on structure and function, Vol. 8: Archaebacteria", I. C. Gunsalus, ed. Academic Press, Inc., Orlando (1985)

[2] M. Kamekura, D. Oesterhelt, R. Wallace, P. Anderson and D. J. Kushner, Lysis of halobacteria in Bacto-peptone by bile acids. *Appl. Environ. Microbiol.* 54: 990 (1988)

[3] O. Kandler and H. Konig, Cell envelope of archaebacteria, *in* "The bacteria: a treatise on structure and function, Vol. 8: Archaebacteria", I. C. Gunsalus, ed. Academic Press, Inc., Orlando (1985)

[4] J. Lechner and M. Sumper, The primary structure of a procaryotic glycoprotein. Cloning and sequencing of the cell surface glycoprotein gene of halobacteria. *J. Biol. Chem.* 262: 9724 (1987)

[5] M. Kessel, I. Wildhaber, S. Cohen and W. Baumeister, Three-dimensional structure of the regular surface glycoprotein layer of *Halobacterium volcanii* from the Dead Sea. *EMBO J.* 7: 1549 (1988)

[6] M. Kamekura, P. Zhou and D. J. Kushner, Protein turnover in *Halobacterium cutirubrum* and other microorganisms that live in extreme environments. *System. Appl. Microbiol.* 7:330 (1986)

[7] F. Rodríguez-Valera, F. Ruiz-Berraquero and A. Ramos-Cormenzana, Isolation of extremely halophilic bacteria able to grow in defined inorganic media with single carbon sources. *J. Gen. Microbiol.* 119: 535 (1980)

[8] M. Kamekura and Y. Seno, A halophilic extracellular protease from a halophilic archaebacterium strain 172 P1. *Biochem. Cell Biol.* 68: 352 (1990)

[9] K. F. Jarrell and G. D. Sprott, Formation and regeneration of *Halobacterium* spheroplasts, *Current Microbiol.* 10: 147 (1984)

[10] S. W. Cline and W. F. Doolittle, Efficient transfection of the archaebacterium *Halobacterium halobium. J. Bacteriol.* 169: 1341 (1987).

AN ARCHAEBACTERIAL ANTIGEN USED TO STUDY IMMUNOLOGICAL

HUMORAL RESPONSE TO C-*MYC* ONCOGENE PRODUCT

Kamel Ben-Mahrez[1], Irène Sorokine[1], Dominique Thierry[2], Toshiyuki Kawasumi[1], Shunsuke Ishii[3], Remy Salmon[4] and Masamichi Kohiyama[1]

[1]Institut Jacques Monod
Université Paris VII
2 Place Jussieu
75251 Paris Cedex 05
France

[2]CEA, IPSN, DPS, SHR
 B.P. 6
92260 Fontenay-aux-Roses Cedex
France

[3]Riken
Tsukuba Life Science Center
3-1-1 Koyadai
Tsukuba-shi, Ibaraki 305
Japan

[4]Institut Curie
26 rue d'Ulm
75231 Paris Cedex 05
France

ABSTRACT

We have previously shown that it is possible to detect by Western blot analysis antibodies against the hyman c-*myc* oncogene product in the serum of some cancer patients using as antigen an 84 kD protein obtained from *Halobacterium halobium* [1]. To study whether the presence of circulating anti-c-*myc* antibodies might be more strongly associated with a specific type of cancer, sera of 28 colorectal cancer patients at the beginning of their treatment at the "Institut Curie" were tested. We observed a significantly higher level of positivity (18 out of 28) than **in** other cancer patient sera.

This conclusion obtained with an archaebacterial antigen was further consolidated by experiments in which we used the purified c-*myc* protein produced in *E. coli* by genetic engineering.

MATERIALS AND METHODS

Strain and medium

Halobacterium halobium CCM 2090, used in the present work, is in the Czekoslovakia Collection of Microorganisms. The strain was grown in Sehgal and Gibbons' complex medium [2] at 37ºC with vigorous shaking and cells were harvested at a concentration of 10^8/ml.

Human sera

Blood samples from cancer patients and healthy individuals were obtained with informed consent. Sera were stored at -20ºC prior to test.

Antibodies

The monoclonal anti-human-c-*myc* antibody (MAB, DCM905) was purchased from Cambridge Research Biochemical Ltd. Anti-IgG antisera linked to peroxidase were from Miles Laboratories.

Human c-*myc* synthetic peptides

Two different peptides were chemically synthesized by Toray Research Center (Japan). The E peptide corresponds to amino acids 305 to 318 and the F peptide to the 19 carboxyterminal amino acids of the human c-*myc* protein.

Purification of human c-*myc* protein

The human c-*myc* cDNA (gift of Dr. Date) was inserted into T7 expression vector [3] downstream of a ribosome binding site and its accompanying ATG initiation codon to generate pAR2106 *myc*. The hybrid protein encoded by this gene contains the first 15 amino acid residues derived from the vector, fused to the C-terminal 439 amino acid residues of the complete c-*myc* protein. The T7 RNA polymerase gene is under the control of the *lac* promoter which is derepressed by IPTG addition. *E. coli* cells carrying pAR2106 *myc* plasmid were resuspended in buffer A (50 mM Tris-HCl pH 8, 0.5 mM EDTA, 0.4 M NaCl, 5 mM $MgCl_2$, 5% glycerol, 0.1 mM DTT and 0.1% PMSF). After addition of lysozyme, the lysate was centrifuged 10 minutes at 12,000 rpm and the pellet was resuspended in buffer B (50 mM Tris-HCl pH 7.5, 1% triton,

6 M urea, 0.1 mM PMSF and 0.1 mM DTT). After centrifuging 2 hours at 45,000 rpm, the supernatant was dialysed overnight against buffer C (10 mM Tris-HCl pH 7, 1 mM EDTA and 1 mM DTT). The dilysate was then centrifuged 10 minutes at 15,000 rpm, the supernatant was withdrawn and passed through a column of DNA-cellulose. The expressed c-*myc* protein was eluted from the column with buffer C containing 0.5 M NaCl.

Western blot analysis

84 kD or human c-*myc* preparations were applied to sodium dodecyl sulfate polyacrylamide gels, transferred onto Millipore nitrocellulose membranes and used in Western blotting analysis as described [1].

INTRODUCTION

The existence of tumor-specific antigens would represent a qualitative difference between a malignant and a normal cell and the antigen would be clinically useful in tumor diagnosis and therapy. A peculiar class of these antigens may be activated cellular oncogenes [4, 5]. To investigate the immunogenicity of cellular oncogenes in humans, we looked for circulating anti-c-*myc* oncogene product antibodies in sera of patients suffering from cancer as well as healthy individuals.

We have previously shown that an extreme halophilic archaebacterium, *Halobacterium halobium* possesses an 84 kD protein which is recognized by anti-human-c-*myc* antiserum. A polyclonal antiserum raised against the 84 kD protein recognizes the c-*myc* protein obtained from nuclei of human HL-60 promyelocytic leukaemia cell line. Using this archaebacterial protein as antigen in Western blot analysis we found that some sera from cancer patients contained anti-c-*myc* antibodies [1]. In this report, we present comparative study on seropositivity of colorectal cancer patients carried out either by using the archaebacterial 84 kD protein or the human c-*myc* protein produced in *E. coli* by genetic engineering.

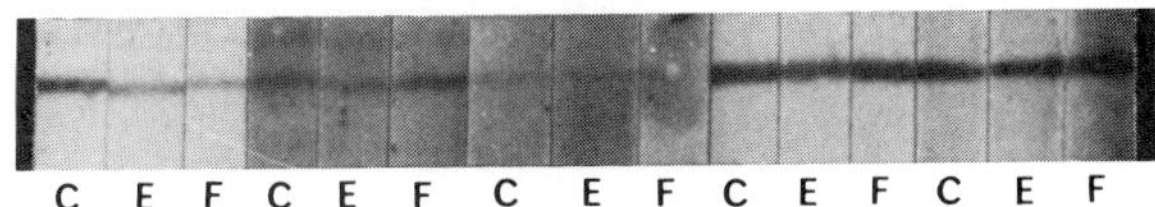

Figure 1.   Western blot analysis of colorectal cancer patient sera using the 84kD protein. Sera (1/20 dilution) were tested as described in Materials and Methods, antibodies were detected by anti-IgG linked to peroxidase. (C) Control, without preincubation. (E) Sera preincubated with peptide E (1 µg/ml). (F) Sera preincubated with peptide F (1 µg/ml).

RECOGNITION OF THE 84 kD PROTEIN BY HUMAN SERA

The 84 kD protein was purified as previously described [1]. The preparation was applied on sodium dodecyl sulfate-polyacrylamide gel, transferred onto nitrocellulose membrane and used as antigen in Western blot analysis to look for circulating anti-c-*myc* antibodies in human sera. Among 214 sera of patients suffering from various cancers 27 recognized the 84 kD protein. Some examples are shown in Figure 1. In contrast, among 62 sera of healthy donors only one sample was positive. Thus, percentage of seropositivity is significantly higher in cancer patient sera than in healthy donors. On the other hand, the percentage varied according to type of cancer.

Evidence that the positive sera contain antibodies against epitopes of the c-*myc* protein was obtained by competition experiments between the 84 kD protein and two different synthetic peptides corresponding to amino acid sequences of the c-*myc* protein. Positive sera were pre-incubated with either of the two peptides and then tested by Western blot analysis using the 84 kD protein as antigen. Among 12 positive sera assayed by this technical approach, one serum was neutralized by the peptide E (corresponding to the sequence 305-318 of the c-*myc* protein), 2 sera were neutralized by the peptide F (sequence 421-439) and 3 were neutralized by both E and F. Some examples are shown in Figure 1. These results demonstrate that the positive sera contain antibodies against epitopes contained in the amino acid sequences E and F of the c-*myc* protein.

CIRCULATING ANTIBODIES IN COLORECTAL CANCER PATIENTS

In order to know whether the presence of circulating anti-*myc* antibodies might be associated with a specific type of cancer, sera of 28 colorectal cancer patients were tested by Western blot using the 84 kD protein. We observed a significantly higher level of positivity (18 out of 28) than other cancer patient sera.

To reconfirm these results with c-*myc* protein, human c-*myc* cDNA was inserted into T7 vector and expressed in *E. coli* and the expressed protein was purified. SDS-polyacrylamide gel electrophoresis revealed two bands of Mr 67,000 and 62,000 (Figure 2A). The Western blot analysis showed that the Mr 67,000 was recognized by anti-human c-*myc* MAB (specific to C-terminal region). However with polyclonal antibodies raised against the 67 kD polypeptide, the two bands were recognized (Figure 2A). This indicates that the Mr 67,000 polypeptide is the human c-*myc* gene product and that the Mr 62,000 is a proteolysis product.

The c-*myc* preparation was used as antigen to look for anti-c-*myc* antibodies in human sera (Figure 2B). Among 44 sera from colorectal cancer patients, 25 recognized the c-*myc* protein, the frequency of seropositivity being practically the same with the archaebacterial 84 kD protein. However, 8 out of 46 healthy individuals' sera reacted with the human c-*myc* gene product, which was almost ten times higher than with archaebacterial protein. These results confirm the presence of circulating anti-c-*myc* antibodies in colorectal cancer patients at high frequency (57%), comparable to that observed with the 84 kD protein (64%). However, in the case of healthy individuals, 17% of tested sera contained anti-c-*myc* antibodies most of which were negative with the 84 kD protein. This suggests that these positive normal sera recognize epitopes which are not present in the archaebacterial protein.

DISCUSSION

Using the archaebacterial 84 kD protein as antigen, Western blot analysis showed that 64% of colorectal cancer patient sera contained anti-*myc* antibodies. Similar percentage (57%) was obtained using the human c-*myc* protein as antigen. Interesting difference between the two antigens was observed in the case of sera from healthy individuals; the result obtained with the c-*myc* protein was 17% of seropositivity and 1.6% with the 84 kD protein. This suggests that the epitopes which can be used as tumor markers exist in both antigens but that the epitopes recognized by normal sera are hardly present in the archaebacterial protein.

The c-*myc* gene was initially identified by its homology with the avian myelocytomatosis virus v-*myc* oncogene [6]. The product of human c-*myc* gene is a

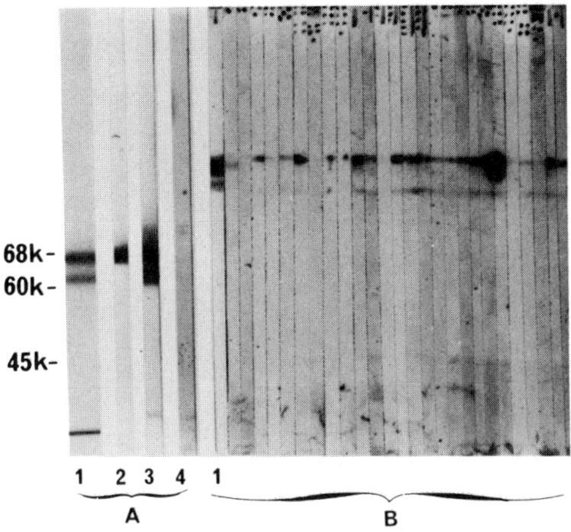

Figure 2. Western blot analysis of colorectal cancer patient sera using the human c-*myc* protein. A - 2 μg of purified *myc* protein were analyzed by coomassie blue-SDS-7% polyacrylamide gel (Lane 1). After transfer onto nitrocellulose membrane, the *myc* protein was incubated with anti-c-*myc* MAB (Lane 2), or with polyclonal anti-*myc*-antibodies (Lane 3) or with control antiserum (Lane 4). B - After transfer onto nitrocellulose membrane, the *myc* protein was used to test sera from colorectal cancer patients. Lane 1: control with polyclonal anti-*myc* antibodies.

nuclear phosphoprotein of 439 amino acids having DNA binding activity [7] and probably plays a role in DNA replication [8] or in transcription. The c-*myc* gene product has been implicated in the control of cell proliferation and tumorigenesis. In fact, levels of c-*myc* mRNA increase in response to mitogenic signals [9], c-*myc* gene is overexpressed and/or rearranged in a variety of tumors [10] and is capable of transforming cells in culture and inducing tumors in animals [11, 12].

In the case of colorectal cancers, c-*myc* oncogene is overexpressed in a majority of colon tumors [13]. Relationship between this overexpression and the presence of circulating anti-c-*myc* antibodies we observed, remains to be analysed.

REFERENCES

[1]   K Ben-Mahrez, D. Thierry, I. Sorokine, A. Danna-Muller and M. Kohiyama, Detection of circulating antibodies against c-*myc* protein in cancer patient sera, *Br. J. Cancer* 57: 529 (1988)

[2]   S. N. Sehgal and N. E. Gibbons, Effect of some metal ions on the growth of *Halobacterium cutirubrum, Can. J. Microbiol.* 6: 165 (1960)

[3]   F. W. Studier and B. A. Moffatt, Use of bacteriophase T7 RNA polymerase to direct selective high-level expression of cloned genes, *J. Mol. Biol.* 189: 113 (1986)

[4]   A. L. Schlechter, D. F. Stern, L. Vaidyamathan, S. J. Decker, J. A. Drebin, M. I. Greene and R. A. Weinberg, The *neu* oncogene: an *erb*-B-related gene encoding a 185000-Mr tumor antigen, *Nature* 312:513 (1984)

[5]   J. A. Drebin, V. C. Link, R. A. Weinberg and M. I. Greene, Inhibition of tumor growth by a monoclonal antibody reactive with an oncogene-encoded tumor antigen, *Proc. Natl. Acad. Sci. USA* 83: 9129 (1986)

[6]   T. Graf and D. Stehelin, Avian Leukemia viruses. Oncogenes and genome structure, *Biochim. Biophys. Acta.* 651: 245 (1982)

[7]   K. Bister and H. W. Jansen, Oncogenes in retroviruses and cells: biochemistry and molecular genetics, *Adv. Cancer Res.* 47: 99 (1986)

[8]   G. P. Studzinski, Z. S. Brelvi, S. C. Feldman and R. A. Watt, Participation of c-*myc* protein in DNA synthesis of human cells, *Science* 234: 467 (1986)

[9]   K. Kelly, B. H. Cochrane, C. D. Stiles and P. Leder, Cell-specific regulation of the c-*myc* gene by lymphocyte mitogens and platelet-derived growth factor, *Cell* 35: 603 (1983)

[10]   J. M. Bishop, The molecular genetics of cancer, *Science* 235: 305 (1987)

[11]   H. Land, L. F. Parada and R. A. Weinberg, Tumorigenic conversion of primary embryo fibroblasts requires at least two cooperating oncogenes, *Nature* 304: 596 (1983)

[12]   A. Leder, P. K. Pattengale, A. Kuo, T. A. Stewart and P. Leder, Consequences of widespread deregulation of the c-*myc* gene in transgenic mice: multiple neoplasms and normal development, *Cell* 45: 485 (1986)

[13]   M. D. Erisman, P. G. Rothberg, R. E. Diehl, C. C. Morce, J. M. Spandorfer and S. M. Astrin, Deregulation of c-*myc* gene expression in human colon carcinoma is not accompanied by amplification or rearrangement of the gene, *Mol. Cell. Biol.* 5: 1969 (1985)

BIOPOLYMER PRODUCTION BY *HALOFERAX MEDITERRANEI*

F. Rodriguez-Valera, J.A. Garcia Lillo, Josefa Antón
and Inmaculada Meseguer

Departamento de Genética Molecular y Microbiologia
Universidad de Alicante
Campus de S. Juan, Apdo. 374
Alicante, Spain

ABSTRACT

The halobacterium *H. mediterranei* is an organism of biotechnological
potential due to its capacity as polymer producer. This microorganism accumulates
poly-ß-hydroxybutyrate as intracellular granules in very large amounts. PHB is an
interesting biopolymer for its possible use as biodegradable thermoplastic. The
conditions of culture for PHB production have been studied and optimized in batch
as well as in continuous culture. Phosphate limitation is an essential condition for PHB
accumulation in important quantities. Glucose and starch were the carbon sources
giving the highest productivities. Under favourable circumstances in batch culture a
production of 6.5 g/l PHB was reached, being 67% of the total biomass dry weight, and
with a yield over the carbon source of 0.33 gg^{-1}. There is evidence indicating that the
PHB produced by *H. mediterranei* is a copolymer with monomers of more than four
carbon atoms, which is favourable for the manipulation of the physical properties of
the polymer. *H. mediterranei* accumulates another polymer extracellularly. It is a
sulfated heteropolysaccharide that has very good rheological properties which are very
resistant to environmental stress including, as may be expected, salinity. High
productivities of this polysaccharide require sugars as carbon source, a condition that
also favours PHB accumulation.

INTRODUCTION

The genus *Haloferax* belongs to the Halobacteriales, and is characterized by
its rapid growth rate and physiological versatility. The species *Haloferax mediterranei*
accumulates two polymers of biotechnological interest. One is poly-β-hydroxybutyrate
(PHB) [1], a carbon reserve material, stored by many eubacteria, which in some cases
can serve as biological substitutes for oil-derived plastics such as polypropylene or

General and Applied Aspects of Halophilic Microorganisms
Edited by F. Rodriguez-Valera, Plenum Press, New York, 1991

under these conditions, since the cytoplasm appears extremely reduced in relation to the volume occupied by the granules.

Production in continuous culture has also been studied, in the conditions which are optimal for batch production and at different dilution rates (D). Under these polyethylene, with the advantages of being biodegradable, therefore not damaging the environment, and biocompatible, and thus suitable for numerous applications in surgery and pharmacology. In fact a copolymer of β-hydroxybutyrate (HB) and β-hydroxyvalerate (HV) is being produced by ICI Ltd. under the trade name Biopol, using the eubacterium *Alcaligenes eutrophus* grown with glucose as carbon source. With a fed batch system maintained in sterile conditions they reach 80% of the biomass dry weight as polymer, with yields around 0.33 gg^{-1} with respect to the carbon source. The copolymerization of hydoxyvalerate is achieved supplying propionic acid to the culture that acts as a precursor, the copolymer with 10-20% HV has better physical properties and is more susceptible to be processed by conventional plastic machinery [2,3].

The other polymer accumulated by *H. mediterranei* is a highly sulfated hetero polysaccharide that is excreted into the medium. The composition and rheological properties of this polymer have been studied previously, showing characteristics which could be of interest for certain biotechnological applications [4].

H. mediterranei could offer specific advantages for the production of either or both polymers. Considering its peculiar growth conditions and high growth rate, cultures of this organism are virtually non-contaminable, which would permit much simpler and cheaper plant production designs than those necessary for use with conventional eubacterial producers. In particular, the scale of the production facility could be vastly increased without additional cost to maintain a sterile process; and large-scale productivity is one of the necessary conditions for biological plastics to be able to replace some of the usual applications of chemically produced plastics.

PRODUCTION OF PHB

PHB accumulation by *H. mediterranei* follows a pattern similar to that found in many eubacteria, starting during the logarithmic phase and reaching a peak at the beginning of the stationary phase when the carbon source reaches its minimum (Figure 1). Culture conditions have been optimized [5] in batch culture using different carbon sources (glucose, starch, sucrose, glutamate, butyrate, acetate, glycerol and methanol), temperatures (30, 38, 45, 50 or 55°C), pH (6.5 or 7.2), salt concentration (15, 20, 25 or 30% SW salts), and oxygen transfer rates (OTR) (0.02, 0.07 or 0.40 mmol O_2/l hr). Different concentrations of the carbon source (glucose 1, 2, 5, or 10%), nitrogen source (ClNH 0.40, 0.20, 0.1, 0.05, 0.025%) and phosphorus source (KH_2PO_4 0.03, 0.015, 0.0075, 0.00375 or 0.0009375%) were also assayed. PHB was quantified by the Law and Slepecky method [6]. All these experiments showed that the two major factors affecting the amount and yield of PHB accumulated by *H. mediterranei* were the nature and concentration of the carbon source and the concentration of the phosphorus source; the others had similar effects on PHB accumulation as on total biomass, i.e. the

374

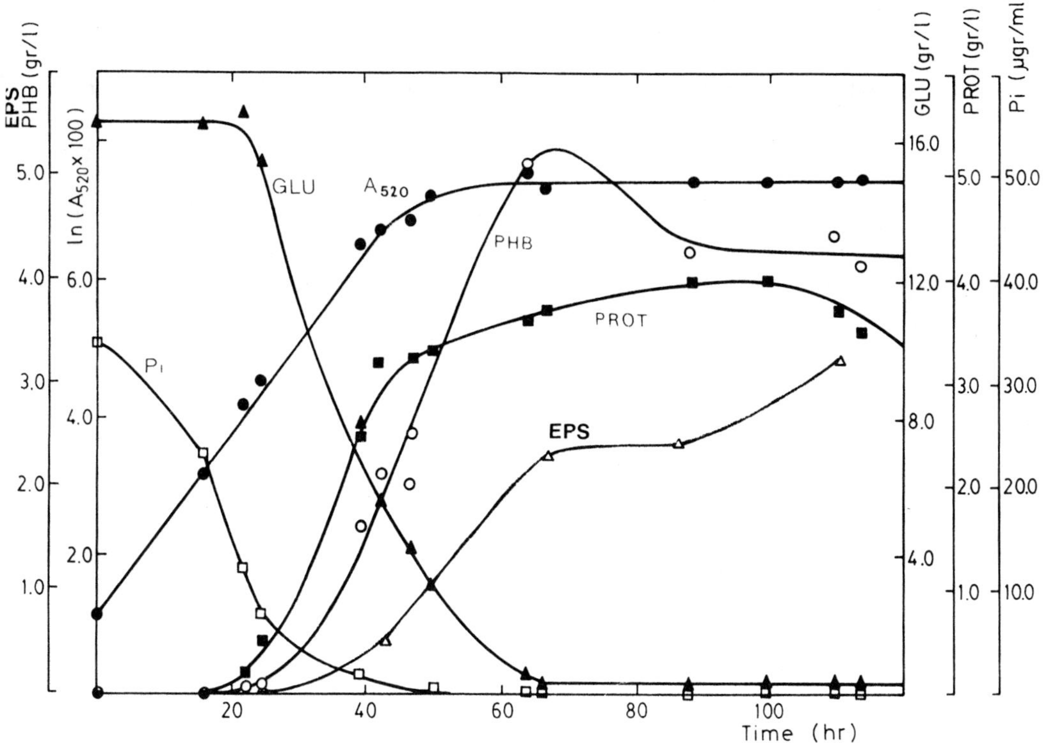

Figure 1. Batch fermentation production curves of *H. mediterranei* grown on 2% glucose at 45°C : (•) Optical density; (▲) glucose concentration; (□) phosphate concentration; (■) protein concentration; (Δ) EPS; and (o) PHB concentration.

optimal values for PHB accumulation were those that gave the highest biomass production. With regard to the carbon source only glucose and its polymer starch gave good results. Surprisingly, a glucose disaccharide sucrose gave very poor results, but growth in general was much poorer with this substrate. The phosphate concentration greatly influenced the PHB accumulated, the best results being obtained with a rather low concentration, 0.00375%. Phosphate limitation is known to stimulate PHB accumulation in eubacteria also [7]. However, nitrogen limitation or oxygen availability, which stimulate polymer accumulation in some eubacteria, did not show any effect in this case except for the general effect on biomass production when changed from optimal conditions.

The medium and culture conditions that gave better values for PHB production were as follows: starch 2% (w/v), NH_4Cl 0.2% (w/v), KH_2PO_4 0.00375% (w/v), marine salts as in SW mixture [8] 25% (w/v), temperature 45°C, pH 7.2, OTR 0.40 mmol O_2/l hr. The results under those conditions appear in Table 1, compared with similar parameters for other prospective PHB producers.

Table 1. Production of poly-β-hydroxyalcanoates by organisms most commonly used.

Microorganism	Substrate	PHA Content (% Dry Weight)	Maximum Production (g/l)	Y_{PHA} (g.g^{-1})	Reference
Alcaligenes eutrophus H16	H_2/CO_2/air	80	-	0.50	[9]
Alcaligenes eutrophus NCIB 11599	Glucose	70	-	0.33	[10]
Haloferax mediterranei R4	Starch	67	6.48	0.32	[5]
Haloferax mediterranei R4	Glucose	60	5.13	0.26	[5]
Pseudomonas Sp. K.	Methanol	67	-	0.18	[11]
Pseudomonas cepacia	Fructose	50	2.50	0.12	[12]
Pseudomonas oleovorans	Octanoate	30	0.35	0.21	[13]
Rhodospirillum rubrum	Butyrate	40	1.53	0.08	[14]
Azotobacter beijerinckii	Glucose	70	0.75	0.04	[15]

Microscopic examination of samples from cultures under optimal PHB production conditions showed the accumulation of large amounts of big granules, similar to those described in eubacteria. Figure 2a shows cells in which the cytoplasm is virtually filled with granules, in the same way as that occurring with *Alcaligenes eutrophus* and other highly efficient PHB-producing eubacteria. These micrographs also show that the proportion of PHB to total biomass must be reaching the maximum conditions PHB production was always much lower than in batch culture, the highest concentration found was at a D of 0.04 h^{-1}, giving 1.62 g/l which represented about 46% of the dry weight of biomass and a yield around 0.16 gg^{-1}. However, the

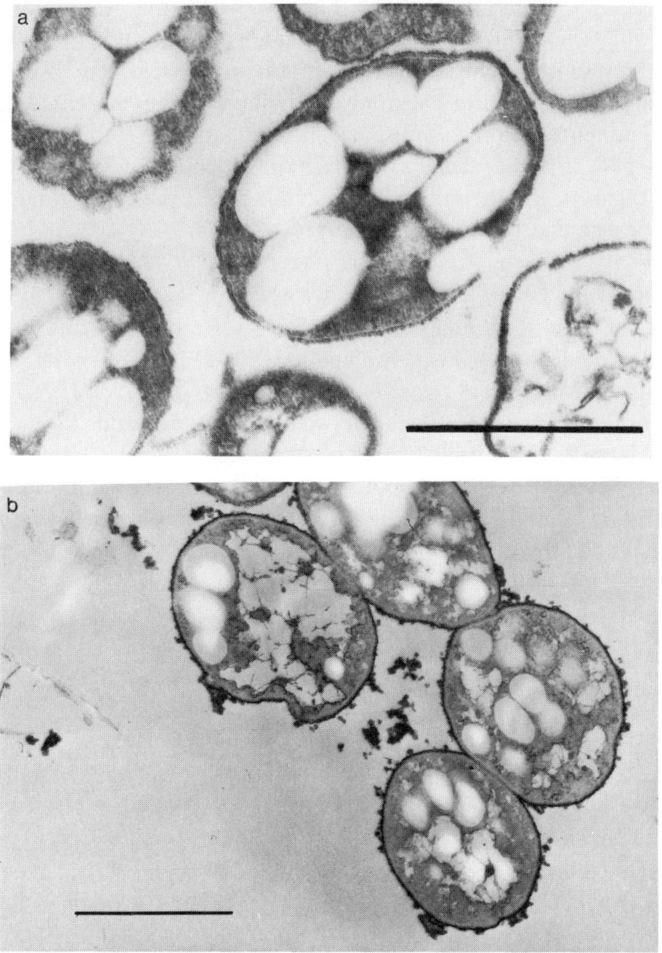

Figure 2. (a) Electron micrograph of *H. mediterranei* grown under optimal conditions for PHB production.

      (b) Electron micrograph of a thin section of ruthenium red stained cells of *H. mediterranei*. The bar represents 1 μm. The EPS appears as an electron dense surrounding layer of irregular thickness.

concentrations of carbon source and phosphate, a critical factor for good productivity, had not been optimized in these experiments.

An important condition for the use of an organism for production in continous culture is the stability of the strain, since very often non-producing mutants are favoured in these growth conditions, and displace the wild producing strain. Moreover, some halobacteria are known to have high genomic instability. To test whether that was the case with PHB production by *H. mediterranei* we set a chemostat and maintained it in the steady state for three months with a $D = 0.13 \ h^{-1}$, which roughly corresponds to 360 generations [5]. Afterwards the production of the strain isolated from the fermenter was evaluated in batch under optimal conditions with 2% starch as carbon source. The total PHB production, yield, etc. were not notably different from those of the original strain . This apparently remarkable stability makes the organism eminently suitable for use in continuous culture for industrial production. On the other hand, in a chemostat with alternating cycles of carbon source starvation, which would theoretically favour overproducing strains, the strain tested at the end of the experiment also had the same productivity and yield.

The polymer produced by *H. mediterranei* grown in batch in the optimal medium described above has been analyzed by 13C-NMR and by diferential scanning calorimetry, also determining the melting point (120.98-145.97ºC). All these data point out that the substance is in fact a copolymer containing 7.5-17.7% HV. This was rather unexpected considering that no precursors were present and adds to the advantages of this organism as producer of biological polyesters. Moreover, data from gel permeation chromatography showed very high molecular weights for the samples obtained under optimal conditions, and this parameter is essential for easy thermal processing (A. Webb, personal comunication).

PRODUCTION OF EPS

Apart from the intracellular accumulation of PHB, *H. mediterranei* also produces a polymer extracellularly, which has been identified as an exopolysaccharide (EPS). This EPS is responsible for the mucous appearance of the colonies, and, as can be observed in Figure 2b, appears as a loose envelope which can be strongly coloured using ruthenium red, a specific stain of acidic polysaccharides.

Chemical analysis of the EPS accumulated in culture media (Table 2) revealed that it consists of a highly sulfated anionic exopolysaccharide [4] and that its

Table 2. Composition of the EPS produced by *Haloferax mediterranei*

	Complex medium	Defined medium
Total carbohydrates	51.65 ± 4.03	17.70 ± 2.62
Hexosamines	1.81 ± 0.69	2.13 ± 0.72
Uronic acids	1.89 ± 0.50	0.62 ± 0.06
Proteins	5.09 ± 0.02	15.34 ± 1.55
Sulfate	5.95 ± 0.64	5.58 ± 0.42
Ashes	7.36 ± 0.0	22.87 ± 0.73
Volatiles	8.25 ± 1.18	11.76± 0.08
Neutralsugars	glucose-gal-man (3:5.15)	glucose-galactose (1:11)

composition varies in accordance with the culture medium. Highly anionic sulfated compounds are often found as structural components in halobacteria [16].

Most industrial uses of microbial polysaccharides are based on their properties to modify the rheological behaviour of acuous systems. Rheological analysis showed that solutions of the EPS are highly viscous and pseudoplastic, that viscosity increases notably with concentration and is stable under changing conditions of pH, temperature and salinity [4]. Since these properties make the *H. mediterranei* EPS interesting from the point of view of biotechnological applications, we proceeded to make a study of optimal production conditions.

We studied the effect of culture conditions (stirring, aeration, pH and temperature) and media composition (NH_4Cl, KH_2PH_4, carbon source, glucose, %SW) on EPS production in batch culture. *H. mediterranei* produces EPS in almost all conditions which permit growth of the microorganism, although the same conditions which give greatest biomass also yield the highest amount of EPS, reaching concentrations of up to 3.2 g/l. Only the effect of stirring-aeration was anomalous in the sense that EPS production was favoured by a lower rate of aeration than that optimal for growth. In regard to culture media composition, the presence of glucose was necessary to achieve good EPS yields, and in this case the effect of salt concentration was interesting, as EPS production increased with increasing salt concentrations up to 30% SW, when it reduced drastically although growth continued.

We have also studied EPS production in a continuous culture system with feedback, which promised to be particularly interesting since it achieves high concentrations of biomass. As could be expected, EPS yield and production were greater than those obtained with batch cultures of *H. mediterranei*.

CONCLUSIONS

H. mediterranei is an interesting organism for PHA production that could presumably carry out the process under less controled conditions than *A. eutrophus* for example. Still, some drawbacks of the process would have to be taken care of to design a competitive production plant based on the halobacteria, such as the cost of the salt that has to be used in the medium. The EPS itself can be considered as a disadvantage, as shown in Figure 1 the production of this polymer is simultaneous and parallel to the accumulation of PHB, sometimes giving a joint yield of 0.55 gg^{-1}. The viscosity of the medium is drastically increased by this polymer. The EPS could be exploited as a by-product of the PHA production, however, if no economical value can be found for this compound, the use of non-polysaccharide-producing mutants may be convenient.

REFERENCES

[1]   R. Fernández-Castillo, F. Rodríguez-Valera, J. González-Ramos and F. Ruiz-Berraquero, Accumulation of poly-(ß-hydroxybutyrate) by halobacteria, *Appl. Environ. Microbiol.*, 51:214 (1986)

[2]   P. A. Holmes, Applications of PHB-a microbially produced biodegradable thermoplastic. *Phys. Technol.* , 16: 32 (1985)

[3]   D. Byrom, Polymer synthesis by microorganisms: technology and economics, *TIBTECH*, 5:246 (1987)

[4]   J. Antón, I. Meseguer and F. Rodríguez-Valera, Production of an extracellular polysaccharide by Haloferax mediterranei, *Appl. Environ. Microbiol.* 54:2381 (1988)

[5]   J. G. Lillo and F. Rodriguez-Valera, Effects of Culture conditions on poly(ß-Hydroxybutyric acid) production by *Haloferax mediterranei*, *Appl. Environ. Microbiol.* 56: (1990)

[6]   J. H. Law and R. A. Slepecky, Assay of poly-ß-hydroxybutyric acid, . 82:33 (1961)

[7]   J. C. Tsai, S. L. Aladegbami and G. R. Vela, Phosphate limited culture of Azotobacter vinelandii, *J. Bacteriol.* , 139:638 (1979)

[8]   F. Rodríguez-Valera, G. Juez and D. J. Kushner, Halobacterium mediterranei spec. nov. , a new carbohydrate utilizing extreme halophile, *System. Appl. Microbiol.* , 4:369 (1983)

[9]   S. H. Collins, Choice of substrate in polyhydroxybutyrate synthesis, *in* "Carbon Substrates in Biotechnology" J. D. Stowell, A. J. Beardsmore, C. W. Keevil and J. R. Woodward (eds.), Irl Press, Washington, D. C. (1987)

[10]   P. A. Holmes, L. F. Wright and S. H. Collins, Betahydroxybutyrate polymers, European Patent Application 52459 (1981)

[11]   T. Suzuki, T. Yamane, and S. Shimizu, Kinetics and effect of nitrogen source feeding on production of poly-ß-hydroxybutyric acid by fed batch culture, *Appl. Microbiol. Biotechnol.* 24:366. (1986)

[12]   B. A. Ramsay, J. A. Ramsay and D. G. Cooper, Production of poly-β-hydroxyalkanoic acid by *Pseudomonas cepacia*, *Appl. Environ. Microbiol.* 55:584 (1989)

[13]   H. Brandl, R. A. Gross, R. W. Lenz and R. C. Fuller, *Pseudomonas oleovorans* as a source of poly-(ß-hydroxyalkanoates) for potential applications as biodegradable polyesters. *Appl. Environ. Microbiol.* , 54:1977 (1988)

[14]   H. Brandl, R. A. Gross, E. J. Knee Jr. , R. W. Lenz and R. C. Fuller, The ability of the phototropic bacterium *Rhodospiriullum rubrum* to produce various poly(ß-hydroxyalkanoates): potential sources for biodegradable polyesters, *Int. J. Biol. Macromol.* 11:1 (1989)

[15]   E. A. Dawes and P. J. Senior, The role and regulation of energy reserve polymers in microorganisms, *Adv. Microbiol. Physiol.* , 10:135 (1973)

[16]   F. Wieland, The cell surfaces of halobacteria. *in* "Halophilic bacteria", Vol. II, F. Rodríguez-Valera (ed.), CRC Press, Boca Raton, Florida (1988)

HALOPHILES IN THAI FISH SAUCE (NAM PLA)

Chaufah Thongthai and Prasert Suntinanalert

Department of Biotechnology
Faculty of Science
Mahidol University
Rama 6 Road, Bangkok 10400, Thailand

ABSTRACT

Halobacterium sp. and *Halococcus* sp. were isolated and reported for the first time in fermenting Thai fish sauce, fish residue and salt grain specimens. Other genera of bacteria including the moderate halophiles and halotolerants were also isolated and characterized. The salt responses and hydrolytic enzyme activities including gelatinolytic, caseinolytic, amylolytic and lipolytic activities of the halophilic bacteria were discussed in relation to their possible roles in fish sauce fermentation.

INTRODUCTION

Fish sauce is a food condiment widely consumed among the Southeast Asian population. It is also known as Nam Pla (Thailand and Lao), Patis (Phillipines) and Nuoc Mam (Vietnam). The clear brown liquid tastes salty, has a meaty and cheesey aroma [1] and about 25 to 50 ml per person per day is consumed (estimated to supply up to 7.5% of daily dietary protein [2]. Its vitamin B12 content which constitutes about 1 μg in daily consumption was sufficient to protect against megaloblastic anaemia [3]. One litre of undiluted first grade Nam Pla at pH 5.8 and S.G. 1.21 contains 26.4 g total nitrogen, 3.7 g ammoniacal nitrogen, 22.7 g organic nitrogen, and 270 g sodium chloride (Department of Health, Ministry of Industry, Bangkok, Thailand). To date fish sauce fermentation remains essentially traditional. Small, fresh fish of *Stolephorus* or *Sardinella* species are mixed together with marine salt to saturation, packed into cubical concrete vats in which the air space is replaced by salt saturated brine, and natural fermentation is allowed to proceed either in the sun or in the shade for up to 12 months. After this, the first grade fluid is filtered and bottled. Re-extraction with hot brine gives a lower grade of fish sauce.

General and Applied Aspects of Halophilic Microorganisms
Edited by F. Rodriguez-Valera, Plenum Press, New York, 1991

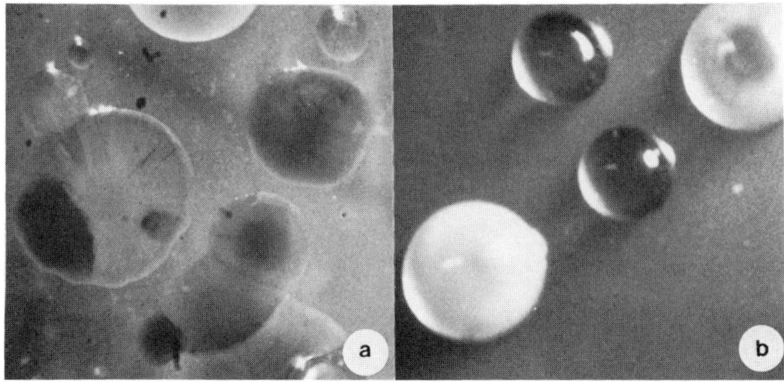

Figure 1. Typical appearance of halophiles cultivated from actively fermenting fish sauce specimens: a) primary culture and b) sub-culture of *Halobacterium*. Both on HMA, 37ºC, aerobically, 14 days. Pictures were taken using X4 objective under a stereomicroscope. Note the sectorial variation of pigment production or opacity.

MICROBIAL PROFILE OF NAM PLA

The early literature reported isolations of bacterial species [4,5] and fungal species [4] from sporadic fish sauce specimens, but none reported isolation of *Halobacterium*. The consistent and quantitatively significant presence of extremely halophilic, red to orange-pigmented, rod-shaped bacteria in Nam Pla during the first month of fermentation was reported by Thongthai and Siriwongpairat in 1978 [6]. Such organisms were sequentially quantitated in a single batch of traditionally fermented Nam Pla to peak at 7×10^8 per ml, three weeks after the onset of fermentation [7]. Identification of the bacterial isolates (Figure 1) according to the criteria set out in Bergey's Manual, 8th edition [3] and Colwell et al. [9] gave *Halobacterium salinarium*. Some characteristics are shown in Table 1 contrasted with those from other bacterial isolates. A total of 128 fermenting liquid specimens, 53 fish residue specimens, 20 marine salt specimens, and 3 rock salt specimens were used in this study. All bacteriological media employed were from Difco Laboratories, Detroit, Michigan, U.S.A. The halobacterium medium agar (HMA) was composed essentially after medium 213 of the ATCC Catalogue of Strains I, 12th Edition, 1976 [10]. From 241 bacterial strains representing isolates of specimens from fish sauce factories in 11 geographical localities in Thailand, four common bacterial genera were found. *Halobacterium*, *Halococcus*, *Bacillus* and coryneform group. Quantitatively, *Halobacterium* and *Halococcus* were uniformly the most significant in number. Typical pigmented colonies showed sectorial variations in shades and opacity, reflecting the high mutation rate in pigment or vacuole production [11, 12]. Other bacterial genera found were similar to those reported by earlier investigators [4, 5]. They included *Micrococcus*, *Staphylococcus*, *Streptococcus*, *Bacillus* and coryneforms. *Pseudomonas* and *Proteus* were encountered only rarely.

Table 1

Testing media	Gram positive cocci				Gram positive rods		Gram negative rods.		
	Micrococcus	Staphylococcus	Streptococcus	Halococcus	Bacillus	Coryneform	Proteus	Pseudomonas	Halobacterium
	%+	%+	%+	%+	%+	%+	%+	%+	%+
Pigmentation	LY 73	WH/Y 100	cl 100	OR 100	G 75	G,Y,OR 75	G 100	G,B 83	OR 95
Colony type	B 95	B 100	A 100	B 100	B 40	B 48	B 100	B 100	A 100
Catalase	+ 100	+ 100	- 0	+ 100	+ 100	+ 92	+ 100	+ 100	+ 100
Oxidase	∓ 36	- 0	- 0	+ 100	± 60	± 63	- 0	+ 100	+ 98
Glucose O/F	-/- 95	+/+ 100	+/+ 70	ND	-/- 83	-/- 96	+/+ 100	+/- 100	ND
Nitrate reduced	∓ 36	± 72	- 0	+ 92	∓ 42	∓ 43	+ 100	- 0	- 4
Indole produced	- 0	- 0	- 0	+ 100	- 2	- 0	- 0	- 0	+ 94
Sodium thiosulfate H$_2$S	∓ 27	- 4	± 57	+ 100	∓ 25	∓ 36	+ 100	- 0	+ 100
Motility	- 0	- 0	- 0	- 0	∓ 35	± 57	+ 100	+ 100	+ 100
Spore-formed	- 0	- 0	- 0	- 0	+ 100	- 0	- 0	- 0	- 0
Anaerobic growth on TGA	- 0	+ 100	+ 100	- 0	∓ 44	∓ 13	+ 100	∓ 33	- 0
Glucose	± 50	+ 96	+ 100	- 0	± 71	∓ 46	+ 100	+ 100	- 6
Mannitol	∓ 18	± 88	+ 90	- 0	± 35	∓ 24	- 0	- 0	- 0
Lactose	- 0	± 56	+ 100	- 0	- 8	- 9	- 0	- 0	- 0
Urease				- 0			+ 100	- 0	- 0
L.D.	ND	ND	ND	∓ 17	ND	ND	- 0	∓ 17	- 8
O.D.				± 67			- 0	NG	± 69
ONPG				∓ 17			- 0	- 0	- 2
Total organisms (241)	22	25	10	12	48	67	3	6	48

%+ = percentage of isolates positive   ND = Not done,   NG = No growth
+   = 90% or more of strains positive   LY = Light yellow   G = Gray
-   = 90% or more of strains negative   WH = White   B = Grown
±   = Majority of strains positive   cl = Colorless   OR = Orange red
∓   = Majority of strains negative   A = Colony: circular, convex, translucent, entire edge.
     B = Colony: circular, convex, opaque, entire edge.

SALT RANGE

All the 241 isolated organisms were incubated aerobically at 37ºC in HMA containing varying amounts of NaCl, and observed daily for up to 14 days. Their differing NaCl requirements or tolerance are shown in Table 2. While 81% of the *Halobacterium* isolates grew strictly in the presence of between 20 to 30% NaCl, the majority of the *Halococcus* isolates (59%) were cultivable in the presence of NaCl from 15% upward. The extreme salt dependence of the cellular morphology of halophilic bacterium strain OR is demonstrated in Figure 2. This normally rod-shape bacterium was observed microscopically in liquids containing 25% (Figure 2a), 20% (b) and 15% NaCl (c). It appeared round when suspended in 10% (d) and 5% NaCl solution (e) and eventually lysed to an amorphous mass of presumably membranous remnants in distilled water (f). Smears were fixed and desalted simultaneously using Dussault's technique [13] before Gram staining. Other bacterial isolates were either halotolerant

Table 2

%NaCl in HMA	gram positive cocci				gram positive rods		gram negative rods.		
	Micrococcus	Staphylococcus	Streptococcus	Halococcus	Bacillus	Coryneform	Proteus	Pseudomonas	Halobacterium
0	5%							17%	
5			70%		29%	8%	100%		
10	27%	4%			27%	24%			
15	5%	20%			25%	18%			
20	18%	44%			11%	21% 2%			
25	5%	8%			2%	5%			
30	9% 27%	8% 16%	30%	8% 59% 33%	4% 2%	20% 2%		83%	6% 13% 81%
Total organisms (241)	22	25	10	12	48	67	3	6	48

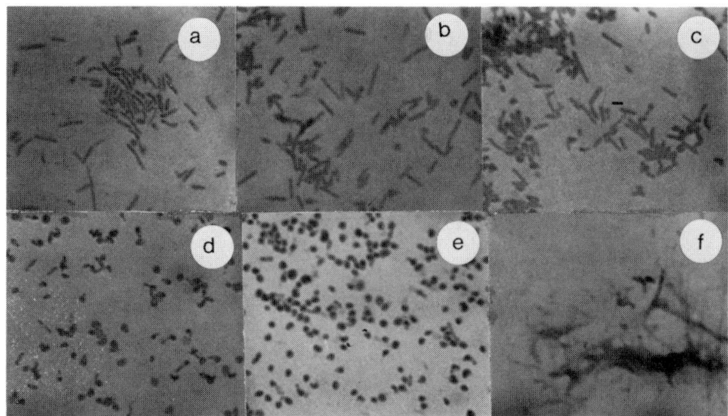

Figure 2. Microscopic appearance of *Halobacterium* strain OR suspended in distilled water containing NaCl at a) 25%, b) 20%, c) 15%, d) 10%, e) 5% and f) 0% concentrations. Smears were fixed and desalted simultaneously using Dussalt's technique [13] before Gram staining.

or moderately halophilic, since they showed growth of wider salt ranges and gave better growth in the presence of less salt. These included isolates of *Micrococcus*, *Staphylococcus*, *Streptococcus*, *Bacillus* and coryneforms (Table 2).

Table 3

Primary Isolation medium	No. of Isolates	Hydrolytic Activity (% of total no.)			
		gelatinase	Caseinase	Amylase	Lipase
TSA	95	75	71	54	30
Bacillus*	42	81	74	52	47
Coryneform**	29	97	86	72	21
TSA+10% NaCl	64	66	48	23	20
Coryneform*	30	90	80	33	14
Staphylococcus**	17	53	24	6	35
HMA	82	62	17	6	7
Halobacterium*	48	63	21	6	6
Halococcus**	12	75	0	0	25

*, ** represent the most and second most frequently isolated genera.

GELATINOLYTIC, CASEINOLYTIC, AMYLOLYTIC AND LIPOLYTIC ACTIVITIES

Since Nam Pla is the resultant product from the liquefaction of fish protein and its constituents, studies concerning the proteolytic activities of the major bacterial strains isolated from fermenting Nam Pla specimens will help us to understand the role of these microorganisms in the fermentation of Nam Pla.

All bacteria were grown in appropriate broth media and were inoculated using a multipoint inoculator onto agar plates, each of which contained 20 ml of the appropriate media. The halotolerant bacteria were tested for the degradation of starch, gelatin, casein and lipid by standard methods using 0.2% starch [14], 0.4% gelatin, 1.5% casein [[15] and spirit blue agar [16]. The halophilic cultures were assayed similarly with the respective media containing either 10% or 25% NaCl. Medium 73 which contained either 1% gelatin [17] or 0.8% skim milk [18] was used to demonstrate proteolytic activities. Starch was added at 0.2% concentration to HMA medium for the determination of amylolytic action. Degradation of lipid was also tested in spirit blue agar containing salt supplement 1 [10] and NaCl as required. The media were incubated at 35ºC for an appropriate number of days before flooding with suitable reagents for final resolution. The degree of hydrolysis was expressed as a ratio of the diameter of the hydrolytic zone divided by the diameter of the colony. Halophilic cultures for comparative studies included *Halobacterium halobium* NRC 34020, and *Halococcus morrhuae* NRC 14034 which were generously supplied by Dr. I. J. McDonald (the National Research Council of Canada, Ottawa, Canada). The *Halobacterium* OR which was originally isolated by one of the authors (C.T.) in 1975 from a fermenting Nam Pla specimen was also included in this investigation.

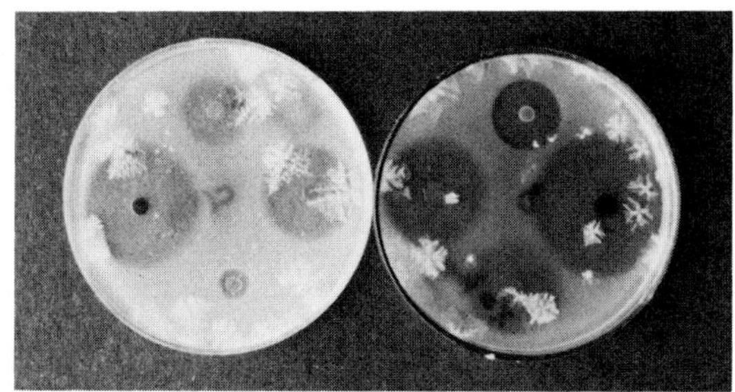

Figure 3.   Gelatinolytic clear zones exhibited by *Halobacterium* isolates from Nam Pla. Both red pigmented and colourless isolates hydrolyzed gelatin in Medium 73 in the presence of 25% NaCl. Note that NaCl has crystallized on the dried surface of the plates.

The percentage of strains showing various hydrolytic activities are shown in Table 3. The most frequently isolated genera in the three salt concentrations of primary isolation conditions, i.e. 0.5% (TSA), 10%, and 25% were respectively the *Bacillus* (44.2%), coryneform (46.8%) and *Halobacterium* (58.5%). The rests of the isolates were varying proportions of *Micrococcus, Streptococcus, Halococcus* and in a few instances *Proteus* and *Pseudomonas*.

The percentage of isolates of each genus which showed activity of either gelatin or casein or both were *Bacillus* (67-81%), coryneform (63-97%) and *Halobacterium* (63%). Only a small fraction (21%) of the *Halobacterium* strains showed both enzymatic activities consistently, and for these the clear zone with gelatin was larger than that with casein. Interestingly, strains of *Bacillus* and the coryneforms showed both amylolytic and lipolytic capabilities, while the *Halobacterium* strains rarely showed such versatility. Figure 3 showed representative strains of *Halobacterium* with strong sodium chloride dependent gelatinolytic activities. Hydrolysis of gelatin was maximal after 7 to 10 days of inclubation, at 35ºC and in the presence of 25% NaCl. About 75% of *Halococcus* strains behaved similarly but none could degrade casein. None of the *Proteus* or *Pseudomonas* strains showed any gelatinolytic or caseinolytic activities.

DISCUSSION

Owing to limitations either in the nature of the specimens or the primary isolation conditions, the presence of *Halobacterium* in fish sauce was heretofore unreported. Extended incubation for up to 14 days together in media supplemented with 25% NaCl made the isolation of these truly extreme halophiles possible. The optimal salt requirement of 25% NaCl for growth (Table 2) and for proteolytic activities (Figure 4); the drastic change in cellular morphology at concentrations of NaCl below 15% and complete lysis in distilled water (Figure 2); the characteristic ability of primary cultures to produce colourless and vacuole-less mutants [12] (Figure

1a and b); and a number of other taxonomic characteristics (Table 1) suggested strongly that these extremely halophilic isolates in fermenting fish sauce were *Halobacterium* strains. However, to conclusively prove the point, pertinent data on the cell wall lipid and DNA content are required. The significantly high concentration of *Halobacterium* during the third week of fermentation [7] and the high gelatinolytic activities (Figure 3) suggested a role in the hydrolysis of fish protein in the Nam Pla fermentation process. However, because the bulk of the protease activity upon fermentation is seen during the first week [19] it is unlikely that Halobacterial proteases are responsible for active proteolysis of fish muscle at the initial phase. It is likely that the halotolerant and moderately halophilic isolates which possessed a multitude of hydrolytic activities are more important at that time. However, as fermentation advances and peptidyl nitrogen accumulates [20], *Halobacterium* species may be relevant as producers of salt resistant endopeptidases. Although *Halobacterium* is often incriminated as a spoilage bacterium in the salted fish and hide industries [21] a batch of Nam Pla which contained high viable counts of *Halobacterium* during the early part of fermentation did not spoil. On the contrary, the aroma of the finished product of this batch was superior to that of a parallel batch which contained lower viable counts. The role of fish sauce aroma formation using controlled cultures and Co^{60} irradiated whole fish homogenates is presently under investigation.

ACKNOWLEDGEMENTS

The authors would like to thank the International Foundation for Science for financial support, the owners of many Thai fish sauce factories for genuine cooperation and T. W. Flegel for help with the manuscript.

REFERENCES

[1]   J. Dougan and G. E. Howard, *J. Sci. Fd. Agric.* 26: 887 (1975).

[2]   K. Amano, *in* "Fish in Nutrition", Fishing of News (Books) Ltd., London (1962).

[3]   S. Areekul, R. Therawibul, and D. Matrakul, *Southeast Asian J. Trop. Med. and Pub. Hlth.* 5: 461 (1974).

[4]   E. V. Crisan and A. Sand, *Appl. Microbiol.* 29: 106 (1975).

[5]   P. Saisithi, B. Kasemsarn, J. Liston and A. M. Dollar, *J. Fd. Sci.* 31: 105 (1966).

[6]   C. Thongthai and M. Siriwongpairat, *J. Sci. Soc. Thailand.* 4: 73 (1978).

[7]   C. Thongthai and M. Siriwongpairat, *Abstr., Symp. Sci. Technol. Devel. Nrth. Thd.* (1978) C6: 86

[8]   R. E. Buchanan and N. E. Gibbons, "Bergey's Manual of Determinative Bacteriology", 8th edition, Williams and Wilkins Co., Baltimore (1974).

[9]   R. R. Colwell, C. D. Litchfield, R. H. Vreeland, L. A. Kiefer and N. E. Gibbons, *Int. J. Syst. Bacteriol.,* 29: 106 (1975).

[10]   R. L. Gherna and H. D. Hatt, "ATCC Catalogue of Strain I", 12th edition, Rockville, Maryland (1976).

[11]   F. Pfeifer, *in* "Halophilic Bacteria" Vol. II, CRC Press Inc., Boca Raton, Florida (1988).

[12]   F. Pfeifer, G. Weidinger and W. Goebel, *J. Bacteriol.* 145: 375 (1981).

[13]   H. P. Dussault, *J. Bacteriol.* 70: 484 (1955)

[14]   A. J. Holding and J. G. Collee *in* "Methods in Microbiology", Vol. 6A, Academic Press, New York (1971).

[15]   T. L. Pitt and D. Dey. *J. Appl. Bacteriol.* 33687 (1970).

[16]   M. P. Starr, *Science*, 93: 333 (1941)

[17]   P. Norerg and B. V. Hofsten, *J. Gen. Microbiol.* 55: 251 (1969).

[18]   N. E. Gibbons, *Can. J. Microbiol.* 3: 249 (1957).

[19]   C. Thongthai, W. Panbangred, C. Khoprasert and S. Dhaveetiyanond. *Proc. Unesco Regnl. Wrkshp, Bundung, Indonesia,* (1986)

[20]   C. Thongthai and H. Okada, *in* "Microbial Utilization of Renewable Resources" Vol. 1, International Center of Cooperative Research and Development in Microbial Engineering, Osaka, Japan (1980).

[21]   J. M. Shewan, *J. Appl. Bacteriol.* 34: 299 (1971).

LIQUID FUEL (OIL) FROM HALOPHILIC ALGAE: A

RENEWABLE SOURCE OF NON-POLLUTING ENERGY

Ben-Zion Ginzburg

Plant Biophysical Laboratory
Botany Department
The Hebrew University
Jerusalem
Israel

ABSTRACT

A system was developed for producing high yields of a biomass rich in proteins and low in polymeric carbohydrates from the halophilic alga *Dunaliella* grown in saline water. It was found that the major nutrients (C, N) could be recycled. On a global scale the use of fuel obtained from biomass need not lead to an increase of CO_2 in the atmosphere. For the efficient conversion of biomass into hydrocarbon by pyrolysis, the most important component is protein. Pyrolysis of biomass was shown to be exothermic, thus enabling the recycling of an important fraction of the thermal energy.

Liquid fuel in the form of a mixture of hydrocarbons has been produced from a renewable source of the green microalga *Dunaliella* [1]. Many of the constituent processes have been tested, at least in the laboratory, and a start was made at upscaling some of them. The project is now inactive, for lack of funds.

Oil is the only form of liquid fuel. Its use has spread over the entire globe and it has been the most important factor in shaping the affluent life-style of the 20th century. It has also made possible the fighting of wars on a world-wide scale. One of the reasons that oil has such power is because it was - and still is - very cheap. Technologically, it is unrivalled among other forms of energy (including nuclear energy). It is a very densely stored form of energy per unit of volume (Table 1 [2]). It is the most readily available form of energy for mobile transport (cars, aeroplanes, etc.). It is easily and relatively cleanly, cheaply and safely transported over long distances. Sociologically, its production requires high skill, and does not demand a large unskilled labour force, and/or dangerous working conditions. It can be one of the least polluting forms of energy if used judiciously.

General and Applied Aspects of Halophilic Microorganisms
Edited by F. Rodriguez-Valera, Plenum Press, New York, 1991

Table 1. Energy Storage Density [2]

Form of energy storage	Energy density KWh/m^3
Pumped water storage	1.5
Pumped air (optimum pressure)	3.0
Rock bed	10.0
Hot water (natural underground)	35.0
Fly wheel	50.0
Batteries (lead acid)	60.0
Eutectic salts	100.0
Hydrogen	2000.0
Gasoline	8000.0

It took many millions of centuries to accumulate the reserves of oil which we know to exist, but it has taken no more than one century to bring these reserves close to exhaustion. The course of the twentieth century has been determined by the uncontrolled use of apparently unlimited quantities of oil. The course of the next century and beyond is dependent on the development of alternative sources of energy, alternative that is to natural oil, which is unrenewable.

The development of a renewable source of energy is a real challenge, and is dependent, in our view, on the production of sufficient amounts of biomass, by capturing sunlight, as was done by nature in the past. Oil is consumed now to the tune of $2-3.10^9$ tons annually. Thus to produce this amount of oil, it is necessary to produce not less than twice the amount of biomass. At present the total amount of biomass produced annually by agricultural methods is about $2-3.10^9$ tons. Thus, the challenge is to double or more the annual output of biomass and will require an increase in yield per unit of area (by one order of magnitude, say) and a new niche in the biosphere. We believe that the <u>saline</u> environment could provide such a niche. Surprisingly enough, it can be very productive indeed. As will be discussed below, a very good candidate for source of biomass is the halophilic alga *Dunaliella*, which has already been tested experimentally, both for quality of biomass and for productivity.

Two considerations should be taken into account when discussing the merits of various components of the biomass as sources for hydrocarbons: materials and energetics. The maximum amount of hydrocarbons that can be obtained from any component of biomass is limited by the percentage of C and H therein (Table 2). The second consideration is energetic: the heat of formation of hydrocarbon is 11 KCal/gram, while that of carbohydrate is only 4 KCal/gram (Table 2). Thus carbohydrates have not more than one-third of the energy content of hydrocarbons. Proteins have half and lipids two-thirds the energy content of hydrocarbons. Thus on theoretical grounds, carbyhydrates are the least desirable components from both the material and energetic points of view. Lipids seem indeed to be preferable to carbohydrates and to proteins.

Table 2. Conversion of biomass into hydrocarbons.

	Percentage Weight				Heat of formation
	C	H	O	N	KCal/gram
Carbohydrates	40	7	53	-	4.0
Proteins	50	10	24	16	5.5
Lipids	70	12	18	-	8.0
Hydrocarbons	88	12	-	-	11.0

When we started this project we thought that we needed to maximise the lipid content of the biomass. Experimental results have changed our view altogether. To our surprise, we found that it is difficult to convert lipids into hydrocarbons by means of a thermal reaction. Furthermore, a high lipid content occurs only in the very old algal cultures (or stressed ones) which therefore have to be left to grow for a very long time. A long period of growth is uneconomical and it is undesirable to use old cultures for production purposes. At the same time we found that high-quality hydrocarbons could be obtained from the pyrolysis of proteins. This has been one of our most important discoveries and when combined with the observation that young, vigorously-growing algal cultures have a high protein content, drew us to the conclusion that the protein component of the biomass is the most important raw material for oil production.

There are several aspects of biomass production for oil. The choice of a halophilic alga as source of biomass has several advantages. The alga can grow in saline

Table 3. Growth of Dunaliella parva 19/9

	Yield	Doubling time
A. In field experiments	Up to 15 g/day/m^2	1–3 days
"Simple system"	50 Tons/Year/Hectare	
B. Partly field experiments, mainly laboratory work. "Elaborate system"	Up to 50 g/day/m^2 150 Tons/Year/Hectare	12–24 hours
C. "Sophisticated" chemostat in the laboratory.		1.5–2 hours.

environments and does not have to compete with other agricultural systems. Its composition is very suitable for conversion to oil, and last but not least, its growth potential was found to be very high (Table 3). It was found by extrapolation from field conditions that the yield was 50 tons per hectare per year. In a more elaborate system, on a medium scale, the yield was three times higher. In a chemostat in the laboratory, the growth rate was such that the doubling-time was less than two hours. By assuming a day length of 14 hours, a potential increase in biomass of 100-fold per day may be possible. It is a challenge to close the technological gap between the growth potential of the laboratory system and that of the practical system in the "energy farm". Let us note that all these growth experiments were done with wild types of alga; we are still comparable to early farmers. Thus, there is enormous scope for improvement by conventional genetical methods. Several unusual biological properties of *Dunaliella* make it a very suitable candidate for genetic engineering. For instance, the cells have no cellulose cell-wall and should therefore be open to the introduction of foreign DNA.

The water source for the "energy farms" should be ocean water, which is available on a large scale. The total area needed is of the order of 1-2 million square kilometers, and is of a type unsuitable for agriculture or other human uses. Suitable regions are available in different parts of the world; thus there need be no monopoly on this resource. The climate required is not restricted to any special type, since different species grow optimally at different temperatures. The most important environmental factor is day-length rather than light-intensity. Several unusual physical and chemical properties of brines are important for the technology and economy of the whole process, e.g. brines influence the process of obtaining the CO_2 from the atmosphere, and the hydrodynamics of the growth system.

Nutrients. 90% of the final product, i.e. the hydrocarbons, is composed of carbon. The remaining 10% is hydrogen, the source of which is water. The only realistic source for the enormous quantities of carbon needed is atmospheric CO_2, present at the very low concentration of 0.03%. Some species of *Dunaliella* have been shown to be able to concentrate CO_2 from the atmosphere at reasonable rates. The mechanism of uptake is biological, very likely coupled to light energy. As is well known, burning fuel is an important source of CO_2 in the atmosphere. Thus, as the energy "farms" are planned as open ponds supplied with CO_2 from air, they present an opportunity to recirculate CO_2 on a global scale. A renewable source of energy of this kind could also serve to maintain the future human ecosystem in a steady state, as is needed for human survival.

The other important nutrient is nitrogen which composes 10-13% of the dry-weight of the biomass of growing cultures of *Dunaliella*. From a practical point of view, nitrogen is a very important component of the input of modern agriculture, and requires more than 50% of the maintenance investment in money and energy.

As will be shown later, a nitrogenous fertilizer is a major by-product of the conversion of the biomass into oil in our process. It is important to remove all

392

the nitrogen from the product to get high-quality oil. It has emerged that the nitrogen thus removed is in a form that can be re-introduced directly into the ponds. In consequence, nitrogen is virtually <u>recirculated</u> in the course of the whole process. This is very important technologically, economically and ecologically because of the large amounts of nitrogen needed. Nitrogen in modern agriculture is a troublesome source of pollution. Recirculation of nitrogen has already been demonstrated experimentally in our system. Phosphorus, another essential nutrient, is absorbed by *Dunaliella* cells from relatively low concentrations in the medium and can be accumulated in the cells to quite high concentrations [3]. There is a reasonably good chance that a large proportion of the P can be recycled. The other essential nutrients occur at sufficiently high concentrations in ocean water for growth to occur. Ocean water needs adding to the growth ponds in a constant flow to replace water lost by evaporation. Thus it is planned to have at least two re-cycling systems in the ecosystem of the oil farms, a carbon cycle on a global scale, and a nitrogen cycle locally.

Conversion of biomass into oil. Surprisingly enough, of all the components in the plant biomass, proteins were found to be the most suitable as a source of oil. During the thermal treatment (pyrolysis) of amino-acids and/or proteins, deamination occurs at a temperature range of 150-300°C and is followed by decarboxcylation. Simultaneously, various molecular rearrangements such as polymerisation and hydrogenation also occur [4] and yield short chain hydrocarbons as products. Dry pyrolysis, accompanied by more intensive polymerization and carbonisation, yields asphaltine, a very viscous mixture of hydrocarbons, as partial product. In wet pyrolysis, i.e. when water is present with the reactants, only short-chain hydro-carbons are obtained. Thus the wet process is more efficient from the material balance point of view, and of course, the quality of the product is superior. It is

Table 4. Pyrolysis by autoclaving [4]

Reactant	Organic Products	Inorganic Products
Phenylalanine Dry powder	Tolune, Ethylbenzene, Asphalt-like material, Carbonised material.	Ammonium carbonate
Phenylalanine In water.	Toluene, Ethylbenzene, only	Ammonium carbonate
Protein fraction of *Dunaliella parva* in NaCl solution	Toluene and many light hydrocarbons	Ammonium carbonate

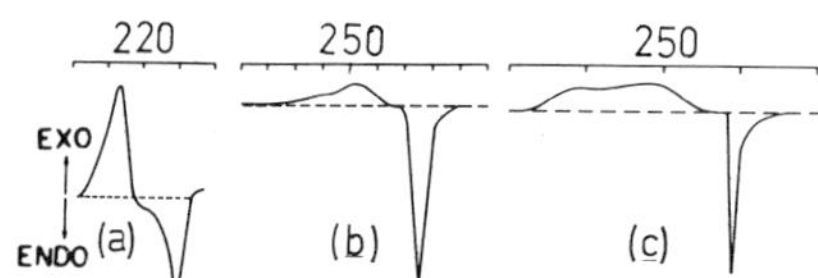

Figure 1.   Thermograms of phenylalanine (a), copolymer of glutamic phenylalanine (b), bovine serum albumin (c), in differential scanning calorimeter. Figures refer to temperature in Centigrade.

suggested that the source of hydrogen for the wet process is water, whereas in dry pyrolysis H comes from the reactants themselves. In the autoclave, which serves as a laboratory version of an industrial reactor, hydrocarbons were obtained in the organic phase and ammonium carbonate in the aqueous phase (Table 4). The energetics of pyrolysis in the autoclave could be followed by means of a Differential Scanning Calorimeter (DSC). In Fig. 1a,b,c, thermograms of pyrolysis of phenylalanine, a copolymer of phenylalanine and glutamic acid and of the protein bovine serum albumin (BSA) are depicted. In the first case, we obtained toluene and ethylbenzene at a ratio of 2:1 respectively; ammonium carbonate was obtained in stoichiometric amounts. Two thermal events can be seen. The first is exothermic with a net value of 11.0 KCal/mole and comes to a peak at 200°C. It is followed by an endothermic event of 9.5 KCal/mole at 240°C. Overall, the thermal event is

Table 5. Energetics of pyrolysis by Differential Scanning Calorimeter [4]

	KCal/mole		
	Exothermic events	Endothermic events	Total thermal events
Phenylalanine	11.0	-9.5	1.5
Copolymer: Phenylalanine- Glutamic acid	5.0	-7.0	-2.0
Protein: Bovine Serum Albumin	8.6	-2.0	6.6

slightly exothermic.In Figure 1b the thermal events occurring during the pyrolysis
of the copolymer are depicted. Qualitatively, they are similar to the thermal events
occurring during the pyrolysis of pure phenylalanine, though they occur at higher
temperatures. In the exothermic event, enthalpy is about half that of the pure
phenylalanine, whereas the endothermic phase is similar. Fig. 1c shows the trace of
pyrolysis of BSA, a model for proteins. The exothermic event occurs over a wide
temperaturerange, with a peak at 235°C, and enthalpy of 8.6 KCal/mole. A very
small endothermic event follows, with a peak at 276°C and enthalpy of 2.0 K-
Cal/mole. To sum up, the pyrolysis of aminoacids and proteins combines endothermic and exothermic processes, which on the whole almost balance each other. The
net overall reaction is slightly exothermic (Table 5). This is a most surprising and
very significant finding since it means that by clever use of heat exchangers, the
heat input in conversion of biomass into oil is very low, since conversion takes
place at the expense of materials which were originally synthesized by means of
energy captured from the sun.

ACKNOWLEDGEMENTS

This project was initiated by the late Professor M. R. Bloch and myself in
1973. Prof. Bloch's inspiration, unusual imagination, creative thinking and industrial
experience were the cornerstone of the project. Many research workers, and other
colleagues have contributed to the development of the project. Dr. Margaret Ginzburg
and Professors N. Garty and Y. Sasson have led teams in this project and played an
important part in the development of the concepts presented in this paper.

REFERENCES

[1]   Report to the Ministry of Energy, Govt. of Israel (1978).
[2]   T. H. Moss, Quart. Rev. Biophys. 10:99 (1977).
[3]   M. Ginzburg and B. Z. Ginzburg., J. Exp. Bot. 36:701   (1985).
[4]   A. Asserin, N. Garty, M. R. Bloch, and B. Z. Ginzburg, in preparation.

INDEX

Halotolerant strains, 12
Heavy-metal resistant strains, 177
Heavy metals, 173
Heteropolysaccharide, 373
History, 63
HMGCoA reductase, 273
Hydrocarbon, 389
Hydrophilic, 236
Hydrophobic, 236
Hydrophobic bonds, 64
Hyocholic acid, 360
Hyodeoxycholic acid, 360

IBMX, 151
Immobilized halophilic nuclease H, 341
In vitro protein synthesis, 64
Induction can take place at low oxygen
    concentrations, 130
Induction of NAR activity, 131
Initiation factors, 66
Insertion elements, 285
Intercellular and intracellular spaces,
    218
Intracellular, 217
Intracellular ions, 218
Ionophores, 165
ISH, 305
ISH-elements, 289
ISH26, 305
ISH27 elements, 290

Killer-toxins, 157

Light-induced pH changes, 163
Lipoic acid, 121
Liquid fuel, 389
Lithocholic acid, 360
Lot, 67
Lot's wife, 67

Macrorestriction fragments, 296
Malic dehydrogenase, 64
Mapping, 272
Marine bacteria, 97
Marine fungi, 107
Membrane-derived oligosaccharides
(MDO), 244
Membrane vesicles, 139
Membranes, 207
Menaquinones, 202
Methyl-accepting proteins, 152
Methylation, 152

Mevalonate pathway, 195
Mevinolin, 273
Microbial mats, 33-36, 40-43
Micrococcus varians subsp. *halophilus*,
    341
Minor circular DNA, 277
Moderate halophiles, 97
Moderately halophilic bacteria, 45
Modes of action, 159
Multienzyme, membrane-bound system,
    195
Museum cultures, 15
Mutant strain L-07, 139
Mutations, 289
Mutual repulsion, 63

Na^+/H^+ antiporter, 97, 162
NA+ PUMP, 97
NADH-quinone-acceptor
    oxidoreductase, 97
Nam Pla, 381, 382, 384-387
^{23}Na-NMR, 219
Natronobacteria, 73
Natronobacterium, 73, 191, 199
Natronobacterium pharaonis, 295
Natronococcus, 191, 199
Navicula, 10
NH_4^+ ions, 66
NMR, 218
NMR-visible, 219
Nomenclature, 7
Novobiocin, 333-336
Number of chromosomes, 295
Numerical taxonomic, 53
Nuoc Mam, 381

2-oxoacid dehydrogenase complexes,
    126
2-oxoacid: ferredoxin oxidoreductases,
    126
Oscillator, 151
Organic Lake, 9, 11
  biota, 10
Osmolarity, 225
Osmoregulate, 221
Osmoregulation, 89
Osmotic adaptation, 243
Osmotic stress, 243
Outer membrane, 226
Oxygen tension, 206

P-NMR, 165

Secondary metabolites, 33, 34, 40
Sensory light, 139, 145
Sensory photosystems, 150
Sensory pigment, 140
Sensory pigments, 140
Sensory rhodopsin I, 150
Sensory rhodopsin I (SR-I), 140
Sensory rhodopsin II, 150
Sensory rhodopsin II (SR-II), 140
Sensory signal, 153
Shift reagent, 219
Shuttle vectors, 273
Signal integration, 152
Signal lifetime, 152, 153
sn-2,3-diphytanylglycerol, 191
sn-2,3-diphytanylglycerol diether, 192
Sodium, 117
Sodium acetate, 139, 144
Sodium carbonate, 116
Sodium sulfate, 116
Solar salterns, 26, 33, 67
Spheroplasts, 364
Splicing, 256
SR-I, 140
SR-II, 140
Stability, 133
Stereological analysis, 109
Structures of archaeol phospholipids and
glycolipids, 192
Structures of isoprenyl precursors of the
diphytanyl ether, 196
Substrate cycles, 111
Subterranean brines, 54
Subtilisin, 359
Succinic acid, 244
Sulfated diglycosyl archaeol (S-DGA-1),
192
Sulfated diglycosyl-archaeol, 191
Sulfated diglycosyl, 193
Sulfated tetraglycosylarchaeol
(S-TeGA), 192
Sulfated triglycosyl-archaeol, 191
Sulfide, 33-35, 38-42

Taurocholic acid, 359
Taxonomy, 3, 45
Temperature,
  minimum, 11
Temporal light gradients, 150
Tetrahydropyrimidine, 351
TGA, 193
The First Halophile, 67
Thermoplasma, 126
T_{Min}, 12
"Tolerant", 222
Topology, 295, 298
Toxic action of Cl⁻ ions, 65
Transcription, 259, 323, 324, 333, 335,
  336
Transformation, 273
Transport mechanisms, 234
Transposition, 289
Triglycosylarchaeol, 191
  (S-TGA-1), 192
  (TGA-2), 192
Trypsin, 364
Tryptophan, 274
Tungstate, 130
Turgor pressure, 221

Uncouplers, 165
Ursodeoxycholic acid, 360
UV-A, 140
UV-B, 139, 140

V-*myc* like protein, 313
V-*myc* oncogene product, 313
Vacuole, 109
Vibrio alginolyticus, 97, 138-2
Vibrio costicola, 97, 218

Water permeability, 208
Water space, 222
Western blot analysis, 314, 367

X-ray microanalysis, 106